METHODS IN MOLECULAR BIOLOGY

Series Editor
John M. Walker
School of Life Sciences
University of Hertfordshire
Hatfield, Hertfordshire, AL10 9AB, UK

For further volumes:
http://www.springer.com/series/7651

Polyadenylation

Methods and Protocols

Edited by

Joanna Rorbach

Mitochondrial Genetics Group, MRC Mitochondrial Biology Unit, Cambridge, UK

Agnieszka J. Bobrowicz

Mitochondrial Research Group, Institute for Ageing and Health, The Medical School, Newcastle University, Newcastle upon Tyne, UK

Humana Press

Editors
Joanna Rorbach
Mitochondrial Genetics Group
MRC Mitochondrial Biology Unit
Cambridge, UK

Agnieszka J. Bobrowicz
Mitochondrial Research Group
Institute for Ageing and Health
The Medical School
Newcastle University
Newcastle upon Tyne, UK

ISSN 1064-3745 ISSN 1940-6029 (electronic)
ISBN 978-1-4939-6055-2 ISBN 978-1-62703-971-0 (eBook)
DOI 10.1007/978-1-62703-971-0
Springer New York Heidelberg Dordrecht London

Printed on acid-free paper

Humana Press is a brand of Springer
Springer is part of Springer Science+Business Media (www.springer.com)

Preface

Across the diverse biological systems a poly(A) tail functions as a *cis*-acting element involved in a range of processes, from modulating mRNA transport, termination of transcription and translational efficiency to acting as the major determinant, either positive or negative, of messenger RNA stability. Polyadenylation appears to play contrasting roles in prokaryotic and eukaryotic gene expression. In bacteria poly(A) tails stimulate RNA degradation suggesting that this is the ancestral function of polyadenylation. In fact, degradation of polyadenylated species in prokaryotes is so rapid that 3′ polyadenylation was thought to be a specifically eukaryotic feature for some time after its discovery. Polyadenylation of nascent transcripts is an essential step in the maturation of nearly all eukaryotic messenger RNAs and plays a role in the export of the mature transcripts into the cytosol from the nucleus. In the cytosol of eukaryotes, polyadenylation functions in stabilizing the transcript (in cooperation with a cytosolic poly(A)-binding protein, PABPC) and promotes initiation of translation. In the process of RNA quality control in the nucleus, polyadenylation of the ribosomal and small nucleolar RNAs by the polyadenylation complex TRAMP stimulates the degradation of these transcripts by the nuclear exosome, a multisubunit complex of 3′–5′ exonucleases. That points to the ancestral, degradation-stimulating role of poly(A) tails, maintained in the eukaryotic nucleus and modified to assist translation in the cytosol. The length of the poly(A) extension is also an important determinant of mRNA fate in eukaryotes and the departure from the standard length can trigger degradation or nuclear retention of the transcript. Hyperadenylation of mRNA in response to the nuclear import of PABPC (which can occur during cellular stress) causes nuclear retention of mRNA, limiting gene expression. Examples presented above point toward polyadenylation as a ubiquitous mechanism across the spectrum of organisms. Therefore its importance in regulation of gene expression is currently extensively studied.

This volume, Polyadenylation: Methods and Protocols, presents perspectives from some of the most active groups in this field. We provide protocols and up-to-date reviews, with a particular focus on the recent advances in the fast-moving area of alternative polyadenylation that only recently has been recognized as a key contributor to the complexity of mammalian gene expression.

In *Part I*, a broad range of all-purpose methods for studying eukaryotic polyadenylation is presented. This includes methods of detection and analysis of polyadenylation status as well as in vitro assays to measure activities and interactions of the different components of polyadenylation machinery. Recent studies point to the abundance of the transcript variants resulting from alternative polyadenylation (APA) of mammalian genes. Alternative polyadenylation is being extensively studied with regards to its role in mRNA stability, miRNA targeting and silencing gene expression, translation, imprinting, cellular proliferation, and differentiation. Some of the methodologies currently being developed to study this fascinating field are presented in *Part I*. *Part II: Bacteria and Organelles* comprises the methods developed specifically to study polyadenylation in bacteria and eukaryotic organelles, mitochondria and chloroplasts. Deadenylation is a common mode of controlling messenger RNA stability and translation. Poly(A)-dependent degradation of mRNA, once thought to be the characteristic of archaea, bacteria, and organelles, is today known to be present in the eukaryotic nucleus and cytosol. The methodology dedicated to the study of this phenomenon is shown in *Part III: Deadenylation and Decay*.

Collectively, the presented protocols illustrate the recent up-to-date knowledge on different aspects of polyadenylation. A review chapter has been included at the beginning of each section to allow for the full scope and perspectives of each of these areas of study to be appreciated.

We are most grateful to the dedicated group of authors for their efforts in contributing to this volume. It is our hope that by combining recent adaptations of the golden standards in the field together with the latest advances in the study of polyadenylation, this volume will become a valuable compendium for both the newcomers and the experienced researchers.

Target audience: Biochemists, molecular biologists, cell biologists.

Cambridge, UK — *Joanna Rorbach*
Newcastle upon Tyne, UK — *Agnieszka J. Bobrowicz*

Contents

Contributors

BASKAR BAKTHAVACHALU • *Center for RNA Biology, The Ohio State University, Columbus, OH, USA; Department of Molecular and Cellular Biochemistry, The Ohio State University, Columbus, OH, USA*

TRAUDE HELENE BEILHARZ • *Department of Biochemistry and Molecular Biology, Monash University Australia, Victoria, VIC, Australia*

AGNIESZKA J. BOBROWICZ • *Mitochondrial Research Group, Institute for Ageing and Health, The Medical School, Newcastle University, Newcastle upon Tyne, UK*

LUKASZ S. BOROWSKI • *Institute of Genetics and Biotechnology, Faculty of Biology, University of Warsaw, Warsaw, Poland; Institute of Biochemistry and Biophysics, Polish Academy of Sciences, Warsaw, Poland*

ÖZGÜL PERSIL ÇETINKOL • *Zentiva Health Products Co., Lüleburgaz, Kırklareli, Turkey*

OLGA COLL • *Gene Regulation, Stem Cells and Cancer Programme, Centre for Genomic Regulation (CRG), Universitat Pompeu Fabra (UPF), Barcelona, Spain*

ASHLEY L. CORNETT • *Department of Biochemistry and Molecular Biology, New Jersey Medical School and The Graduate School of Biomedical Sciences, Rutgers, The State University of New Jersey, Newark, NJ, USA*

EMRAL DEVANY • *Department of Chemistry, Hunter College and Graduate Center, City University of New York, New York, NY, USA*

JULIE A. DOUGHERTY • *Center for RNA Biology, The Ohio State University, Columbus, OH, USA; Department of Molecular and Cellular Biochemistry, The Ohio State University, Columbus, OH, USA; Biomedical Sciences Graduate Program, The Ohio State University, Columbus, OH, USA*

GRANT A. ENCALADE • *Department of Chemistry and Biochemistry, Lamar University, Beaumont, TX, USA*

JAMIE VAN ETTEN • *Department of Biological Chemistry and Genetics Training Program, University of Michigan Medical School, Ann Arbor, MI, USA*

ALEKSANDRA FILIPOVSKA • *Harry Perkins Institute of Medical Research, Centre for Medical Research, The University of Western Australia, WA, Australia; School of Chemistry and Biochemistry, The University of Western Australia, Crawley, WA, Australia*

XIANG-DONG FU • *Department of Cellular and Molecular Medicine, Institute for Genomic Medicine, University of California, San Diego, La Jolla, CA, USA*

DOMINIQUE GAGLIARDI • *Institut de Biologie Moléculaire des Plantes, Centre National de la Recherche Scientifique (CNRS), Université de Strasbourg, Strasbourg Cedex, France*

FÁTIMA GEBAUER • *Gene Regulation, Stem Cells and Cancer Programme, Centre for Genomic Regulation (CRG), Universitat Pompeu Fabra (UPF), Barcelona, Spain*

AARON C. GOLDSTROHM • *Department of Biological Chemistry and Genetics Training Program, University of Michigan Medical School, Ann Arbor, MI, USA*

PETAR N. GROZDANOV • *Department of Cell Biology and Biochemistry, Texas Tech University Health Sciences Center, Lubbock, TX, USA*

J. Andrew Hockert • *Department of Biology, University of the Cumberlands, Williamsburg, KY, USA*
Mainul Hoque • *Department of Biochemistry and Molecular Biology, University of Medicine and Dentistry of New Jersey (UMDNJ)-New Jersey Medical School, Newark, NJ, USA*
Joel Hrit • *Department of Biological Chemistry and Genetics Training Program, University of Michigan Medical School, Ann Arbor, MI, USA*
Jie Huang • *State Key Laboratory of Virology, College of Life Sciences, Wuhan University, Wuhan, Hubei, China*
Amrei Jänicke • *Department of Biochemistry and Molecular Biology, Monash University Australia, Melbourne, VIC, Australia*
Mandy Jeske • *Institute of Biochemistry and Biotechnology, Martin Luther University Halle-Wittenberg, Heidelberg, Germany*
Ge Jin • *Zhengzhou University, Zhengzhou, Henan, China*
Ioannis Kavakiotis • *Department of Informatics, Aristotle University of Thessaloniki, Thessaloniki, Greece*
Frida E. Kleiman • *Department of Chemistry, Hunter College and Graduate Center, City University of New York, New York, NY, USA*
Sidney R. Kushner • *Department of Genetics, University of Georgia, Athens, GA, USA*
Man Chun Lee • *Department of Biochemistry and Molecular Biology, Monash University Australia, Melbourne, VIC, Australia*
Antonio Lembo • *Department of Genetics, Biology and Biochemistry, Molecular Biotechnology Center, University of Turin, Turin, Italy*
Hai-Ri Li • *Department of Cellular and Molecular Medicine, Institute for Genomic Medicine, University of California, San Diego, La Jolla, CA, USA*
Wencheng Li • *Department of Biochemistry and Molecular Biology, University of Medicine and Dentistry of New Jersey (UMDNJ)-New Jersey Medical School, Newark, NJ, USA*
Donny D. Licatalosi • *Center for RNA Molecular Biology, Case Western Reserve University, Cleveland, OH, USA*
Jia-Hui Lin • *Department of Chemistry, National Sun Yat-sen University, Kaohsiung, Taiwan*
Carol S. Lutz • *Department of Biochemistry and Molecular Biology, New Jersey Medical School and The Graduate School of Biomedical Sciences, Rutgers, The State University of New Jersey, Newark, NJ, USA*
Clinton C. MacDonald • *Department of Cell Biology and Biochemistry, Texas Tech University Health Sciences Center, Lubbock, TX, USA*
James L. Manley • *Department of Biological Sciences, Columbia University, New York, NY, USA*
Roshan Mascarenhas • *Department of Pharmacology, The Ohio State University, Columbus, OH, USA*
Michal Minczuk • *Mitochondrial Genetics Group, MRC Mitochondrial Biology Unit, Cambridge, UK*
Bijoy K. Mohanty • *Department of Genetics, University of Georgia, Athens, GA, USA*
Takashi Nagaike • *Biomedical Research Institute, National Institute of Advanced Industrial Science and Technology, Tsukuba-shi, Ibaraki, Japan*
Fatih Ozsolak • *Helicos BioSciences Corporation, Cambridge, MA, USA*

Deepak P. Patil • *Center for RNA Biology, The Ohio State University, Columbus, OH, USA; Department of Molecular and Cellular Biochemistry, The Ohio State University, Columbus, OH, USA*
Sarah Pearce • *Mitochondrial Genetics Group, MRC Mitochondrial Biology Unit, Cambridge, UK*
Adam Petterson • *Department of Biological Chemistry and Genetics Training Program, University of Michigan Medical School, Ann Arbor, MI, USA*
Paolo Provero • *Department of Molecular Biotechnology and Health Sciences, Molecular Biotechnology Center, University of Torino, Torino, Italy ; Center for Translational Genomics and Bioinformatics, San Raffaele Scientific Institute, Milano, Italy*
Oliver Rackham • *Harry Perkins Institute of Medical Research, Centre for Medical Research, The University of Western Australia, Perth, WA, Australia; School of Chemistry and Biochemistry, The University of Western Australia, Crawley, WA, Australia*
Nathan Raynard • *Department of Biological Chemistry and Genetics Training Program, University of Michigan Medical School, Ann Arbor, MI, USA*
Joanna Rorbach • *Mitochondrial Genetics Group, MRC Mitochondrial Biology Unit, Cambridge, UK*
Kamya Sankar • *Department of Biological Chemistry and Genetics Training Program, University of Michigan Medical School, Ann Arbor, MI, USA*
Daniel R. Schoenberg • *Center for RNA Biology, The Ohio State University, Columbus, OH, USA; Department of Molecular and Cellular Biochemistry, The Ohio State University, Columbus, OH, USA*
François M. Sement • *Institut de Biologie Moléculaire des Plantes, Centre National de la Recherche Scientifique (CNRS), Université de Strasbourg, Strasbourg Cedex, France*
Yongsheng Shi • *Department of Microbiology and Molecular Genetics, School of Medicine, University of California, Irvine, Irvine, CA, USA*
Maxim V. Sukhodolets • *Department of Chemistry and Biochemistry, Lamar University, Beaumont, TX, USA*
Thomas J. Sweet • *Center for RNA Molecular Biology, Case Western Reserve University, Cleveland, OH, USA*
Roman J. Szczesny • *Institute of Genetics and Biotechnology, Faculty of Biology, University of Warsaw, Warsaw, Poland; Institute of Biochemistry and Biophysics, Polish Academy of Sciences, Warsaw, Poland*
Claudia Temme • *Institute of Biochemistry and Biotechnology, Martin-Luther-University Halle-Wittenberg, Halle, Germany*
Bin Tian • *Department of Biochemistry and Molecular Biology, University of Medicine and Dentistry of New Jersey (UMDNJ)-New Jersey Medical School, Newark, NJ, USA*
Wei-Lung Tseng • *Department of Chemistry, National Sun Yat-sen University, Kaohsiung, Taiwan; School of Pharmacy, College of Pharmacy, Kaohsiung Medical University, Kaohsiung, Taiwan*
George Tzanis • *Department of Informatics, Aristotle University of Thessaloniki, Thessaloniki, Greece*
Ana Villalba • *Gene Regulation, Stem Cells and Cancer Programme, Centre for Genomic Regulation (CRG), Universitat Pompeu Fabra (UPF), Barcelona, Spain*
Ioannis Vlahavas • *Department of Informatics, Aristotle University of Thessaloniki, Thessaloniki, Greece*
Elmar Wahle • *Institute of Biochemistry and Biotechnology, Martin-Luther-University Halle-Wittenberg, Halle, Germany*

MOTOAKI WAKIYAMA • *Post-transcriptional Control Research Unit, RIKEN Center for Life Science Technologies, Tsurumi-ku, Yokohama, Japan*
CHENGGUO YAO • *Department of Microbiology and Molecular Genetics, School of Medicine, University of California, Irvine, Irvine, CA, USA*
SHIGEYUKI YOKOYAMA • *RIKEN Structural Biology Laboratory, Tsurumi-ku, Yokohama, Japan*
XIAOKAN ZHANG • *Department of Chemistry, Hunter College and Graduate Center, City University of New York, New York, NY, USA*
YU ZHOU • *Department of Cellular and Molecular Medicine, Institute for Genomic Medicine, University of California, San Diego, La Jolla, CA, USA*

Part I

3′-End Processing in Eukaryotes

Chapter 1

3′ End Formation and Regulation of Eukaryotic mRNAs

Thomas J. Sweet and Donny D. Licatalosi

Abstract

The polyadenosine (polyA) "tail" is an essential feature at the 3′ end of nearly all eukaryotic mRNAs. This appendage has roles in many steps in the gene expression pathway and is subject to extensive regulation. Selection of alternative sites for polyA tail addition is a widely used mechanism to generate alternative mRNAs with distinct 3′UTRs that can be subject to distinct forms of posttranscriptional control. One such type of regulation includes cytoplasmic lengthening and shortening of the polyA tail, which is coupled to changes in mRNA translation and decay. Here we present a general overview of 3′ end formation in the nucleus and regulation of the polyA tail in the cytoplasm, with an emphasis on the diverse roles of 3′ end regulation in the control of gene expression in different biological systems.

Key words Alternative polyadenylation, PolyA tail control, Posttranscriptional regulation

1 Introduction

Transcription by RNA polymerase II yields precursor mRNA (pre-mRNA) that undergoes multiple processing events in order to become mature mRNA that can serve as a template for translation in the cytoplasm. Pre-mRNA processing events include the addition of a 7-methyl guanosine "cap" to the pre-mRNA 5′end, the removal of introns and splicing of exons, and maturation of the mRNA 3′ end. With the exception of histone mRNAs, the 3′ ends of all mRNAs are processed by endonucleolytic cleavage and the non-templated addition of a continuous stretch of adenosines [1]. This polyadenosine (polyA) "tail," which has an average length of 250 bases in human cells, is an essential feature of mRNA with important roles in multiple steps of the mRNA life cycle. In the nucleus, 3′ end cleavage and polyadenylation of mRNA is essential for transcription termination, release of mRNA from the site of transcription, and export to the cytoplasm [2]. In the cytoplasm the polyA tail protects mRNA from degradation and enhances mRNA translation [3].

Joanna Rorbach and Agnieszka J. Bobrowicz (eds.), *Polyadenylation: Methods and Protocols*, Methods in Molecular Biology, vol. 1125, DOI 10.1007/978-1-62703-971-0_1,

Posttranscriptional control of gene expression via factors that act on mRNA provides the cell with multiple layers to diversify and fine-tune protein output. This includes alteration of the protein-coding content of an mRNA through alternative processing of pre-mRNA into mRNA [4]. In addition, mature mRNAs can be subject to translational repression, localization to discrete subcellular locations, and changes in the rate of mRNA translation and decay [5, 6]. Posttranscriptional regulation is largely dependent on the binding of specific factors (RNA-binding proteins or miRNAs) to sequences in the mRNA 3′ untranslated region (3′UTR). This includes binding sites for factors that act directly or indirectly to modify the length of polyA tail present on specific mRNAs and therefore affect mRNA translation and stability. Importantly, 3′UTR regulatory sequences can be included or omitted from mature mRNA by selection of an alternative site of mRNA 3′ end formation [7]. Thus, 3′ end formation in the nucleus can impact mRNA fate in the cytoplasm.

In recent years, there has been renewed interest in the 3′end of mRNAs due to the recognition that 3′ ends are subject to extensive regulation (both in the nucleus and cytoplasm) and have important roles in cell- and tissue-specific gene expression programs. In addition, there is increasing evidence that global changes in 3′UTR length are associated with changes in cell proliferation and differentiation and that modification of polyA tail length has important roles in diverse developmental pathways [1, 7]. In this review, we provide an overview of eukaryotic mRNA 3′ end formation and regulation, with the goal of highlighting the prevalence of 3′ end regulation and the importance of the 3′UTR in posttranscriptional regulation of gene expression.

2 Cleavage and Polyadenylation of Eukaryotic mRNA

The addition of a polyA tail to the 3′ end of mRNA occurs via two tightly coupled steps. The first step is endonucleolytic cleavage of the pre-mRNA to generate a free 3′ hydroxyl that is the substrate for the second step, the non-templated addition of adenosines (polyadenylation). Cleavage and polyadenylation are dependent on multiple *cis*-acting elements and their recognition by specific RNA-binding proteins within the multi-subunit complexes that comprise the cleavage and polyadenylation machinery [8]. Here we focus on the *cis* sequences and *trans*-acting factors in mammals. For a more detailed discussion on species-specific similarities and differences in diverse organisms, we refer the reader to the following excellent reviews [9, 10].

In higher eukaryotes, >90 % of sites of cleavage and polyadenylation have a hexanucleotide polyadenylation signal (PAS) approximately 10–30 nucleotides upstream of the site of cleavage and polyadenylation (polyA site). The most common PAS hexamers used in higher eukaryotes are AAUAAA or AUUAAA, but many other variants exist, with over ten different possibilities identified in mouse and humans [8]. A second element, the downstream sequence element (DSE) is situated 10–30 bases downstream of the polyA site and consists of a more variable U-rich or UG-rich sequence. A U-rich upstream sequence element (USE) is generally present upstream of the PAS and frequently consists of UGUA. The *trans*-acting factors necessary for cleavage and polyadenylation include CPSF (cleavage and polyadenylation specificity factor), CstF (cleavage stimulatory factor), CFIm and CFIIm (mammalian cleavage factors I and II, respectively), and PAP (polyA polymerase). The 160 kilodalton (kD) subunit of CPSF (CPSF160) binds the PAS, while CPSF73 functions as the endonuclease. The position of pre-mRNA cleavage site is determined by interactions between CPSF and CstF, the latter of which is bound to the DSE via the 64 kD subunit of CstF (Cstf64). CFIm recognizes UGUA elements upstream of the cleavage site and is believed to regulate the interaction of CPSF with the PAS. The role of CFIIm is not well defined, although it is required along with CPSF, Cstf, CFIm, and PAP for cleavage and polyadenylation in vitro. PolyA-binding protein (PABP) has an important role in 3′ end formation to stabilize the interaction of CPSF to PAP during polyadenylation to ensure PAP processivity until the growing polyA tail has reached an appropriate (species-specific) length [11]. For additional details on the core factors and elements necessary for 3′ end cleavage and polyadenylation, see the following reviews [1, 7–10].

3 mRNA 3′ End Formation Is Coupled to Multiple Steps in the mRNA Life Cycle

Although pre-mRNA cleavage and polyadenylation activity can be reconstituted in vitro using only the factors described above, 3′ end formation in vivo is physically and functionally coupled to multiple steps in the mRNA production pathway, including the RNA polymerase II transcription cycle [2]. Interactions between 3′ end processing factors and the transcription machinery have been described from yeast to human cells [12]. This includes, but is not limited to, the association of CPSF with transcription initiation factor IID [13] and the co-transcriptional recruitment of 3′ end processing factors to the C-terminal domain (CTD) of elongating RNA polymerase II [2]. The CTD is required for proper 3′ end formation in vivo [14] and can stimulate cleavage and polyadenylation when added to in vitro 3′ end processing assays [15]. Conversely, mutation of cleavage and polyadenylation elements results in defects in transcription termination [16]. Connections between 3′ end formation and other steps in the mRNA life cycle have also been described [1]. This includes interactions between cleavage and polyadenylation factors with pre-mRNA splicing factors, including some that have been shown to have splicing-independent functions in the regulation of polyA site utilization [17, 18]. In addition, there are links between subunits of cleavage and polyadenylation complexes and the mRNA quality control and export machinery [19]. It is believed that these interactions couple 3′ end formation to mRNA export from the nucleus and facilitate the degradation of transcripts that are not properly processed or exported. Consistent with physical and functional links between 3′ end processing factors and other proteins involved in mRNA biogenesis and gene regulation, biochemical purification of the cleavage and polyadenylation machinery from human cells revealed ~70 proteins that are associated with the "core" factors described above [20]. In addition to splicing factors and components of the transcription machinery, this ~1 megadalton complex contains a number of unexpected proteins including translation factors suggesting that 3′ end formation of mRNA may be coupled to other regulatory events in the cell.

4 Cytoplasmic Regulation of mRNA 3′ Ends

The polyA tail has an important role in mRNA translation and stability [21]. The interaction of PABP (bound to the polyA tail) with the translation initiation factors (bound to the mRNA 5′ cap) forms a "closed loop" or "circular" mRNA that facilitates translation and protects mRNA from degradation [22]. The formation and stability of this loop is believed to be dependent on the length of the polyA tail [3]. In general, shortening of the tail polyA results in reduced PABP binding and lower levels of translation, whereas polyA tail lengthening is associated with higher translation. A number of posttranscriptional regulatory factors have been described that control the translation of their mRNA targets by directly or indirectly modifying polyA tail length. For example, PUF family proteins (present in genomes from yeast to humans) recognize specific 3′UTR sequences to recruit deadenylases and repress translation [23]. Similarly, AU-rich elements (AREs, classically defined as AUUUA, although other variants are known to exist) comprise a class of posttranscriptional regulatory sequences in 3′UTRs that are bound by ARE-binding proteins that recruit deadenylases such as polyA ribonuclease (PARN) or the CCR4-NOT complex [24].

Although deadenylation can be coupled to mRNA decay, in some instances, deadenylation is associated with translational repression and mRNA stabilization [5]. Such regulation is believed to be essential for the proper control of mRNAs whose translation must be restricted to discrete subcellular locations and/or activated in response to specific intra- or extracellular cues. Cytoplasmic lengthening of the polyA tail promotes the translation of specific mRNAs and has been shown to be dependent on cytoplasmic polyadenylation elements (CPEs) in the 3′UTR [3]. CPEs are typically (but not limited to) UUUUAU or UUUUAAU and are bound by CPE-binding proteins (CPEBs). Interestingly, CPEBs can have dual roles in polyA tail length regulation. Depending on the location and number of CPEs in a 3′UTR, as well as the posttranslational phosphorylation status of CPEBs, CPEBs can positively or negatively impact translation by recruiting translational regulators, polyA polymerases such as Gld-2 to lengthen the polyA tail, or deadenylases such as PARN to shorten the polyA tail [25]. Such dual control of polyA tail length by the same factor binding to the same element(s) highlights the difficulty in predicting roles for RNA-binding proteins based solely on the presence of their cognate binding elements and the importance of using biochemical approaches to monitor protein-RNA interactions, posttranslational modifications, and polyA tail length.

5 Fine-Tuning Gene Expression Through Alternative Polyadenylation

Bioinformatic analyses of ESTs, and more recently, global RNA profiling approaches, have revealed that the majority of human genes (>80 %) yield multiple mRNA isoforms with alternative 3′UTRs due to differences in the position of 3′ end cleavage and polyadenylation [1, 7]. Since posttranscriptional regulatory sequences are contained within 3′UTRs, alternatively polyadenylated mRNA variants generated from the same gene are likely to be bound by different combinations of *trans*-acting factors (proteins and miRNAs) that can affect mRNA localization, translation, stability, and decay. Thus changes in the position of cleavage and polyadenylation have the potential to significantly impact downstream events in the life cycle of the mRNA by including or excluding posttranscriptional regulatory sequences in the mRNA's 3′UTR [26].

The most common type of alternative polyadenylation is that in which multiple polyA sites are arranged in tandem in the same 3′terminal exon [1]. In such cases, selection of an alternative site for cleavage and polyadenylation alters the 3′UTR sequence with no impact on protein-coding sequence. An excellent example of how such regulation can impact gene expression (particularly mRNA localization and translation) comes from the study of the mRNA encoding brain-derived neurotrophic factor (BDNF). BDNF is an important signaling molecule in the brain implicated in neuronal survival, circuit development, and synaptic plasticity and through these functions is also genetically linked to neurological and psychiatric disorders. Importantly, neurons produce distinct forms of BDNF mRNA [27]. BDNF mRNA generated using a proximal polyA site (shorter 3′UTR) is not localized and is translated constitutively. In contrast, selection of a distal polyA site includes 3′UTR sequences that confer localization of BDNF mRNA to dendrites. Additionally, the long 3′UTR isoform of BDNF is not translated normally, but is induced upon signaling. Thus, cell-specific differences in the relative amounts of long and short 3′UTR isoforms of BDNF may confer different levels of BDNF responsiveness in neurons.

A second class of alternative polyadenylation events includes changes in polyA site selection that are associated with alternative splicing events and therefore have the potential to change both coding potential and 3′UTR sequence. A classic example of this type of regulation is the alternative processing of IgM pre-mRNA that is associated with the switch from a membrane-bound form of IgM (which acts as a receptor on B-cells) to a secreted form of IgM that can perform the various functions of soluble immunoglobulins [28]. In resting B-cells, a distal polyA site is selected that gives rise to an mRNA bearing sequences that encode IgM with a C-terminal membrane association domain. In active B-cells, a proximal polyA site is chosen that causes loss of the membrane association motifs and thus generates a secreted form of IgM.

Alternative processing of pre-mRNA encoding the transcription factor CREM-tau represents an example of the ability of alternative polyadenylation to modulate gene expression at multiple levels. CREM is a transcription factor critical for proper male germ cell development (spermatogenesis). CREM pre-mRNA is subject to complex RNA processing, producing various isoforms of the transcription factor with various activities ranging from repressor to activator [29, 30]. In early stages of spermatogenesis, the CREM gene yields a transcription factor that functions as a repressor. However, at a specific point in spermatogenesis, alternative promoter usage and alternative splicing generate CREM mRNA isoforms that encode a protein with transcriptional activation activity. This switch is coupled with selection of a polyA site that generates a short 3′UTR that lacks instability elements, and thus becomes stabilized, and thus leads to a large accumulation of CREM-tau activator. This switch is critical for activator activity of CREM-tau and proper progression through male germ cell development.

6 Global Programs of 3′ End Regulation

It is becoming increasingly apparent that many developmental processes exhibit distinct signatures of alternatively polyadenylated mRNAs and that global reprogramming of 3′UTRs has important functions in cell proliferation, differentiation, and tissue development (for reviews, *see* refs. 1, 7). Early embryogenesis highlights the intersection of regulated 3′ end formation and polyA tail length control [31–33]. Very early after fertilization of oocytes, transcription is silent, and maternally derived mRNAs with long 3′UTRs containing CPEs are positively regulated by CPEBs to drive early embryogenesis. Upon transcriptional activation of the zygotic genome, factors are expressed that bind to maternal mRNAs 3′UTRs and mediate their degradation. In addition, the newly expressed genes generally have shorter 3′UTRs and thus lack these negative regulatory sites. In general, these genes are involved in the rapid cell divisions in the early embryo.

A second example which illustrates how global reprogramming of 3′UTRs could be driving cellular processes is provided by the study of T-cell activation [34]. In quiescent T-cells, genes required for activation express mRNAs with long 3′UTRs, many of which have sequences recognized by miRNAs; thus these mRNAs are poorly translated and readily degraded. While the protein products are being made at low levels, the mRNAs encoding activation factors are still being produced, resulting in a T-cell "primed" for rapid activation. Upon stimulation of T-cells, alternative polyA site selection results in a general shortening of 3′UTRs in important regulatory genes. This 3′UTR shortening removes miRNA binding sites that negatively regulate translation and mRNA abundance, thus allowing higher levels of expression of the encoded proteins. Interestingly, global analyses have revealed tissue-specific biases in 3′UTR length, with the most notable examples being brain and testis that generally express mRNAs with long and short 3′UTRs, respectively [35–38]. The functional significance of these tissue-specific differences remains unclear.

While it is clear that alternative polyadenylation and cytoplasmic control of polyA tail length are widely used mechanisms of posttranscriptional gene regulation [3, 26], our understanding of how these processes are controlled remains limited. In some cases, tissue-specific auxiliary factors are believed to control the access of the polyA machinery to alternative polyA sites. For example, the neuron-specific RNA-binding protein Nova2 can bind to sequences that overlap with the PAS or DSE of proximal polyA sites to repress 3′ end formation presumably by competition with CPSF and CstF, respectively [39]. In contrast, Nova2 binding to sequences adjacent to (but not overlapping with) core cleavage and polyadenylation sequences is associated with Nova2-dependent selection of alternative polyA sites, possibly through antagonizing negative auxiliary factors. Modulation of core components of the cleavage and polyadenylation machinery is also believed to be a mechanism of polyA site regulation. For example, in resting B-cells, lower levels of CstF are thought to be one factor favoring distal polyA site usage and thus expression of membrane-bound IgM (*see* example above). Activation of B-cells is accompanied by increased expression of CstF, which is thought to be one factor favoring proximal polyA site usage, and thus production of secreted IgM [28]. While more recent studies have shown that regulation of this processing switch involves additional *cis*-acting sequences and *trans*-acting factors [40], it is clear that regulation by CstF levels is part of the picture. The coupling of 3′ end processing and regulation to multiple steps in the mRNA life cycle and the multifunctionality of RNA-binding proteins suggests that 3′ ends are most likely subject to combinatorial control by a number of factors that act at multiple steps in the biogenesis and metabolism of mRNA. Interestingly, a recent study demonstrated a role for PABP in regulation of alternative polyadenylation [41], while a second report found evidence that CPEB can regulate alternative splicing and polyadenylation in addition to regulating polyA tail length [42]; thus these proteins can coordinate multiple layers of posttranscriptional gene regulation.

7 Conclusion

The examples described in this chapter were selected to illustrate the significance of the mRNA 3′ end (both the position of polyadenylation and length of polyA tail) in posttranscriptional control of gene expression. Moving forward, a better understanding of how networks of mRNAs are co-regulated at their 3′ends will provide new insights into mechanisms of gene regulation in different cells and stages of development. New global profiling methods (e.g., HITS-CLIP and bioinformatic analyses) combined with traditional in vitro biochemical assays provide a powerful approach to explore the roles of putative regulatory factors in polyA site regulation and polyA tail length control [43]. Considering the impact of the mRNA 3′ end on gene output, the use of gene expression profiling approaches that do not incorporate polyA site information is likely to yield an incomplete picture of the mRNA landscape of the cell and preclude the identification of 3′ end regulatory events that may impact gene expression pathways. Fortunately, a number of new deep sequencing methods have been developed in recent years that are capable of generating comprehensive measurements of alternative 3′UTR expression in a global manner [1]. These tools have shown that alternative 3′ end processing is a widespread occurrence with central roles in many important processes, such as cellular proliferation, differentiation, T-cell activation, and development of many tissues [26]. A major challenge for the future will be to move beyond global descriptions of regulated mRNA 3′ ends and towards a better understanding of the functional significance of 3′ end regulation in different cellular contexts. Combining global profiling studies of polyA sites with analyses of mRNA polyA tail length, translation, stability, and localization will help better understand the impact of 3′ end regulation on the expression of networks of genes important for different biological processes.

References

1. Shi Y (2012) Alternative polyadenylation: new insights from global analyses. RNA 18:2105–2117
2. Perales R, Bentley D (2009) "Cotranscriptionality": the transcription elongation complex as a nexus for nuclear transactions. Mol Cell 36:178–191
3. Weill L, Belloc E, Bava FA et al (2012) Translational control by changes in poly(A) tail length: recycling mRNAs. Nat Struct Mol Biol 19:577–585
4. Nilsen TW, Graveley BR (2010) Expansion of the eukaryotic proteome by alternative splicing. Nature 463:457–463
5. Villalba A, Coll O, Gebauer F (2011) Cytoplasmic polyadenylation and translational control. Curr Opin Genet Dev 21:452–457
6. Martin KC, Ephrussi A (2009) mRNA localization: gene expression in the spatial dimension. Cell 136:719–730
7. Tian B, Manley JL (2013) Alternative cleavage and polyadenylation: the long and short of it. Trends Biochem Sci 38:312–320
8. Tian B, Graber JH (2012) Signals for pre-mRNA cleavage and polyadenylation. Wiley Interdiscip Rev RNA 3:385–396
9. Mandel CR, Bai Y, Tong L (2008) Protein factors in pre-mRNA 3′-end processing. Cell Mol Life Sci 65:1099–1122
10. Zhao J, Hyman L, Moore C (1999) Formation of mRNA 3′ ends in eukaryotes: mechanism, regulation, and interrelationships with other steps in mRNA synthesis. Microbiol Mol Biol Rev 63:405–445
11. Kuhn U, Gundel M, Knoth A et al (2009) Poly(A) tail length is controlled by the nuclear poly(A)-binding protein regulating the interaction between poly(A) polymerase and the cleavage and polyadenylation specificity factor. J Biol Chem 284:22803–22814
12. Hsin JP, Manley JL (2012) The RNA polymerase II CTD coordinates transcription and RNA processing. Genes Dev 26:2119–2137
13. Dantonel JC, Murthy KG, Manley JL et al (1997) Transcription factor TFIID recruits factor CPSF for formation of 3′ end of mRNA. Nature 389:399–402
14. McCracken S, Fong N, Yankulov K et al (1997) The C-terminal domain of RNA polymerase II couples mRNA processing to transcription. Nature 385:357–361
15. Hirose Y, Manley JL (1998) RNA polymerase II is an essential mRNA polyadenylation factor. Nature 395:93–96
16. Logan J, Falck-Pedersen E, Darnell JEJ et al (1987) A poly(A) addition site and a downstream termination region are required for efficient cessation of transcription by RNA polymerase II in the mouse beta maj-globin gene. Proc Natl Acad Sci U S A 84: 8306–8310
17. Kaida D, Berg MG, Younis I et al (2010) U1 snRNP protects pre-mRNAs from premature cleavage and polyadenylation. Nature 468: 664–668
18. Berg MG, Singh LN, Younis I et al (2012) U1 snRNP determines mRNA length and regulates isoform expression. Cell 150:53–64
19. Schmid M, Jensen TH (2010) Nuclear quality control of RNA polymerase II transcripts. Wiley Interdiscip Rev RNA 1:474–485
20. Shi Y, Di Giammartino DC, Taylor D et al (2009) Molecular architecture of the human pre-mRNA 3′ processing complex. Mol Cell 33:365–376
21. Eckmann CR, Rammelt C, Wahle E (2011) Control of poly(A) tail length. Wiley Interdiscip Rev RNA 2:348–361
22. Derry MC, Yanagiya A, Martineau Y et al (2006) Regulation of poly(A)-binding protein through PABP-interacting proteins. Cold Spring Harb Symp Quant Biol 71:537–543
23. Miller MA, Olivas WM (2011) Roles of Puf proteins in mRNA degradation and translation. Wiley Interdiscip Rev RNA 2:471–492
24. Barreau C, Paillard L, Osborne HB (2005) AU-rich elements and associated factors: are there unifying principles? Nucleic Acids Res 33:7138–7150
25. Kim JH, Richter JD (2006) Opposing polymerase-deadenylase activities regulate cytoplasmic polyadenylation. Mol Cell 24: 173–183
26. Di Giammartino DC, Nishida K, Manley JL (2011) Mechanisms and consequences of alternative polyadenylation. Mol Cell 43:853–866
27. Lau AG, Irier HA, Gu J et al (2010) Distinct 3′UTRs differentially regulate activity-dependent translation of brain-derived neurotrophic factor (BDNF). Proc Natl Acad Sci U S A 107:15945–15950
28. Takagaki Y, Seipelt RL, Peterson ML et al (1996) The polyadenylation factor CstF-64 regulates alternative processing of IgM heavy chain pre-mRNA during B cell differentiation. Cell 87:941–952
29. Foulkes NS, Mellstrom B, Benusiglio E et al (1992) Developmental switch of CREM function during spermatogenesis: from antagonist to activator. Nature 355:80–84
30. Foulkes NS, Schlotter F, Pevet P et al (1993) Pituitary hormone FSH directs the CREM functional switch during spermatogenesis. Nature 362:264–267
31. Wang QT, Piotrowska K, Ciemerych MA et al (2004) A genome-wide study of gene activity reveals developmental signaling pathways in

the preimplantation mouse embryo. Dev Cell 6:133–144
32. Belloc E, Pique M, Mendez R (2008) Sequential waves of polyadenylation and deadenylation define a translation circuit that drives meiotic progression. Biochem Soc Trans 36:665–670
33. Pique M, Lopez JM, Foissac S et al (2008) A combinatorial code for CPE-mediated translational control. Cell 132:434–448
34. Sandberg R, Neilson JR, Sarma A et al (2008) Proliferating cells express mRNAs with shortened 3′ untranslated regions and fewer microRNA target sites. Science 320:1643–1647
35. Smibert P, Miura P, Westholm JO et al (2012) Global patterns of tissue-specific alternative polyadenylation in Drosophila. Cell Rep 1:277–289
36. Miura P, Shenker S, Andreu-Agullo C et al (2013) Widespread and extensive lengthening of 3′ UTRs in the mammalian brain. Genome Res 23:812–825
37. Macdonald CC, McMahon KW (2010) Tissue-specific mechanisms of alternative polyadenylation: testis, brain, and beyond. Wiley Interdiscip Rev RNA 1:494–501
38. Zhang H, Lee JY, Tian B (2005) Biased alternative polyadenylation in human tissues. Genome Biol 6:R100
39. Licatalosi DD, Mele A, Fak JJ et al (2008) HITS-CLIP yields genome-wide insights into brain alternative RNA processing. Nature 456:464–469
40. Peterson ML (2011) Immunoglobulin heavy chain gene regulation through polyadenylation and splicing competition. Wiley Interdiscip Rev RNA 2:92–105
41. Jenal M, Elkon R, Loayza-Puch F et al (2012) The poly(A)-binding protein nuclear 1 suppresses alternative cleavage and polyadenylation sites. Cell 149:538–553
42. Bava FA, Eliscovich C, Ferreira PG et al (2013) CPEB1 coordinates alternative 3′-UTR formation with translational regulation. Nature 495:121–125
43. Licatalosi DD, Darnell RB (2010) RNA processing and its regulation: global insights into biological networks. Nat Rev Genet 11:75–87

Chapter 2

Retracted: Poly(A) Polymerase-Based Poly(A) Length Assay

Deepak P. Patil, Baskar Bakthavachalu, and Daniel R. Schoenberg

Abstract

mRNA polyadenylation functions in nuclear export, translation, and stability. We describe an efficient protocol designed to assess poly(A) tail length that is based on 3′ tailing by yeast poly(A) polymerase and product analysis to single-nucleotide resolution by capillary electrophoresis.

Key words Poly(A) tail length assay, Adenylation, Yeast poly(A) polymerase, Capillary electrophoresis

1 Introduction

Assays designed to determine the length of mRNA poly(A) tail have been helpful in understanding several processes in RNA biology, including mRNA transport, translation, stability, and turnover. Recent contributions include unraveling the mechanism of miRNA-mediated control of gene expression [1], discovery of *cis-acting* elements that support translation despite the presence of a short poly(A) tail [2], and identification of stable transcripts with short poly(A) tails [3].

Until recently most assays for quantifying poly(A) tail length used Northern blotting after treating RNA with RNase H + oligo(dT) to remove poly(A) or a PCR approach in which the length of the poly(A) tail was determined by PCR after hybridization with modified forms of oligo(dT) [4]. These gel- and PCR-based assays were time consuming, depended on the use of radioactively labeled probes and required significant optimization. Here we describe a facile method for quantifying poly(A) length that is based on using poly(A) polymerase (PAP) to tag the 3′ end of the poly(A) tail. When integrated with capillary electrophoresis this assay yields information on poly(A) tail length with single-nucleotide resolution. This approach is outlined in Fig. 1.

Joanna Rorbach and Agnieszka J. Bobrowicz (eds.), *Polyadenylation: Methods and Protocols*, Methods in Molecular Biology, vol. 1125, DOI 10.1007/978-1-62703-971-0_2, © Springer Science+Business Media New York 2014

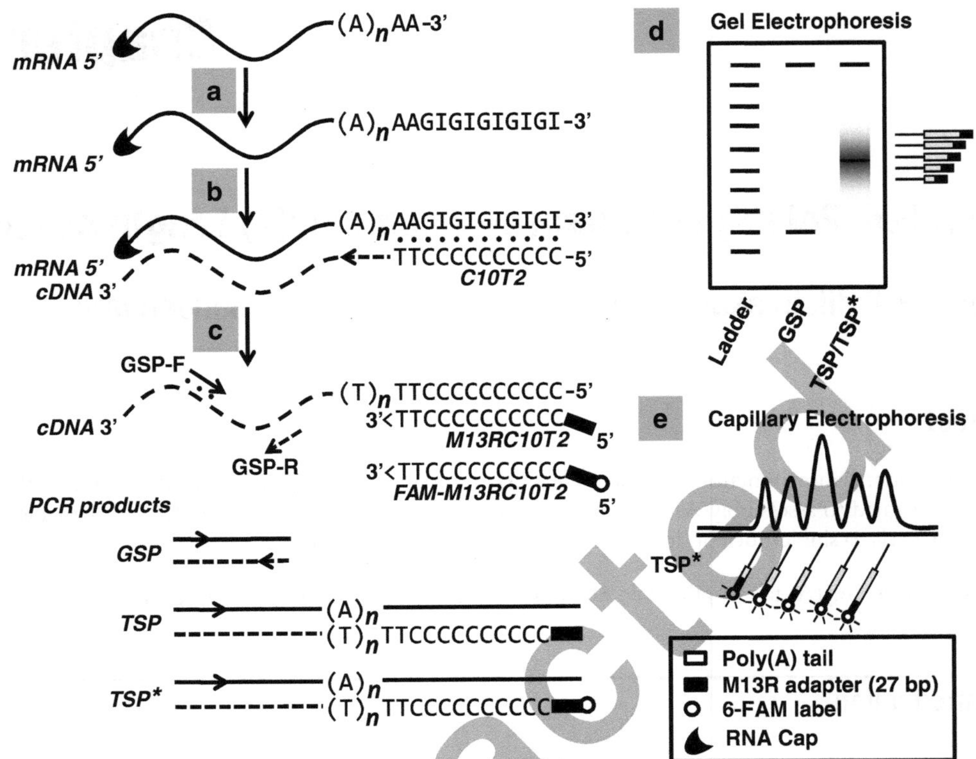

Fig. 1 Overview of the PAP-based poly(A) length assay. (**a**) Polyadenylated mRNA is tagged at the 3′ end with GMP and IMP residues by PAP (n = number of adenosines). (**b**) The newly synthesized 3′ GI-tag on mRNA is used for cDNA synthesis using a complementary primer (C10T2). (**c**) PCR amplification of cDNA using gene-specific primers (GSP-F and -R) yields information on the expression of mRNA (GSP product). Tail-specific PCR, using the GSP-F and M13C10T2 primers, yields information on the length of A-tail in the form of tail-specific product (TSP). Alternatively, the use of a fluorescently labeled FAM-M13C10T2 primer generates a labeled PCR product (TSP*) that is later analyzed by capillary electrophoresis. (**d**) Gel electrophoresis resolves GSP as a discrete band and TSP/TSP* as a smear. Smear distribution is proportional to the length of A-tail. (**e**) Analysis of TSP* by capillary electrophoresis coupled with fluorescence detection yields an exact measure of the A-tail in the form of an electropherogram at a single-nucleotide resolution

In the first step yeast PAP is used to tag the 3′ end of the polyadenylated tail of the mRNA with a random polymer of 12–14 guanosine and inosine residues [5]. The limited processivity of yeast PAP with these nucleotides restricts the length of polymer added, and random insertion of inosine bases prevents the formation of self-paired G-tetrad structures that may inhibit the downstream steps [6]. The newly synthesized GI-tag positions a complementary primer (C10T2) that is used for cDNA synthesis to the 3′-end of the poly(A) tail. PCR amplification of cDNA using a gene-specific forward primer and GI-tag-specific C10T2 primer yields a product that on gel electrophoresis appears as a smear for long poly(A) [7] and a tight band for short poly(A) tail. Higher resolution is obtained by using fluorescently labeled primers and separating PCR products by capillary electrophoresis [1]. Fluorescent labeling of gene-specific forward primer yields comparatively cleaner results than the labeled C10T2 primer; however, the use of fluorescently labeled C10T2 primer facilitates the analysis of multiple transcripts.

2 Materials

Prepare all the solutions in RNase-free water (*see* **Note 1**). Carry out all the steps in an RNase-free workspace. Follow waste disposal as per the regulations. Use sterile deionized water (resistance of 18 MΩ-cm at 25 °C) for PCR and DNA electrophoresis.

2.1 GI Tailing

1. RNase-free water (*see* **Note 1**).
2. PAP: Yeast Poly(A) polymerase (600 U/μL) (Affymetrix, *see* **Note 2**). Store at −20 °C.
3. 5× PAP buffer (Affymetrix, supplied with the enzyme, *see* **Note 3**): 100 mM Tris–HCl pH 7.0, 3.0 mM $MnCl_2$, 0.1 mM EDTA,1 mM DTT, 500 μg/mL acetylated BSA, 50 % glycerol. Store at −20 °C.
4. 10× Guanosine 5′-triphosphate and inosine 5′-triphosphate (GTP–ITP) mix (5 mM each): Mix, and dilute the commercially available 100 mM GTP and ITP stocks to 5 mM each in RNase-free water (*see* **Note 4**). Store at −20 °C in small aliquots, and avoid multiple freeze–thaws.
5. Ribonuclease Inhibitor (RNaseOUT, Life Technologies). Store at −20 °C (*see* **Note 5**).
6. 10 mM Ethylenediaminetetraacetic acid (EDTA): Dilute 1:50 times from a 0.5 M stock of EDTA.
7. RNA cleanup column (RNA Clean & Concentrator-5, Zymo Research, *see* **Note 6**).

2.2 cDNA Synthesis

1. C10T2 DNA primer (5′ CCC CCC CCC CTT 3′): 10 μM stock. Store at −20 °C.
2. Superscript II (200 U/μL, Life Technologies): Store at −20 °C (*see* **Note 7**).
3. 5× FSB (supplied with Superscript II, Life Technologies): 250 mM Tris–HCl, pH 8.3 at room temperature, 375 mM KCl, 15 mM $MgCl_2$. Store at −20 °C (*see* **Note 8**).

4. 0.1 M Dithiothreitol (DTT, supplied with Superscript II, Life Technologies): Store at −20 °C.
5. 10 mM dNTP stock: Mix of 10 mM each dNTP. Store at −20 °C.

2.3 PCR

1. Taq DNA polymerase (5 U/μL, NEB) (*see* **Note 9**).
2. 10× ThermoPol Reaction Buffer (supplied with Taq DNA polymerase, NEB): 200 mM Tris–HCl, 100 mM $(NH_4)_2SO_4$, 100 mM KCl, 20 mM $MgSO_4$, 1 % Triton X-100, pH 8.8 at 25 °C.
3. Oligonucleotide primers M13RC10T2 (5′ CA GGA AAC AGC TAT GAC CCC CCC CCC CTT 3′) or FAM-M13RC10T2 (5′ 6-FAM-CA GGA AAC AGC TAT GAC CCC CCC CCC CTT 3′), GSP-F and GSP-R: 10 μM stock each. GSP-F and -R are transcript-specific forward and reverse primers designed upstream to the predicted polyadenylation site in the gene (Fig. 1, *see* **Note 10**).
4. 10× dNTP stock (2 mM each): Dilute the commercially available 10 mM stocks of dNTP by mixing with water 4:1. Store at −20 °C in aliquots.
5. PCR cleanup column (DNA Clean & Concentrator™-5-Capped Columns, Zymo Research, *see* **Note 11**).

2.4 Electrophoresis

1. Agarose: Molecular biology-grade agarose.
2. 5× TBE: Dissolve 54 g of Tris base, 27.5 g of boric acid, and 20 mL of 0.5 M EDTA (pH 8.0) in 800 mL of water and make up the volume to 1 L. Autoclave for 15 min at 15 psi pressure on a liquid cycle, and store at room temperature (*see* **Note 12**).
3. 0.5× TBE: Dilute from the 5× TBE stock in water 1:10.
4. DNA loading dye: EZ-VISION® THREE 6× DNA loading dye (Amresco LLC, *see* **Note 13**).
5. Capillary electrophoresis: ABI 3730 DNA Analyzer (Applied Biosystems, *see* **Note 14**).

3 Methods

Handle the reagents and carry out all the steps in an RNase/nuclease-free environment. Perform all the incubations on a PCR thermal cycler. Thaw the reagents and RNA on ice. During additions, maintain all the reactions on ice. Before each incubation, mix the reaction by gentle pipetting followed by a quick centrifugation step (12,000 × *g* for 5 s) to pool the contents.

3.1 GI Tailing of RNA

1. In a sterile 0.2 mL PCR tube, set up the GI tailing reaction as follows (*see* **Note 15** for the recommended controls):

Components	Volume (in μL)
DNA-free total RNA (*see* **Note 16**)	*x* (up to 5 μg)
RNase-free water	14.5 − *x*
Heat at 65 °C for 5 min followed by rapid cooling on ice	
5× PAP buffer	5.0
10× GTP–ITP mix	2.5
RNase inhibitor (RNaseOUT, 40 U/μL)	1.0
PAP (600 U/μL)	2.0

2. Mix and incubate at 37 °C for 60 min.
3. Add 25 μL of RNase-free water, and purify the RNA using RNA cleanup column (for example RNA Clean & Concentrator-5, Zymo Research, *see* **Note 6**). At the end, elute the GI-tailed RNA with 10 μL of RNase-free water.

3.2 cDNA Synthesis

1. In a sterile 0.2 mL PCR tube, mix 9 μL of the GI-tailed RNA with 1 μL of C10T2 primer (10 μM), and heat the mix at 65 °C for 5 min followed by gradual cooling to 42 °C and a continuous incubation for 5 min.
2. Prepare the following reverse transcription (RT) mix in 0.2 mL PCR tube (*see* **Note 17** for the recommended controls):

Components	Volume (in μL)
5× FSB buffer	4.0
0.1 mM DTT	2.0
10 mM dNTPs	1.0
RNase inhibitor (RNaseOUT, 40 U/μL)	1.0
Superscript II (200 U/μL)	1.0
RNase-free water	1.0

3. Transfer the "RT" mix to the denatured RNA–primer mix at 42 °C.
4. Mix, and incubate at 42 °C for an additional 50 min followed by heating at 94 °C for 5 min.
5. After cooling to room temperature, dilute the reaction by adding 20 μL of PCR-grade water (*see* **Note 18**).

3.3 PCR

1. As outlined in Fig. 1 and Subheading 2.3, design and synthesize the primers using the RNA/gene sequence of interest (*see* **Note 10**).
2. In a sterile 0.2 mL PCR tube, set up PCR as follows (TSR = tail-specific reaction, and GSR = gene-specific reaction; a separate TSR is used for the FAM-labeled primer, *see* **Note 19**):

	Volumes in µL	
Components	**TSR**	**GSR**
10× Taq polymerase buffer	5.0	5.0
10× dNTPs	5.0	5.0
10 µM GSP-F primer	1.0	1.0
10 µM GSP-R primer	–	1.0
10 µM M13RC10T2/FAM-M13RC10T2 primer	1.0	–
Diluted cDNA	2.0	2.0
Taq DNA polymerase (5 U/µL)	0. 5	0.5
PCR-grade water	to 50	to 50

3. Mix, and place the tube in a thermocycler. Amplify the cDNA using the cycling conditions given in Table 1 (*see* **Note 20**).

Table 1
PCR cycling conditions

Step no.	Time/temperature	Remark(s)
1	3 min at 95 °C	Hold
2	10 s at 95 °C	30 cycles for GSR and 15–20 for FAM-TSR[c]
3	30 s at 60°C[a]	
4	1 min at 72 °C[b]	
5	1 min at 72 °C	Hold
6	4 °C	Hold

[a]Annealing temperature may require optimization
[b]Set polymerization time to 1 min/kb
[c]Low cycles ensure linearity of the reaction in the FAM-labeled TSR. The number of cycles also depends upon the expression level of the transcript

3.4 Electrophoresis

Gel electrophoresis

1. Electrophorese appropriate size markers and 5 μL of each reaction on a 2 % agarose gel in 0.5× TBE buffer (*see* **Notes 13** and **21**).
2. Stain the gel, if required (*see* **Note 13**), and capture a gel image using a gel documentation system (refer to Fig. 2 for a representative image).

Capillary electrophoresis

1. To remove unused labeled primers and their dimers, purify 25 μL of the FAM-labeled TSR reaction using PCR cleanup column (for example, DNA Clean & Concentrator™-5-Capped Columns, Zymo Research) according to the manufacturer's instructions (*see* **Note 11**). Elute the product with 25 μL of water.
2. Mix 1, 1:2, 1:4, and 1:8 dilutions of the purified product with an appropriate size standard and analyze on ABI 3730 DNA Analyzer following the standard amplified fragment length polymorphism (AFLP) assay (*see* **Note 14**). A representative electropherogram is in Fig. 3.

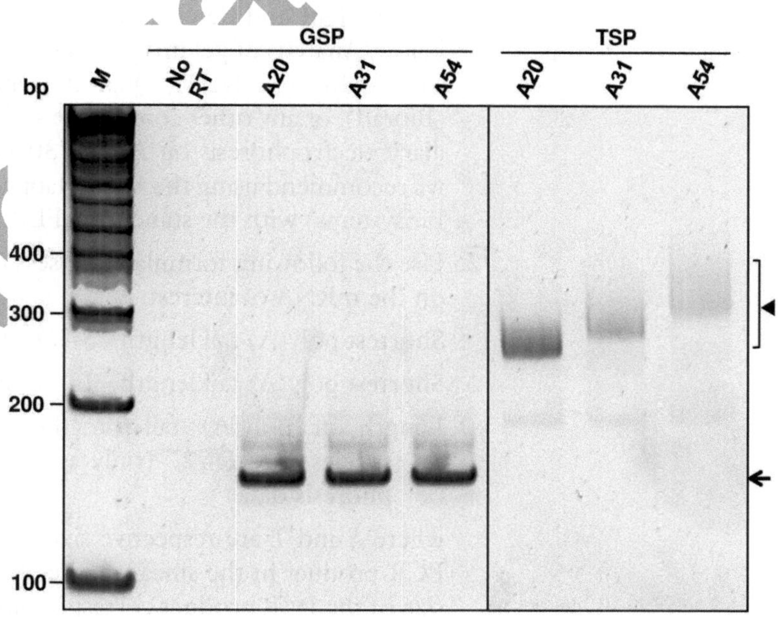

Fig. 2 Assessment of poly(A) tail length of in vitro-transcribed luciferase mRNAs with varying length poly(A) tails. Lanes A20, A31, and A58 are luciferase mRNAs with the stated number of adenosines at their 3′ terminus. Note that the TSP lanes show smears proportional to poly(A) tail length (*shaded wedge*=TSP products, *arrow*=GSP products, "*No RT*" is no reverse transcriptase reaction, *M*=DNA ladder, *bp*=base pairs)

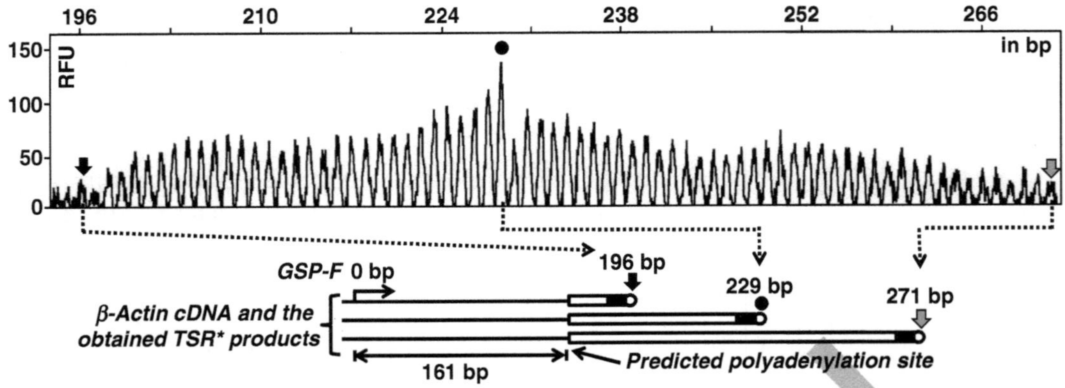

Fig. 3 Determination of β-actin mRNA poly(A) tail length by capillary electrophoresis. FAM-labeled TSP* products were separated by capillary electrophoresis, and the products are shown as relative fluorescence as a function of product size. Each peak represents a single-nucleotide difference in poly(A) tail length. The GSP forward primer (TTT GAA TGA TGA GCC TTC GTG CCC) anneals 161 nt upstream of the predicted poly(A) site. The smallest product (*dark arrow*) is 196 bp and the longest is 271 bp (*gray arrow*). Using the formula from Subheading 3.5, the length of the poly(A) tail on β-actin mRNA from HEK293 cells spans 8–83 nucleotides. For legend information refer to Fig. 1

3.5 Data Analysis

1. Analyze the gel image or data from capillary electrophoresis for the TSR and determine the variables indicated in **step 2** below. Analysis of gel image can be performed using software, such as ImageJ (NIH), Quantity One 1-D Analysis Software (Biorad), or any other compatible software. For data from capillary electrophoresis on ABI 3730 DNA analyzer (.fsa files), we recommend using the GeneMapper® v4.1 package (Applied Biosystems) with the standard AFLP analysis protocol.
2. Use the following formulae to assess the length of poly(A) tail on the mRNA of interest:

   Shortest poly(A) tail length = $S - D - 27$

   Shortest poly(A) tail length = $L - D - 27$

   Length of poly(A) tail on the most abundant mRNA species = $M - D - 27$ (only applicable to capillary electrophoresis data)

   where S and L are respective sizes of the smallest and largest PCR product in the smear or the electropherogram. M is the size of the PCR product corresponding to the highest peak on the electropherogram. D is the distance between the GSP forward primer and the end of the last exon or the predicted polyadenylation site in the gene sequence. It includes the length of the primer-binding site. In the formulae, number "27" is the length of the adapter in M13RC10T2 without the last two T's. All the variables are in base pairs or nucleotides.

4 Notes

1. To prepare RNase-free water, add 100 μL of diethylpyrocarbonate (DEPC) per liter of glass-distilled water. Mix until globules disappear, and let stand overnight. Autoclave for 30 min at 15 psi, and store in aliquots at –20 °C in RNase-free tubes. Note that DEPC reacts with primary amines such as Tris, and as such commercially available concentrated solutions of Tris should be added to the autoclaved DEPC-treated water.
2. Use of other polymerases such as *E. coli* PAP [8] is not recommended.
3. 5× PAP buffer can be made from stock reagents of higher concentrations in RNase-free water (*see* **Note 1**).
4. Commercially available sodium salts of GTP and ITP can also be used to make stock solutions in RNase-free water. Do not include the GTP/ITP mix with RNA in the denaturation step. GTP is unstable at high temperatures.
5. Other commercially available RNase inhibitors such as SuperaseIn and RNasin may also be used.
6. Glycogen/LiCl precipitation is a cheap alternative to column-based purification; however, these procedures are susceptible to a considerable RNA loss. Chilling the contents during the precipitation step at –20 °C for 1 h improves yield.
7. Use of alternative reverse transcriptase may require optimization (*see* **Note 16**).
8. FSB buffer may also be made from stock reagents in RNase-free water (*see* **Note 1**).
9. A change in the source of thermophilic DNA polymerase may require optimization of PCR conditions.
10. Considering the PCR mix composition, use standard parameters for designing the primers (length = 20–40 nucleotides, %GC = 40–60, and melting temperature = 65–75 °C). Avoid primers showing a tendency to form secondary structures and dimers. We also recommend a verification step for sequence specificity. This may be done by a transcriptome-level comparison of primer sequences using the BLAST tool at the GenBank database. In addition, the distance between the binding site of the forward primer and the polyadenylation site is of considerable importance. This influences the choice of gel separation and the method of analysis. For TSP products (tail-specific products), we recommend an amplicon size below 1,000 bp including the length of the A-tail. For fluorescent labeling of TSP products, 6-FAM labeling of GSP-F primer instead of M13RC10T2 yields much cleaner data; however, it may not be

cost effective when analyzing multiple transcripts. In addition, apart from 6-FAM, the users may also use compatible fluorophores such as ROX™, TAMRA, and PET®. While designing the labeled primers, avoid labeling of bases that show proximal-base quenching effect, e.g., 6-FAM to a G base.

11. This is an optional step. A PCR mix can be directly analyzed by gel or capillary electrophoresis. PCR cleanup may be necessary to avoid interference due to salts and residual primers in capillary electrophoresis. We highly recommend the cleanup columns from Zymo Research Corporation because their design efficiently reduces carryover residual ethanol in the eluates. Contaminating ethanol, being lighter, causes loss of PCR product during loading of wells in the gel electrophoresis step.
12. 5× TBE stock is more stable at room temperature than 10× stock and does not have the risk of precipitation from being too concentrated.
13. DNA loading dye used here pre-stains the PCR products. Users may also use other available staining methods involving ethidium bromide and nontoxic stains, such as SYBR GOLD. The fluorescently labeled TSP product may not be visible on gel due to a low-cycle PCR.
14. For capillary electrophoresis, we recommend the use of ABI 3730 DNA analyzer (Applied Biosystems). The manufacturer recommends using GeneScan Size Standards (Applied Biosystems), which are available in various size ranges. Use size standards that are appropriate for the expected size of the FAM-labeled TSP products.
15. The use of "no-RNA" and "no-PAP" controls is a good practice for GI tailing reaction.
16. The amount of RNA input can be empirically determined by the abundance of transcript of interest. DNase I treatment of input RNA is also highly recommended.
17. To rule out amplification from DNA contamination, we recommend a "no-RT" control for all the gene-specific reactions.
18. Quantitative PCR for housekeeping transcripts, such as actin and GAPDH, may be used to assess the efficiency of cDNA synthesis. However, amplification in "no-PAP control" may be seen due to nonspecific priming. We have seen a 21-fold amplification of β-actin transcripts in "PAP-treated" over "no-PAP" control reactions ($\Delta Cq > 4$). To avoid nonspecific priming, cDNA synthesis may be performed at higher temperature with a thermally stable reverse transcriptase.
19. "No-template" control is highly recommended.

20. In certain cases, an optimization of PCR conditions may be essential, especially, the annealing temperature.
21. Percentage and choice of gel (agarose vs. polyacrylamide) depend upon the size of the expected products. Note that the Bioanalyzer system (Agilent Inc.) can also be used as an alternative to gel or capillary electrophoresis.

Acknowledgements

We thank Ms. Trinh Tat (Nilsen lab at the Center for RNA Molecular Biology, Case Western Reserve University, Cleveland, OH) for exchange of ideas. We are also grateful to the Plant-Microbe Genomics Facility (the Ohio State University) for technical assistance with capillary electrophoresis. Additionally, we would like to acknowledge the members of the Schoenberg lab for their helpful comments. This work was supported by National Institute of General Medical Science of the National Institutes of Health under award numbers GM084177 and GM038277.

References

1. Bazzini AA, Lee MT, Giraldez AJ (2012) Ribosome profiling shows that miR-430 reduces translation before causing mRNA decay in zebrafish. Science 336:233–237
2. Das Gupta J, Gu H, Chernokalskaya E et al (1998) Identification of two cis-acting elements that independently regulate the length of poly(A) on Xenopus albumin pre-mRNA. RNA 4:766–776
3. Yang L, Duff MO, Graveley BR et al (2011) Genomewide characterization of non-polyadenylated RNAs. Genome Biol 12:R16
4. Murray EL, Schoenberg DR (2008) Assays for determining poly(A) tail length and the polarity of mRNA decay in mammalian cells. Methods Enzymol 448:483–504
5. Martin G, Keller W (1998) Tailing and 3′-end labeling of RNA with yeast poly(A) polymerase and various nucleotides. RNA 4:226–230
6. Burge S, Parkinson GN, Hazel P et al (2006) Quadruplex DNA: sequence, topology and structure. Nucleic Acids Res 34:5402–5415
7. Kusov YY, Shatirishvili G, Dzagurov G et al (2001) A new G-tailing method for the determination of the poly(A) tail length applied to hepatitis A virus RNA. Nucleic Acids Res 29: E57-7
8. Yehudai-Resheff S, Schuster G (2000) Characterization of the E. coli poly(A) polymerase: nucleotide specificity, RNA-binding affinities and RNA structure dependence. Nucleic Acids Res 28:1139–1144

Chapter 3

Using Klenow-Mediated Extension to Measure Poly(A)-Tail Length and Position in the Transcriptome

Man Chun Lee, Amrei Jänicke, and Traude Helene Beilharz

Abstract

The poly(A)-tail that terminates most mRNA and many noncoding RNA is a convenient "hook" to isolate mRNA. However the length of this tail and its position within the primary RNA transcript can also hold diagnostic value for RNA metabolism. In general, mRNA with a long poly(A)-tail is well translated, whereas a short poly(A)-tail can indicate translational silencing. A short poly(A)-tail is also appended to RNA-decay intermediates via the TRAMP complex. A number of approaches have been developed to measure the length and position of the poly(A)-tail. Here, we describe a simple method to "tag" adenylated RNA using the native function of DNA polymerase I to extend an RNA primer on a DNA template in second-strand DNA synthesis. This function can be harnessed as a means to purify, visualize, and quantitate poly(A)-dynamics of individual RNA and the transcriptome en masse.

Key words ePAT, End labeling, Polyadenylation, Poly(A)-tail length, Klenow polymerase, Translational control

1 Introduction

The addition of a poly(A)-tail is the final modification to nascent mRNA prior to its export from the nucleus and the start of its posttranscriptional fate. The global adenylation state of the transcriptome is often used as a diagnostic in RNA metabolism. For example, hyperadenylation is a common phenotype of mRNA export mutants [1, 2] and nuclear mRNA retention in response to stress conditions [3]. Within the cytoplasm, poly(A)-dynamics are important for the control of translation. There is a global correlation between poly(A)-tail length and translation state [4, 5]; and poly(A)-shortening is a central feature of most models of translational silencing (reviewed: ref. 6). Moreover, poly(A)-extension by cytoplasmic poly(A) polymerase to activate translation is a field now gaining importance with the discovery that this pathway is deregulated in cancers to activate oncogenes ([7] and reviewed: ref. 8).

Joanna Rorbach and Agnieszka J. Bobrowicz (eds.), *Polyadenylation: Methods and Protocols*, Methods in Molecular Biology, vol. 1125, DOI 10.1007/978-1-62703-971-0_3, © Springer Science+Business Media New York 2014

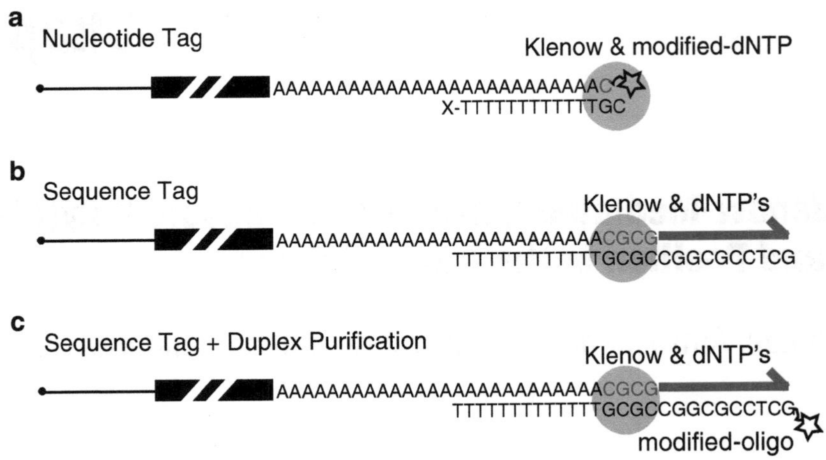

Fig. 1 Using the Klenow fragment of DNA polymerase I to tag adenylated RNA. (**a**) A simple nucleotide tag such as an α-[^{32}P]-dCTP, biotin, or otherwise derivatized nucleotide can be directly incorporated into the 3′ end of the RNA molecule using Klenow polymerase and a template oligonucleotide (x—indicates a 3′ amino group on the oligonucleotide). (**b**) A complex sequence tag is generated when dNTPs and Klenow are added to RNA with pre-annealed template oligo(dT)-template primer. (**c**) Inclusion of a 5′ modification such as biotin in the template oligo can provide a convenient way to isolate the end-extended duplex

This, in addition to cytoplasmic adenylation's well-established role in germ line and neuronal function (recent reviews: refs. 6, 9, 10), means that simple tools to measure poly(A)-tail length are more and more needed. A number of assays have been devised to measure the adenylation state of the transcriptome en masse and the poly(A)-tail associated with individual genes [11–15]. Here we show how the native function of DNA pol I can be recruited to 3′-tag adenylated RNA.

In the process of DNA replication, the lagging strand is synthesized piece by piece by RNA priming and DNA extension [16]. The RNA primer-extension step, that is, extending the 3′ end of RNA with dNTPs on a DNA template to generate Okazaki fragments, was previously exploited by Huang et al. (1996) to 3′-label specific RNA molecules with radionucleotides using the Klenow fragment of DNA polymerase I. In what we called the ePAT assay (for *e*xtension *P*oly(*A*) *T*est), we extend the Klenow extension approach as a tool to measure the length distribution of poly(A)-tails on specific mRNA ([17] and Fig. 1). In this case the 3′ recessed end of the RNA:DNA duplex is filled by Klenow to extend the RNA molecule with regular or derivatized dNTPs. A simple two-base overhang can be used to incorporate a specific radionucleotide (Figs. 1a and 2) or derivatized nucleotide such as Biotin-dCTP. Alternatively, a longer overhang and all four dNTPs can generate a specific sequence tag at the end of the molecule (Fig. 1b). Finally, the inclusion of biotin either into the RNA directly as illustrated in

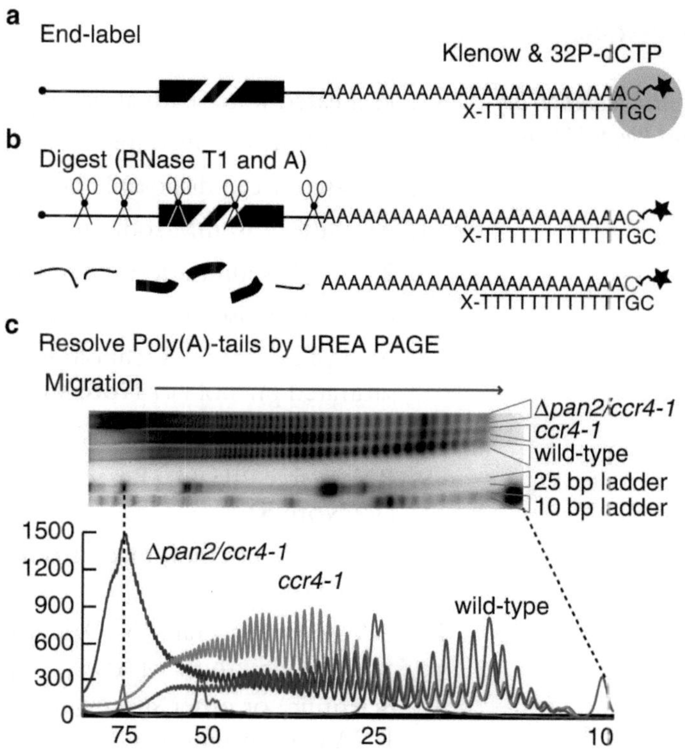

Fig. 2 Klenow-mediated tagging of adenylated RNA for global poly(A)-length measure. (**a**) An oligo-dT primer with a two-base extension is annealed to adenylated RNA. Addition of a modified or derivatized nucleotide (α[^{32}P]-dCTP in this case) results in the covalent incorporation into the 3′ end of the adenylated RNA molecule. (**b**) Nucleolytic digest of the body of the RNA with RNases T1 and A (cleavage of G, U, and C, respectively) leaves only poly(A)-tracts intact. (**c**) The poly(A)-tracts can be visualized and sized with nucleotide resolution using ultrathin 14 % urea-PAGE. Inclusion of 5′ end-labeled DNA ladder (25 bp and 10 bp) allows sizing of the poly(A)-tail lengths in specific samples. Total RNA samples analyzed from wild-type yeast cells or strains mutant in the major cellular deadenylase Ccr4, or the double mutant (Δ*pan2*/*ccr4-1*), have characteristic poly(A)-length distributions [4, 28]. Wild-type cells show an enrichment of short-tailed mRNA, *Ccr4-1* cells accumulate mRNA having tails between 25 and 50 adenosine residues, and the double-mutant cells (*ccr4-1*/Δ*pan2*) have predominately long poly(A)-tails. The radioactive signal was quantitated by phosphorimaging and multi-gauge software analysis

Fig. 1a or by the use of a biotinylated template primer (Fig. 1c) can facilitate the purification of the labeled RNA. The methods in this chapter provide simple approaches toward the purification of total RNA having sufficiently high quality and purity for downstream enzymatic processing for the measure of global adenylation state and the specific poly(A)-tail length and position of genes of interest, using the Klenow-mediated extension of the poly(A)-tail.

2 Materials

2.1 RNA Isolation by the Hot Phenol Method

1. Preheated water bath at 65 °C (*see* **Note 1**).
2. Ice-bath with chilled metal block (*see* **Note 2**).
3. 2 ml screw cap disposable plastic tubes (*see* **Note 3**).
4. AE buffer (50 mM sodium acetate pH 5.2, 10 mM EDTA).
5. 25 % w/v sodium dodecyl sulfate (SDS).
6. Chloroform isoamyl alcohol mixture.
7. Acid phenol (Life Technologies; 15594-047) or other water-saturated phenol (*see* **Note 4**).
8. Isopropanol/2-propanol.
9. RNase-free dH_2O (Baxter water for irrigation; AHF7114) (*see* **Note 5**).
10. 80 % V/V ethanol, made from absolute ethanol and RNase-free dH_2O.
11. 3 M sodium acetate (NaOAc, pH 5.2).
12. Nanodrop 2000 UV–Vis spectrophotometer (Thermo Scientific) or other spectrophotometer suitable for measuring RNA concentration and purity (*see* **Note 6**).

2.2 RNA Isolation Using TRIzol® or Similar Guanidinium Thiocyanate–Phenol-Based Preparation

1. Ice-bath with chilled metal block (*see* **Note 2**).
2. Mini-BeadBeater-8 (Biospec Products) or similar pulverizing device.
3. Rocking platform or rotary mixing device at room temperature.
4. TRIzol® (Life Technologies; 15596-026) or similar commercial or homemade guanidinium thiocyanate–phenol-based preparation.
5. Zirconia beads (Biospec Products; 11079107zx) or similar >1 mm glass beads.
6. Chloroform isoamyl alcohol mixture.
7. Isopropanol/2-propanol.
8. RNase-free dH_2O (Baxter water for irrigation; AHF7114) (*see* **Note 5**).
9. 3 M sodium acetate (NaOAc, pH 5.2).
10. Nanodrop 2000 UV–Vis spectrophotometer (Thermo Scientific) or other spectrophotometer suitable for measuring RNA concentration and purity (*see* **Note 6**).

2.3 Klenow-Mediated RNA 3′ End Labeling

1. ^{32}P work area with appropriate occupational health, safety, and environment precautions and dedicated equipment that includes heat block, cooled benchtop microfuge, high-voltage power pack, and vacuum dryer.

2. Vertical gel running apparatus at least 20 cm in length (preferably longer) with sequencing-width (≥0.4 mm) spacers and gel combs. The front plate should be siliconized.
3. Electrophoresis-grade urea.
4. 40 % acrylamide and bis-acrylamide solution, 37.5:1.
5. α-[^{32}P]-dCTP, 800 Ci/mmol 10 mCi/ml.
6. Klenow fragment (3′→5′ exo-) and NEB buffer 2 (New England Biolabs; M0212L).
7. 50:50 mix of phenol/chloroform prepared fresh from acid phenol and chloroform isoamyl alcohol mixture.
8. Desalting columns such as Micro Bio-Spin 6 Columns (Bio-Rad; 732-6221).
9. 1.5 ml Phase Lock Gel Heavy phase separation tubes (5 Prime; 2302810).
10. $dT_{14}GC$ 3′ blocked template oligonucleotide 5′ CGTTTTT TTTTTTTTT/3AmMO/at 100 μM dissolved in dH_2O.
11. RNase T1 and A purchased as a commercial premix such as Ambion® RNase Cocktail™ (Life Technologies; AM2286). Alternatively, use 1 μl of RNase T1 (100 U/μl; Roche; 10109495001) and 1 μl RNase A (4 μg/μl: Roche; 109169) per reaction.
12. GlycoBlue™ (Life Technologies; AM9516) or similar co-precipitant.
13. tRNA for ballast (10 mg/ml: Roche; 10109495001).
14. 2× RNA gel loading buffer. Made by adding 1/100th volume loading dyes to a solution of 9 parts deionized formamide and 1 part 10× TBE. The loading dyes are preprepared by dissolving 40 mg of each bromophenol blue and xylene cyanol FF in 1 ml of methanol.

2.4 ePAT and TVN-PAT Size Control

1. Thermo cycler with accessible lid. This should be preprogrammed with temperature hold and pause steps to encompass the full reaction cycle.
2. ePAT template oligonucleotide 5′ GCGAGCTCCGCGGCC GCGTTTTTTTTTTTT 3′ at 100 μM dissolved in dH_2O.
3. TVN-PAT 3′RACE oligonucleotide 5′ GCGAGCTCCGCGG CCGCGTTTTTTTTTTTTVN 3′ 100 μM dissolved in dH_2O.
4. SuperScript® III reverse transcriptase and associated buffers (Life Technologies; 18080-085).
5. RNaseOUT™ Recombinant Ribonuclease Inhibitor (Life Technologies; 10777-019).
6. 10 mM dNTP mix suitable for reverse transcription/PCR.
7. Klenow fragment (3′→5′ exo-) (New England Biolabs; M0212L).

2.5 Amplification of ePAT and TVN-PAT cDNA

1. Gene-specific PCR amplification primer. The primers shown in Fig. 2 are S.c TVP38-PAT 5′ GTGTCCACTGATCAA CTACCTG 3′, S.c SNR19-PAT 5′ GGTCGGGATTGTAG TTTGAAG 3′, and H.s GAPDH-PAT 5′ GGACCACCAGCC CCAGCAAG 3′ where S.c and H.s stand for *Saccharomyces cerevisiae* and *Homo sapiens*, respectively (*see* **Note 7**).
2. AmpliTaq Gold® 360 Master Mix (Life Technologies; 4398901) or equivalent hot-start enzyme preparation.
3. High-resolution agarose such as UltraPure™ Agarose 1000 (Life Technologies; 16550100).
4. 10× TBE (1.1 M Tris; 900 mM Borate; 25 mM EDTA; pH 8.3) or 20× SB (sodium borate buffer [18]) gel running buffers (*see* **Note 8**).
5. Horizontal slab-gel running apparatus with gel trays of at least 10 cm in length and narrow (0.75 mm) gel combs (*see* **Note 9**).
6. 100 bp DNA ladder (New England Biolabs; N3231S) diluted to 0.1 μg/μl with dH_2O and orange G gel loading buffer.
7. Orange G gel loading buffer is made by dissolving 0.1 % Orange G (Sigma; O3756) in a 40 % sucrose solution (*see* **Note 10**).
8. Nucleic acid stain such as ethidium bromide or SYBR® Safe DNA gel stain (Life Technologies; S33102) (*see* **Note 11**).
9. High-resolution gel documentation system such as LAS 3000 or FLA-5000 imagers and multi-gauge software (Fujifilm).

3 Methods

The following protocols are currently used in our laboratory. The protocols for RNA extraction are suitable from model organisms having tough external cuticles and cell walls such as fungi, nematode worms, and zebrafish embryos and result in high-quality and purity preparations suitable for downstream enzymatic modification. The protocols applying Klenow-mediated 3′ extension of adenylated RNA provide quick and simple new approaches to measure the poly(A)-tails of the transcriptome in bulk 3′ end-labeling reactions and for individual genes by ePAT.

3.1 The Isolation of Fungal RNA by the Hot Phenol Method

This method is suitable for single cell fungi such as *S. cerevisiae*, *Candida albicans*, and *Cryptococcus neoformans*. The high-temperature extraction removes the normal-phase separation between the phenol/chloroform and aqueous buffers causing maximum disruption of cellular proteins. The volumes here are suitable for 10–100 OD_{600} units of cells; if more RNA is required, increase the reaction volumes proportionately. Approximately

2.5 μg of total RNA can be expected per OD600 unit. It is often convenient to snap freeze washed, whole-cell pellets (in 2 ml screw cap tubes) in liquid nitrogen and store at −80 °C freezer.

1. Remove 2 ml screw cap tubes from −80 °C freezer to a metal block on ice. Immediately add 30 μl of 25 % SDS, 400 μl AE buffer, and 400 μl acid phenol. Shake or vortex vigorously.
2. Transfer tubes to a preheated 65 °C water bath and incubate for 20 min. Vortex every 2–5 min to ensure maximum cell disruption. Ensure that the phases merge.
3. Transfer tubes back to the metal block on ice and incubate for 5 min to precipitate SDS.
4. Centrifuge for 5 min at 16,000 × *g* in a prechilled 4 °C microfuge to pellet the cell debris and excess SDS.
5. Transfer the supernatant, both phases, to a new 1.5 ml tube (screw caps no longer essential).
6. Add 400 μl chloroform IAA and shake or vortex vigorously.
7. Centrifuge for 5 min at 16,000 × *g* in an ambient temperature microfuge.
8. Transfer the aqueous phase (avoiding the white interphase) to a new tube.
9. Add 400 μl Chloroform, shake or vortex vigorously, and centrifuge for 5 min at 16,000 × *g* in an ambient temperature microfuge.
10. Transfer the supernatant (~350 μl when being generous to avoid interphase) to a new tube containing 1/10th volume 3 M NaOAc. If working with a low number of input cells, inclusion of 2 μl of GlycoBlue™ can aid in recovery and visualization of the RNA pellet. Vortex briefly to ensure good mixing of the salts with the RNA.
11. Add an equal volume of isopropanol, vortex briefly, chill for 5 min on ice, and then centrifuge for 10 min at 16,000 × *g* in a precooled 4 °C microfuge.
12. Aspirate the supernatant and flash-spin. Remove remaining liquid with a fine-bore pipette avoiding the clear/white RNA pellet (*see* **Note 12**).
13. Add 1 ml ice-cold 80 % ethanol and centrifuge for 5 min at 16,000 × *g* in a pre-cooled 4 °C microfuge. Remove all liquid as in **step 11** and air-dry (*see* **Note 13**).
14. Suspend the RNA in RNase-free dH_2O. As a rule of thumb, add ~5 μl dH_2O per OD_{600} Unit of starting yeast cells.
15. Measure concentration and purity of RNA by Nanodrop. Adjust concentration as necessary.

3.2 The Isolation of RNA from Metazoan Model Organisms Using TRIzol® or Similar Guanidinium Thiocyanate–Phenol-Based Preparation

To prepare total RNA from organisms having tough outer cuticles such as the nematode worm *Caenorhabditis elegans* [19], it can be useful to include a mechanistic homogenization step to ensure complete extraction of the RNA. In the following protocol, we suggest modifications to the manufacture's specifications to ensure rapid and effective tissue disruption and a final total RNA preparation having high purity. In our hands the following protocol effectively extracts RNA from zebrafish embryos, nematode worms, and mammalian tissue samples. The proportions given are suitable for a packed tissue mass of 10–50 mg.

1. Transfer fresh or frozen tissue to a 2 ml screw cap tube on ice.
2. Add the equivalent of ~200 μl Zirconia beads and 1.5 ml of TRIzol®.
3. Homogenize for 30 s in a mini-8 BeadBeater or equivalent tissue disruptor.
4. Flash-spin to remove TRIzol® from around the lid and check clarity. Incubate for 15 min on a rocking platform/rotating wheel (*see* **Note 14**).
5. Add 300 μl of chloroform and shake or vortex vigorously.
6. Centrifuge for 15 min at 11,000 × *g* in a prechilled 4 °C microfuge.
7. Transfer the clear aqueous phase (~700 μl) to a fresh 1.5 ml tube containing 2 μl of GlycoBlue™ (*see* **Note 15**).
8. Add an equal volume (~700 μl) of isopropanol, vortex briefly to mix, and centrifuge for 15 min at 11,000 × *g* in a prechilled 4 °C microfuge.
9. Aspirate liquid, flash-spin, and carefully remove any remaining liquid.
10. Resuspend the pellet with 90 μl of dH_2O and 10 μl 3 M NaOAc. Ensure the pellet is fully suspended, then add 250 μl absolute ethanol, vortex briefly, and incubate on ice for 5 min.
11. Reprecipitate by centrifugation for 15 min at 11,000 × *g* in a prechilled 4 °C microfuge. Aspirate liquid, flash-spin, and carefully remove any remaining liquid.
12. Wash pellet with 1 ml 80 % ethanol and centrifugation for 5 min at 11,000 × *g* in a 4 °C microfuge. Aspirate liquid, flash-spin, and carefully remove any remaining liquid.
13. Air-dry the pellet, and suspend in 20–50 μl of dH_2O (depending on starting material).
14. Determine the concentration and purity by nanodrop (*see* **Note 16**).

3.3 Klenow-Mediated RNA Tagging for Bulk Poly(A)-Tail Analysis

This assay, illustrated in Fig. 2, can replace traditional assay for bulk poly(A)-tail length measure that involves T4 RNA ligase-mediated attachment of α-[^{32}P] pCp to the 3′ end of RNA molecules. The assay involves radioactive end labeling followed by nucleolytic digestion of all but poly(A)-tracts using a mix of RNase T1 and A which cleavage after G and C or U, respectively ([12, 20] and Fig. 2b). The Klenow-mediated approach significantly simplifies this traditional assay and avoids the need to synthesize α-[^{32}P]-pCp, commercial stocks of which are not always readily available. The method described here is analogous to our previously published method [12] except for the mechanism for incorporation of the [^{32}P]-CTP at the 3′ end of RNA. In our hands, the Klenow approach to end labeling is considerably more efficient than the ligation approach, probably because it is poly(A)-directed rather than labeling all 3′ termini. We have also successfully used this protocol to tag mRNA with the fluorescent nucleotide ChromaTide® OBEA-dCTP (Life Technologies; C21556) and Biotin-C14-dCTP (Life Technologies; 19518-018), albeit such nonradioactive approaches still suffer from a much lower signal/noise ratio than radionucleotides in this assay.

1. In a 1.5 ml tube, assemble 2 μg total RNA, 2 μl $dT_{14}GC$ template primer, and dH_2O to 16 μl.
2. Denature at 80 °C for 5 min and return to room temperature. Flash-spin to collect condensate.
3. Add 2 μl Klenow buffer (NEB 2), 1 μl α[^{32}P]-dCTP, and 1 μl Klenow polymerase. Incubate for 1 h at room temperature, or 30 min at 37 °C. Heat inactivate at 80 °C for 5 min and flash-spin.
4. Remove unincorporated α-[^{32}P]-dCTP using desalting columns according to the manufacturer's instructions.
5. Add 1 μl tRNA ballast, 10 μl digest buffer, and 70 μl dH_2O to the tube containing the end-labeled RNA.
6. Transfer the (~100 μl) RNA mix to a pre-spun phase-lock tube containing 2 μl of RNAse T1/A premix. Vortex briefly and incubate for 2 h at 37 °C.
7. During RNA digestion, prepare a 14 % urea-PAGE sequencing gel and pre-run for ~30 min at a current compatible with the gel size (30 mA or 45 °C for a 20 cm × 20 cm gel).
8. Inactivate the RNases by adding 100 μl phenol/chloroform mix to the phase-lock tube. Shake vigorously, and centrifuge for 5 min at 16,000 × *g* and ambient temperature.
9. Add 100 μl of Chloroform, shake vigorously, and repeat centrifugation.

10. Transfer the aqueous phase to a fresh tube, containing 2 μl GlycoBlue™. Add 250 μl absolute ethanol and incubate on ice for 5 min.
11. Precipitate ^{32}P-labeled poly(A)-tracts by centrifugation for 15 min at 16,000 ×*g* in a 4 °C microfuge. Aspirate liquid, flash-spin, and carefully remove any remaining liquid. Expect a large pale blue (salty) pellet.
12. Wash pellet by addition of 1 ml 80 % ethanol and centrifuge for 5 min at 16,000 ×*g* in a prechilled 4 °C microfuge. Aspirate liquid, flash-spin to carefully remove any remaining liquid, and air-dry the pellet.
13. Suspend the pellet in 4 μl dH_2O and 6 μl 2× RNA loading buffer. Heat to 80 °C for 5 min immediately prior to loading.
14. Load 4 μl of each sample alongside of a size ladder (*see* **Note 17**).
15. Run the gel until the bromophenol blue dye has migrated ~2/3rd down the gel. The high salt concentration will cause some distortion early in the run. Transfer the gel to Whatman paper, dry, and visualize by autoradiography or phosphorimager analysis (*see* **Note 18**).

3.4 ePAT

The ePAT method is a quick and simple approach to the measure of the poly(A)-tail associated with specific RNA (Fig. 3 and [17]). The approach can also efficiently detect alternative polyadenylation and the poly(A)-tail that is added to RNA in the process of decay (Fig. 3d). Here we provide additional information that extends our previously published method [17]. Before starting, determine if you require a spike-in efficiency control. When working with metazoan samples, this can be RNA from a different species such as total RNA from a deadenylase-deficient *S. cerevisiae* strain (Δ*pan2/ccr4-1*). The mRNA from this strain have uniformly long poly(A)-tails and thus provide a useful internal control for the assay. For yeast work, it can be useful to spike-in total mammalian RNA, which typically has longer poly(A)-tails. Alternatively, a synesthetic, adenylated control RNA may be used. The spike-in can also serve as ballast if the total RNA of interest is limiting (less than 100 ng). If a spike-in is used, ensure that it is also used in the TVN-PAT size control reaction (Subheading 3.5). It is possible to reduce the time of both 3′ extension and reverse transcription to 15 min (Fig. 3e) but is critical that temperature is maintained during the addition of the reverse transcriptase (*see* **Note 19**).

1. Preprogram the thermo cycler with an 80 °C lid temperature and the following temperature holds and pauses: 80 °C for 5 min; 37 °C for 1 min || pause || 37 °C for 60 min; 80 °C for 5 min; 55 °C for 1 min || pause || 55 °C for 60 min; 80 °C for 5 min; 4 °C hold (*see* **Note 19**).

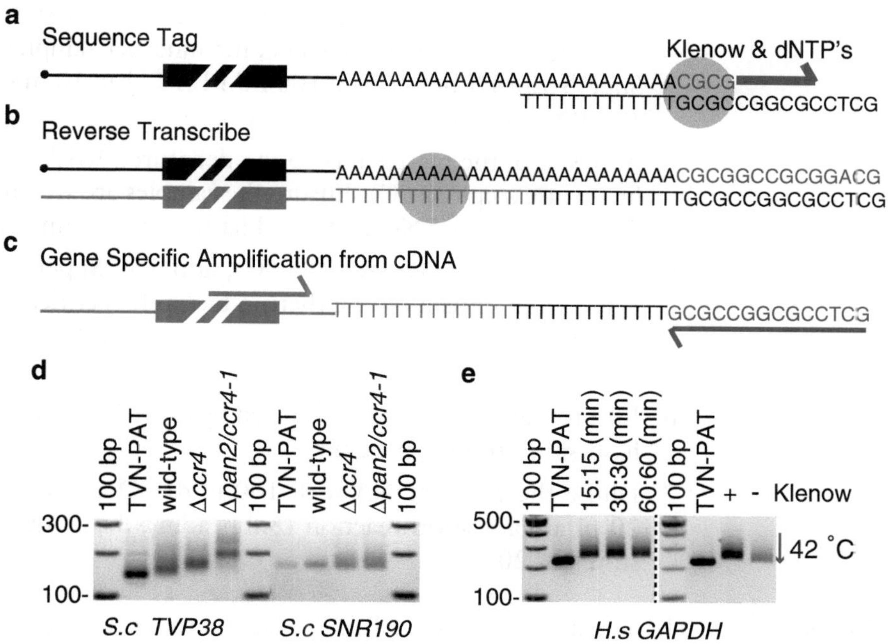

Fig. 3 The ePAT approach to poly(A)-tail length measure. (**a**) Total RNA and a template oligo-dT-anchor primer are assembled in buffers that support both end extension and reverse transcription. In a first step, Klenow polymerase extends the 3′ end with the complimentary sequence to the template. (**b**) Addition of a heat-stable reverse transcriptase at 55 °C means that only the oligo-dT template primer annealed to end-extended RNA has a melt temperature sufficiently high to support cDNA synthesis. In this way internal priming is avoided and reverse transcription is limited to adenylated molecules. (**c**) PCR amplification using a gene-specific forward primer and a universal reverse primer generates products that encompass the full native poly(A)-tail. (**d**) The adenylation state of *TVP38* is analyzed in wild-type yeast and mutants lacking either Ccr4 or both Ccr4 and Pan2. The inclusion of the TVN-PAT reaction provides a useful reference point as it is generally a discrete band, whereas the smear of amplicons generated by the ePAT reactions represents the steady-state poly(A)-tail length distribution of both newly synthesized and aged transcripts. It is important to note that adenylated noncoding and structural RNA are also detected by this approach, and thus the non-coding RNA *SRN190* is also detected, albeit the tramp complex in this case mediates the adenylation during decay rather than in biogenesis. For *SRN190* the increase in abundance and poly(A)-tail length in the mutant presumably represents an indirect result of increased demand on this complex in the absence of normal deadenylation driven decay. Note: The characteristic step upshift of the average tail length of mRNA is not seen in this noncoding RNA. (**e**) The ePAT reaction is very sensitive and also quick; little difference is detected between 15 min extension and reverse transcription (15:15) and the longer (60:60) reaction time. It is critical however that the temperature is maintained during the first minutes of reverse transcription. If the temperature drops even marginally, the template oligonucleotide can prime reverse transcription from anywhere along the length of the poly(A)-tail and accuracy of tail length measure lost. The *red arrow* indicates the downward shift of the amplicons when reverse transcription was at 42 °C and the reaction was performed in the absence of Klenow (Color figure online)

2. In thin-walled PCR strip-tubes, assemble 1 μg total RNA (including spike-in if using); 1 μl the template primer, and dH_2O to a total of 12 μl. Flash-spin, transfer to cycler, and start the program.

3. During the first denaturation step, prepare a master mix for the number of samples ($n+1$) containing per sample: 4 μl 5× SuperScript III buffer, 1 μl 100 mM DTT, 1 μl 10 mM dNTP, 1 μl RNaseOUT, and 1 μl Klenow-exo.
4. Add 8 μl of the master mix to the denatured RNA during the first pause step (37 °C). Ensure the samples are well mixed by flick inversion (*see* **Note 20**) and flash-spin. Resume program.
5. At the second pause step (55 °C), add 1 μl SuperScript III® while tubes remain on the thermo block. It is critical to maintain the temperature at this step to avoid low-stringency mispriming.
6. Ensure the samples are well mixed by rapid flick inversion and flash-spin. Resume program.
7. Dilute the ePAT cDNA with 100 μl dH_2O and use 5 μl per 20 μl amplification reaction (Subheading 3.6). Alternatively, store at −20 °C.

3.5 The TVN-PAT Size Control Reaction

This reaction is a classic 3′ RACE reaction that generates a compatible cDNA that can be used as size control to calculate the proportion of the amplicons that represents the poly(A)-tail (Fig. 3d, e). The VN-variable bases lock reverse transcription to the transition from variable 3′UTR sequence into poly(A)-tail (and unavoidably, internal adenosine homopolymers). The TVN-PAT reaction is an important accompanying assay to ePAT as it provides a means to interpret relative length changes between samples. TVN-PAT takes the place of a "no-tail" control in traditional high-resolution (RNase H) northern blot assays. If a number of samples are used where adenylation site usage or transcript abundance changes are expected, it can be helpful to make a combined stock of mixed total RNAs, to generate a representative amplicon size reference. The final PCR amplicons generated from a TVN-PAT reaction represent the size of the product for mRNA having an invariant 12(A)-tail (*see* **Note 21**).

1. Program the thermo cycler with an 80 °C lid temperature and the following temperature holds and pauses: 80 °C for 5 min; 42 °C for 1 min || pause || 42 °C for 15 min; 48 °C for 15 min; 55 °C for 15 min; 80 °C for 5 min; 4 °C hold.
2. In thin-walled PCR tubes, combine 1 μg total RNA (including spike-in if using); 1 μl of the TVN-PAT oligo, and dH_2O to a total of 12 μl. Flash-spin, transfer to cycler, and start the program.
3. During the first denaturation step, prepare a master mix for the number of samples ($n+1$) containing 4 μl 5× SuperScript III buffer, 1 μl 100 mM DTT, 1 μl 10 mM dNTP, 1 μl RNaseOUT, and 1 μl SuperScript III®. Add 8 μl of this master mix to the reaction during the pause step. Ensure the samples are well mixed by flick inversion and flash-spin. Resume program.

4. Dilute the TVN-PAT cDNA with 200 μl dH_2O and use 5 μl per 20 μl amplification reaction (Subheading 3.6). Alternatively, store at −20 °C (*see* **Note 22**).

3.6 PCR Amplification of ePAT and TVN-PAT cDNA for Poly(A)-Tail Length Analysis

To avoid size bias and ensure linear amplification of all ePAT amplicons, it is important to (a) use a hot-start polymerase and (b) monitor cycle number to catch amplification at the critical point where both the product detection and amplification are linear. This is often only the case over a narrow range (2–3 cycles). As a rule of thumb, if the products are more intense than the ladder, the PCR is over-cycled and bias may occur.

1. Program the thermo cycler: 94 °C for 10 min [94 °C for 20 s; 60°C for 20 s; 72 °C for 20 s]×28 cycles; 72 °C for 1 min; 12 °C hold (*see* **Note 23**).
2. Preprepare a 2 % high-resolution agarose gel in 1× running buffer including nucleic acid stain and store at 4 °C until required (*see* **Note 11**).
3. Prepare a master mix for amplification of the gene of interest in a 20 μl reaction, remembering to include the TVN-PAT cDNA. Per sample ($n+1$), assemble 10 μl 2× AmpliTaq Gold 360 Master Mix, 0.2 μl gene-specific primer, 0.2 μl universal reverse primer (i.e., the ePAT template oligo), and 4.6 μl dH_2O.
4. Pipette 5 μl of each cDNA (ePAT and TVN-PAT) into thin-walled PCR strip-tubes. Add 15 μl of the PCR master mix (**step 3**), mix by flick inversion, and flash-spin.
5. Transfer to thermo cycler and start program.
6. Once the amplification cycle is complete, add 5 μl Orange G gel loading buffer to each PCR reaction, mix, and load 10 μl per well, alongside of 5 μl (0.5 μg) 100 bp DNA ladder.
7. Visualize using a high-resolution gel documentation system aiming to capture the image when there is good size resolution between the TVN-PAT band migration and the distribution of ePAT amplicons. The exact time will depend on the size range of the amplicons, check once Orange G has migrated ~1/2 the gel length.

4 Concluding Remarks and Future Directions

The Klenow-mediated extension of adenylated RNA is a rapid and selective approach to 3′ tagging of adenylated RNA that in our hands surpasses traditional methods in terms of efficiency, resolution, and user-friendliness. At the time of writing, our laboratory is extending the ePAT approach to a genome-wide deep-sequencing application. Our approaches in this area are superficially similar to approaches to 3′ focused RNA-seq that depend on TVN-based [21–25] or RNA ligase-based tagging of RNA [26, 27].

However, a standout difference is that the Klenow approach, being poly(A)-directed, avoids the necessity to poly(A)-select or rRNA-deplete RNA as a first step in RNA-seq. The Klenow extension approach also avoids internal priming and may be less subject to bias because it does not depend on variable bases that can be selectively depleted by abundant RNA in TVN-based approaches. We are thus confident that the extension mediated approach to RNA 3′ tagging will find an important place in the toolkit of methods for the analysis of 3′ UTR dynamics.

5 Notes

1. It is important that the organic and aqueous phases mix completely (i.e., no-phase separation). At temperatures above 65 °C, the phases will merge. In our hands, the better heat transfer of a water bath tends to promote this more quickly than a heat block.
2. The metal block taken from a heat block and placed on ice is a convenient method to rapidly chill samples while keeping them in order.
3. The use of screw cap tubes is a safety precaution when using caustic reagents. In both the hot phenol method (Subheading 3.1) and in the TRIzol® extraction method (Subheading 3.2), considerable pressure is placed on the tube lid. The extra seal strength of a screw cap can avoid unwanted leaks.
4. The lifetime of acid phenol can be extended by storage of aliquots at −20 °C in dark-glass Duran® bottles. We routinely purchase a 500 ml stock and aliquot this into 50 ml working solutions that are stored in the freezer until required. The working stock is stored at 4 °C.
5. A convenient, inexpensive, and consistent source of RNase-free water is to purchase Baxter water for irrigation. We use this for all our work that requires “clean” water, keeping the stock bottle in an RNase and PCR amplicon-free environment away from our general work areas. Aliquot smaller volume (1–10 ml) into new disposable plasticware and discard after having tubes have been opened in the lab environment more than 2–4 times.
6. The nanodrop returns two quality control measures: the 260/280 and the 260/230 absorbance ratios. The latter is an important measure of phenol and/or guanidinium contamination. In the protocols described here, it is normal for both ratios to be above 1.8.
7. To design gene-specific primers for the ePAT and TVN-PAT approaches, it is best to manually choose an 18–24 base region

of high complexity that has a melt temperature of 60 °C or above and is positioned within 100–200 bases from the poly(A)-site. If the poly(A)-site is not known, choose the best site directly upstream of the stop codon. To determine the melt temperature, we use the Sigma-Genosys DNA calculator (http://www.sigma-genosys.com/calc/DNACalc.asp), but note that different calculators often generate significantly different melt temperatures for identical oligos.

8. Both TBE and SB gel running buffers work well when using ethidium bromide as the nucleic acid stain in the gels. However when using SYBR® Safe, SB is the better choice. We find that the SYBR® Safe stain is lost from the gel very early in the run when using TBE as the gel/running buffer. This means the gels have little power to resolve size differences.
9. The use of narrow (0.75 mm) gel combs increases the resolution of the gels and thus makes size estimation more accurate.
10. The migration of bromophenol blue often overlaps the region of interest for ePAT/TVN-PAT amplicons in 2 % agarose gels and leaves an unwanted shadow on the gel images. We thus depend on the faster migration of the Orange G dye, which precedes the migration of the amplicons of interest.
11. Many institutions are moving away from the use of ethidium bromide as the routine gel stain of choice. In our hands SYBR® Safe is an acceptable alternative (*see* also **Note 8**). However, some other gel stains (including SYBR® Gold (Life Technologies) and Gel star (Lonza)) can significantly distort the migration of the DNA amplicons and should thus be avoided in this assay.
12. The term flash-spin is used in these protocols to refer to the rapid 1–2 s spin in a minifuge to collect liquid in the bottom of 1.5 ml or 0.2 ml tubes.
13. Air-drying of RNA is conveniently done by opening tubes ~10 cm under a desk lamp with a normal incandescent globe. This generates a gentle heat and associated updraft that reduces the chance of unwanted material floating into the open tubes.
14. If the TRIzol/tissue homogenate is very hazy, it may indicate that there is too much tissue (relative to the TRIzol). If this is the case, remove the homogenate and split into two to four fresh tubes and add additional TRIzol to ensure effective extraction.
15. The use of GlycoBlue™ can aid in the recovery of dilute RNA solutions resulting from low starting material. It is an especially useful addition when also analyzing small RNA such as microRNA. The addition of a carrier also helps in visualization of the pellet and thus gives confidence when removing all traces of liquid with a pipette.
16. The additional ethanol precipitation step in this protocol fixes the notoriously poor A260/A230 ratios of the TRIzol® family

of reagents. This is likely because removing all traces of liquid at each step significantly reduces the contaminating Guanidinium in the final RNA preparation.

17. Suitable ladders for this experiment can be prepared by T4 polynucleotide kinase-mediated 5′ labeling of commercial ladders (Fig. 3). Alternatively, a ladder can be conveniently prepared by a standard Klenow-mediated end-filling reaction of an MspI digest of pBR233 (NEB: N3032S) using α-[32P] dCTP (thus avoiding the need for additional purchase of γ-[32P]-dNTP).
18. The transfer of large, thin, brittle, and radioactive gels to filter paper is not pleasant and should not be performed by junior personnel. The filter paper does not stick to the gels, as would be expected, and thus peeling the gel off the plate is the only option. We find that the best approach is to separate the gel plates (the gel will stick to the siliconized front plate), cover the gel with a larger sheet of Whatman filter paper, turn it over, and lay it face down on the bench (paper, gel, glass plate). Pull the glass plate toward you until you can prize the gel from the plate (with a spacer or similar thin device) onto the filter paper. Continue to pull the glass plate toward you letting the gel gently fall onto the filter paper. Finally, lift the glass plate up and back, allowing the last part of the gel to fall onto the filter. Once the gel is off the glass and on the filter, cover it with cling film and dry down in a vacuum drier.
19. Maintenance of temperature at the time of SuperScript III addition is critical. If the temperature drops, the advantage of end extension is lost, and the template primer will anneal (and extend from) along the full length of the poly(A)-tract. This is illustrated in Fig. 3e which shows that even in the absence of Klenow, a smear of amplicons is achieved but the length of these is significantly shortened as compared to the + Klenow reaction that has had careful temperature control. Reverse transcription is only locked to the end-extended mRNA if temperature remains high at the critical earliest steps. The solution is to add the reverse transcriptase to the open tubes in the PCR block and to remove tubes only to rapidly flick invert to mix and to flash-spin.
20. The term flick inversion is used here to mean the rapid shake of the wrist to invert the solution from one end of the tube to the other to promote mixing. It is combined with a ~1 s flash-spin to collect all the liquid back to the bottom of the tube.
21. Smeary TVN-PAT reactions can be caused by microheterogeneity of polyadenylation sites. These can obscure the true lengths of the poly(A)-tail. Two discrete TVN-PAT products can indicate either alternative polyadenylation site usage or

internal priming. We note that some 3′UTRs are poorly detected by the TVN-PAT (relative to ePAT) and suspect that this is due to the selective depletion of specific VN-variable combinations by abundant RNA such as adenylated rRNA (decay intermediates) and mRNA encoding ribosomal proteins.

22. The reason to dilute the TVN-PAT reaction more than the ePAT reaction is simply for the visual aesthetic of the gel. Because the TVN-PAT reaction normally gives a tight band and the signal is thus compacted, it "looks" more intense when compared to the diffuse smear of ePAT amplicons.
23. The number of amplification cycles depends on the abundance of the transcript of interest. As a rule of thumb, we usually start with 24 cycles for abundant transcripts, 28 cycles for transcripts of average abundance, and 30 cycles for rare transcripts.

Acknowledgement

We acknowledge members of the Beilharz laboratory for critical discussions. Monash University start-up funds and the Australian Health and Medical Research Council (APP1042851, APP1042848) supported this work, and an Australian Research Fellowship from the Australian Research Council (DP0878224) supported T.H.B.

References

1. Baker KE, Coller J, Parker R (2004) The yeast Apq12 protein affects nucleocytoplasmic mRNA transport. RNA 10:1352–1358
2. Hector RE, Nykamp KR, Dheur S et al (2002) Dual requirement for yeast hnRNP Nab2p in mRNA poly(A) tail length control and nuclear export. EMBO J 21:1800–1810
3. Izawa S, Kita T, Ikeda K et al (2008) Heat shock and ethanol stress provoke distinctly different responses in 3′-processing and nuclear export of HSP mRNA in Saccharomyces cerevisiae. Biochem J 414:111–119
4. Beilharz TH, Preiss T (2007) Widespread use of poly(A) tail length control to accentuate expression of the yeast transcriptome. RNA 13:982–997
5. Lackner DH, Beilharz TH, Marguerat S et al (2007) A network of multiple regulatory layers shapes gene expression in fission yeast. Mol Cell 26:145–155
6. Weill L, Belloc E, Bava FA et al (2012) Translational control by changes in poly(A) tail length: recycling mRNAs. Nat Struct Mol Biol 19:577–585
7. Ortiz-Zapater E, Pineda D, Martinez-Bosch N et al (2012) Key contribution of CPEB4-mediated translational control to cancer progression. Nat Med 18:83–90
8. D'Ambrogio A, Nagaoka K, Richter JD (2013) Translational control of cell growth and malignancy by the CPEBs. Nat Rev Cancer 13:283–290
9. Radford HE, Meijer HA, de Moor CH (2008) Translational control by cytoplasmic polyadenylation in Xenopus oocytes. Biochim Biophys Acta 1779:217–229
10. Richter JD, Klann E (2009) Making synaptic plasticity and memory last: mechanisms of translational regulation. Genes Dev 23:1–11
11. Salles FJ, Richards WG, Strickland S (1999) Assaying the polyadenylation state of mRNAs. Methods 17:38–45
12. Beilharz TH, Preiss T (2009) Transcriptome-wide measurement of mRNA polyadenylation state. Methods 48:294–300
13. Garneau NL, Sokoloski KJ, Opyrchal M et al (2008) The 3′ untranslated region of sindbis virus represses deadenylation of viral transcripts in mosquito and Mammalian cells. J Virol 82:880–892
14. Couttet P, Fromont-Racine M, Steel D et al (1997) Messenger RNA deadenylylation precedes

decapping in mammalian cells. Proc Natl Acad Sci U S A 94:5628–5633
15. Charlesworth A, Cox LL, MacNicol AM (2004) Cytoplasmic polyadenylation element (CPE)- and CPE-binding protein (CPEB)-independent mechanisms regulate early class maternal mRNA translational activation in Xenopus oocytes. J Biol Chem 279:17650–17659
16. Okazaki T, Okazaki R (1969) Mechanism of DNA chain growth. IV. Direction of synthesis of T4 short DNA chains as revealed by exonucleolytic degradation. Proc Natl Acad Sci U S A 64:1242–1248
17. Janicke A, Vancuylenberg J, Boag PR et al (2012) ePAT: a simple method to tag adenylated RNA to measure poly(A)-tail length and other 3′ RACE applications. RNA 18: 1289–1295
18. Brody JR, Kern SE (2004) Sodium boric acid: a Tris-free, cooler conductive medium for DNA electrophoresis. Biotechniques 36:214–216
19. Sengupta MS, Low WY, Patterson JR et al (2012) ifet-1 is a broad scale translational repressor required for normal P granule formation in C. elegans. J Cell Sci 126:850–859
20. Minvielle-Sebastia L, Winsor B, Bonneaud N et al (1991) Mutations in the yeast RNA14 and RNA15 genes result in an abnormal mRNA decay rate; sequence analysis reveals an RNA-binding domain in the RNA15 protein. Mol Cell Biol 11:3075–3087
21. Yoon OK, Brem RB (2010) Noncanonical transcript forms in yeast and their regulation during environmental stress. RNA 16: 1256–1267
22. Shepard PJ, Choi EA, Lu J et al (2011) Complex and dynamic landscape of RNA polyadenylation revealed by PAS-Seq. RNA 17:761–772
23. Beck AH, Weng Z, Witten DM et al (2010) 3′-end sequencing for expression quantification (3SEQ) from archival tumor samples. PLoS One 5:e8768
24. Derti A, Garrett-Engele P, Macisaac KD et al (2012) A quantitative atlas of polyadenylation in five mammals. Genome Res 22:1173–1183
25. Mangone M, Manoharan AP, Thierry-Mieg D et al (2010) The landscape of C. elegans 3′UTRs. Science 329:432–435
26. Ulitsky I, Shkumatava A, Jan CH et al (2012) Extensive alternative polyadenylation during zebrafish development. Genome Res 22: 2054–2066
27. Jan CH, Friedman RC, Ruby JG et al (2011) Formation, regulation and evolution of Caenorhabditis elegans 3′UTRs. Nature 469: 97–101
28. Tucker M, Valencia-Sanchez MA, Staples RR et al (2001) The transcription factor associated Ccr4 and Caf1 proteins are components of the major cytoplasmic mRNA deadenylase in Saccharomyces cerevisiae. Cell 104:377–386

Chapter 4

Detection of Uridylated mRNAs

François M. Sement and Dominique Gagliardi

Abstract

Uridine addition at the 3′ end of RNAs (i.e., uridylation) emerges as a critical posttranscriptional modification promoting RNA degradation. Uridylation has been notably linked to the degradation of small RNAs, correlated with the 5′ shortening of RISC-cleaved transcripts and the degradation of mRNAs. We describe here a method based on 3′ RACE (3′ Rapid Amplification of cDNA End) PCR that has been successfully used to investigate nucleotide addition at the 3′ end of RISC-cleaved transcripts and full-length mRNAs in plants.

Key words Uridylation, 3′RACE PCR, mRNA, Noncanonical poly(A) polymerase, RNA degradation

1 Introduction

RNA posttranscriptional modification by noncanonical poly(A) polymerases (ncPAPs) is a widespread mechanism involved in diverse aspects of RNA metabolism. ncPAPs are defined by opposition to canonical poly(A) polymerases that are responsible for the addition of the long and stabilizing poly(A) tail at the 3′ end of eukaryotic messenger RNAs. These tails range from 70 residues in yeast to 250 in mammals and are covered by poly(A)-binding proteins. Messenger RNA poly(A) tails are crucial for stabilization [1], export from the nucleus [2], and efficient translation [3]. Whereas canonical poly(A) polymerases are only present in eukaryotes, ncPAPs are present in virtually all organisms. They catalyze RNA modifications that play different roles depending on the ncPAP, the nature of the nucleotide extension, the RNA substrate, the cellular compartment, and the organism considered. For instance, polyadenylation by the ncPAP GLD-2 activates the translation of silenced cytosolic mRNAs in *Xenopus* oocytes [4] and neurons [5], while polyadenylation by the ncPAP Trf4 in *Saccharomyces cerevisiae*, targets nuclear RNAs for degradation by the exosome [6]. The short poly(A) tails added by Trf4 serve as landing pads for 3′–5′ exoribonucleases and therefore earmark RNA for

Joanna Rorbach and Agnieszka J. Bobrowicz (eds.), *Polyadenylation: Methods and Protocols*, Methods in Molecular Biology, vol. 1125, DOI 10.1007/978-1-62703-971-0_4, © Springer Science+Business Media New York 2014

degradation. Such a destabilizing role by short poly(A) tails likely corresponds to the ancestral role of polyadenylation since polyadenylation-assisted RNA degradation operates in bacteria [7], several *Archaea* [8], plant organelles [9], and in the nucleus of all eukaryotes investigated so far [10, 11].

ncPAPs are not all poly(A) polymerases sensu stricto. Members of the ncPAP family can also catalyze the addition of other nucleotides, such as uridines, at the 3′ end of an RNA molecule. RNA uridylation by ncPAPs is emerging as a conserved RNA modification controlling RNA stability. Uridylation, like polyadenylation, can have stabilizing or destabilizing effects depending on the enzyme and its substrate. U6 snRNA uridylation is prerequisite for its stabilization and its role in mRNA splicing [12], while uridylation is also involved in the degradation of small RNAs in organisms like *Caenorhabditis elegans* [13] or plants [14, 15] and the degradation of histone mRNAs in human [16] and mRNAs in general in *Schizosaccharomyces pombe, Aspergillus nidulans and Arabidopsis thaliana* [17–20]. The number of uridine residues added at the 3′ end of RNA is often very limited (1–5) and the detection of those residues on diverse RNA targets is a critical step to investigate the role of uridylation in genome expression. This chapter describes a PCR-based method used to detect uridylation of mRNAs. The method can also be used to determine the nature and number of nucleotides posttranscriptionally added at the 3′ end of any RNA of more than 200 nt.

2 Material

2.1 RNA Preparation

1. Refrigerated centrifuge able to spin at 16,000 × *g*.
2. Phenol/chloroform/isoamyl alcohol (25/24/1) saturated with 1 M Tris–HCl, pH 4.
3. 100 % ethanol.
4. 3 M sodium acetate (Na Ac) pH 5.2.
5. RNase-free glycogen 1 μg/μL.
6. FastAP Thermosensitive Alkaline Phosphatase (Thermo Scientific).
7. 10× FastAP buffer (Thermo Scientific).
8. T4 RNA ligase 1 (New England Biolabs).
9. 10× T4 RNA ligase buffer and 10 mM ATP (New England Biolabs).
10. RNA adapter 5′(Pho)CUAGAUGAGACCGUCGACAUGAAUUC(AmC7)3′ 200 μM (*see* **Note 1**).
11. Water bath (37 °C).
12. Heat block (65 °C).

2.2 RNA Clean-Up

1. NucleoSpin® RNA Clean-up (Macherey-Nagel™) (*see* **Note 2**).
2. RNase-free water.

2.3 cDNA Synthesis

1. SuperScript® III First-Strand Synthesis System (Invitrogen™).
2. RNaseOUT (Invitrogen).
3. cDNA synthesis primer (5′-GAATTCATGTCGACGGTC TCA-3′) (reverse primer 1).
4. 10 mM dNTPs mix (10 mM of each dNTP).
5. Thermocycler.

2.4 PCR Amplification of 3′ cDNA Ends (3′ RACE PCR)

1. Adapter-specific reverse primer (5′-CATGTCGACGGTCTC ATCTAG-3′) (reverse primer 2).
2. 2 specific forward primers for each mRNA to be studied (forward primer 1 and forward primer 2) (*see* **Note 3**).
3. GoTaq® Flexi DNA Polymerase (Promega™) (*see* **Note 4**).

2.5 PCR Products Cloning and Sequencing

1. NucleoSpin® Gel and PCR Clean-up (Macherey-Nagel™) (*see* **Note 5**).
2. pGEM®-T Easy Vector System I (Promega™) (*see* **Note 6**).
3. Thermo-competent *E. coli* cells.

3 Method

An overview of the protocol is presented in Fig. 1.

3.1 RNA Preparation

1. Prepare 5–10 μg of total RNA (*see* **Note 7**).
2. Dephosphorylate 5–10 μg of total RNA using FastAP (*see* **Note 8**). Ideally, FastAP treatment is performed on RNA at a concentration of 0.1 μg/μL in the presence of 1 unit of FastAP per microgram of RNA. Incubate 20 min at 37 °C.
3. Add 1 volume of phenol/chloroform/isoamyl alcohol (25/24/1) and vortex for 15 s.
4. Centrifuge at 16,000 × *g* at 4 °C for at least 2 min.
5. Transfer the aqueous phase (upper phase) in a new 1.5 mL tube.
6. Ethanol precipitate the dephosphorylated RNA by adding 0.1 volume of 3 M Na Ac pH 5.2 and 2.5 volume of 100 % ethanol. Vortex and store the samples at least 10 min at −20 °C.
7. Pellet RNA by spinning the samples in a centrifuge at 16,000 × *g* for 10 min at 4 °C.
8. Discard supernatant. Wash pellet by adding 1 mL 70 % ethanol.

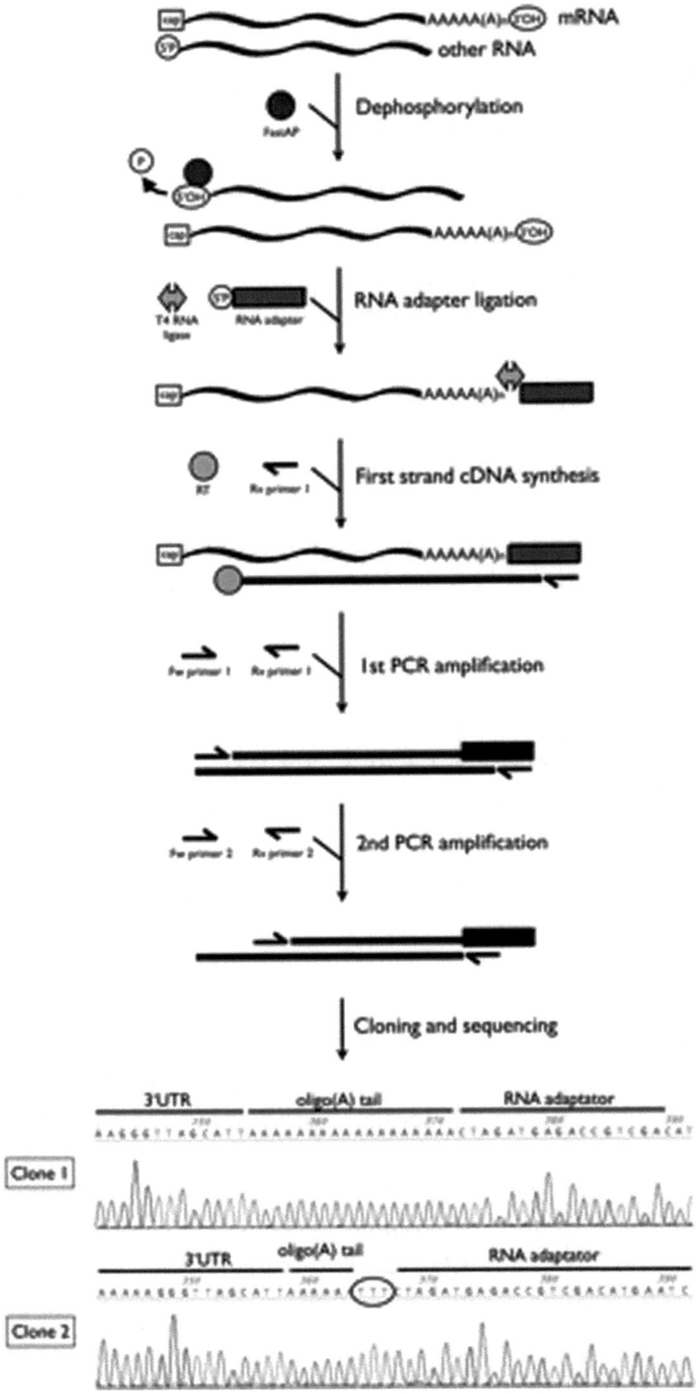

Fig. 1 Flowchart steps for detecting uridylated RNAs. After a phosphatase treatment, total RNA is ligated with an RNA adapter at the 3′ end of every RNA molecule (Subheading 3.1 in the text). After elimination of the RNA adapter excess (Subheading 3.2 in the text), cDNA synthesis by the reverse transcriptase (RT) is initiated from this RNA adapter using reverse primer 1 (rv primer 1) (Subheading 3.3 in the text). The mRNA target 3′ regions are then PCR amplified by two successive rounds of PCR using two gene-specific forward primers (fw primers 1 and 2) and two adapter-specific primers (rv primers 1 and 2) (Subheading 3.4 in the text). PCR products are then cloned and sequenced (Subheading 3.5 in the text). Two examples of sequencing results obtained from the same mRNA target in the same experiment are shown. mRNA 3′ UTR, oligo(A) tail, and the sequence corresponding to the RNA adapter are indicated. Added nucleotides at the end of clone 2 oligo(A) tail are *circled*

9. Spin the samples at 16,000 × *g* for 3 min at 4 °C and discard supernatant.
10. Dissolve the RNA pellet in 40 μL H_2O and determine the RNA concentration by measuring absorbance at 260 nm (*see* **Note 9**).
11. Transfer 5 μg of RNA in a new 1.5 mL tube for each sample (*see* **Note 10**).
12. Add 0.5 μL of 200 μM RNA adapter and water up to 39 μL.
13. After a 3 min denaturation step at 65 °C, cool the samples on ice for at least 1 min.
14. Add 5 μL 10× T4 RNA ligase buffer, 5 μL 10 mM ATP, and 1 μL (10 units) of T4 RNA ligase 1.
15. Incubate for 1 h at 37 °C.
16. Add water up to 100 μL.

3.2 RNA Clean-Up (Macherey-Nagel™ RNA Clean-up Kit)

1. To 100 μL samples, add 600 μL of buffer RA1:ethanol 1:1 mix and mix by vortexing.
2. Load the samples on labeled columns and spin 30 s at 8,000 × *g*.
3. Discard flow-through, add 700 μL of buffer RA3 (wash buffer), and spin 30 s at 8,000 × *g*.
4. Discard flow-through and put the column back in the collection tube. Add 350 μL of buffer RA3 and spin 2 min at 8,000 × *g*.
5. Transfer the columns to new 1.5 mL tubes (avoid any contact between the columns and wash buffer). Open the lid of the column and allow the membrane to dry for 3 min.
6. Add 60 μL of RNase-free water at the center of the column.
7. Spin at 8,000 × *g* for 1 min and put the 60 μL eluate back at the center of the column and spin again 1 min at 8,000 × *g* (*see* **Note 11**).
8. Determine the RNA concentration by measuring absorbance at 260 nm (*see* **Note 9**).
9. Ethanol precipitate the maximum common amount of RNA: add 0.1 volume of 3 M Na Ac pH 5.2, 1 μg glycogen (*see* **Note 12**), and 2.5 volume ethanol. Shortly vortex the samples and store them at −20 °C for 30–60 min.
10. Pellet RNA by spinning the samples in a centrifuge at max speed (16,000 × *g*) for 10 min at 4 °C.
11. Discard supernatant. Wash pellets by adding 1 mL 70 % ethanol.
12. Spin the samples at 16,000 × *g* for 3 min at 4 °C.
13. Repeat **steps 11** and **12** once.
14. Discard supernatants and air-dry the RNA pellets. Resuspend pellets in 10 μL of RNase-free water. The RNA is now ready for reverse transcription.

3.3 cDNA Synthesis (SuperScript® III First-Strand Synthesis System) (See Note 13)

1. Transfer the desired amount of RNA in thermocycler-compatible tubes.
2. For one 20 μL reaction, add 2 μL of 10 μM reverse primer 1, 1 μL of 10 mM dNTPs mix, and water up to 13 μL. Transfer samples in a thermocycler.
3. Denature samples 5 min at 65 °C.
4. Cool down samples to 0 °C for 1 min.
5. During **steps 3** and **4**, prepare a master mix containing, per sample, 4 μL of 5× cDNA First-Strand buffer, 1 μL of 0.1 M DTT, 1 μL of RNaseOUT (40 U) (Invitrogen), and 1 μL of SuperScript III (200 U) (Invitrogen).
6. After samples have cooled down, take out the samples from the thermocycler and put them on ice. Add 7 μL of master mix to the 13 μL samples. Mix by pipetting up and down.
7. Incubate cDNA reactions for 1 h at 50 °C in a thermocycler.
8. Inactivate SuperScript III reverse transcriptase by incubating samples 15 min at 70 °C.
9. cDNAs are ready for PCR (*see* **Note 14**).

3.4 PCR Amplification of 3′ cDNA Ends (3′ RACE PCR)

1. PCR amplify your target by using forward primer 1 and reverse primer 1 (*see* **Note 2**). Use 1 μL of cDNA as PCR template (*see* **Note 15**). In general, the conditions (*see* **Note 16**) used for this PCR are:

   Step 1: 94 °C for 30 s

   Step 2: 94 °C for 20 s

   Step 3: 50 °C for 20 s

   Step 4: 72 °C for 20 s (Repeat **steps 2–4** twenty-nine times)

   Step 5: 72 °C for 30 s
2. Use 1 μL of the previous PCR (*see* **Note 17**) as the template for a second PCR reaction. For this second reaction, use a forward gene-specific primer (forward primer 2) located downstream of the one used in the first PCR reaction and the adapter-specific reverse primer (reverse primer 2). In general, the conditions (*see* **Note 16**) used for this PCR are the same than for the first PCR reaction.
3. The results of these PCR can be visualized on a 2–2.5 % agarose gel stained with ethidium bromide.

3.5 PCR Products Cloning and Sequencing (NucleoSpin Gel) and PCR Clean-up (Macherey-Nagel™) (See Note 5) and pGEM®-T Easy Vector System I (Promega) (See Note 6)

1. Add water to the PCR reaction up to 100 μL.
2. Add 200 μL of 1/4 NT buffer (*see* **Note 18**) and perform the PCR clean-up with the NucleoSpin® Gel and PCR Clean-up kit according to the manufacturer's recommendation and elute the purified PCR products with 15 μL of elution buffer.
3. PCR products are cloned in pGEM®-T Easy vector using pGEM®-T Easy Vector System I according to the manufacturer's recommendation. Briefly, 25 to 100 ng of purified PCR product are ligated to 25 ng of linearized pGEM®-T Easy vector, in the presence of 2.5 U of T4 DNA ligase in a final reaction volume of 5 μL. The reaction is incubated at room temperature for 1 h. The reaction is then transformed in *E. coli* thermo-competent cells and plated on LB agar plates supplemented with ampicillin, X-Gal, and IPTG (*see* **Note 6**).
4. Ampicillin-resistant colonies are then used to purify plasmid DNA for sequencing analysis.

4 Notes

1. An incomplete XbaI restriction site has been incorporated at the 5′ end of the RNA adapter. If the last nucleotide of the target mRNA is a uridine, the cDNA synthesis from the RNA adapter will reconstitute an XbaI restriction site that can be use to rapidly test the presence of uridines at the 3′ end of the studied RNA by restriction analysis of the PCR products. This is especially helpful when studying small RNA uridylation.
2. Silica-based RNA purification columns are designed to retain only 200 nt or larger nucleic acid fragments. This step allows eliminating excess of RNA adapter at step in Subheading 3.2. Do not perform this step if your RNA targets are small RNAs.
3. For the design of gene-specific forward primers: forward primer 1 is located at 500–600 nt upstream the poly(A) site and forward primer 2, 300 nt upstream of the poly(A) site.
4. We routinely use GoTaq® Flexi DNA Polymerase but any Taq polymerase can be used.
5. NucleoSpin® Gel and PCR Clean-up can be replaced by any kit able to eliminate primers or primer dimers from a PCR reaction.
6. pGEM®-T Easy Vector System I can be replaced by any T/A-based cloning system. When using pGEM®-T Easy Vector System I, the bacteria transformed with the ligation reaction have to be plated on LB agar supplemented with 100 μg/mL ampicillin. Before plating, add 100 μl of 0.1 M isopropyl β-D-1-thiogalactopyranoside (IPTG) and 40 μl 2 % 5-Bromo-4-chloro-3-indolyl β-D-galactoside (X-Gal) (in DMSO) to a

LB/antibiotic plate. The colonies transformed with empty vectors will develop a blue color on the LB plate.

7. Total RNA can be isolated by TRIzol® (Invitrogen) or TRI Reagent® (Sigma) extraction according to the manufacturer's recommendation or by silica column-based extraction (e.g., NucleoSpin® RNA Plant kit (Macherey-Nagel) or RNeasy Mini kit (Qiagen)) according to the manufacturer's recommendation.
8. The phosphatase treatment avoids the formation of RNA concatemers. Removal of all 5′ phosphate groups from total RNA allows ligating only the RNA adapter 5′ phosphate ends to RNA 3′OH ends.
9. We use a nanodrop device that allows UV spectrum collection from 220 to 300 nm. The absorbance at 260 nm allows determining the RNA concentration. The different ratios A260/A280 and A260/A230 are representative of protein and salt contamination, respectively.
10. Use 5 μg or the maximum common amount of RNA between the different samples.
11. This step maximizes the amount of RNA retrieved from the columns.
12. Glycogen can be omitted if you are working with more than 1 μg of RNA.
13. The maximum amount of total RNA for a 20 μL reverse transcription reaction is 5 μg with the SuperScript III First-Strand synthesis kit (Invitrogen™). However, we do not recommend the use of more than 2.5 μg of RNA per 20 μL reaction. The reaction volume can be raised to 40 μL, but we recommend performing two 20 μL reactions instead of one 40 μL reaction.
14. RNase H treatment of the synthetized cDNA is optional.
15. You can compare 10× diluted or undiluted cDNA.
16. These are general conditions for 3′RACE PCRs. Annealing temperature, elongation time, template amounts, and cycle number may have to be adjusted from one target to the other. We usually perform nested 3′RACE PCR. We amplify 600 bp and 300 bp fragments in the first and the second PCR rounds, respectively.
17. Undiluted first PCR reaction or 10–100 fold dilution can be tested for optimal amplification.
18. NT buffer dilutions are useful to optimize the size of the fragments you want to recover from the PCR reaction (*see* NucleoSpin® Gel and PCR Clean-up manual for more information).

Acknowledgment

This work was supported by Centre National de la Recherche Scientifique (France).

References

1. Bernstein P, Peltz SW, Ross J (1989) The poly(A)-poly(A)-binding protein complex is a major determinant of mRNA stability in vitro. Mol Cell Biol 9:659–670
2. Brodsky AS, Silver PA (2000) Pre-mRNA processing factors are required for nuclear export. RNA 6:1737–1749
3. Wells SE, Hillner PE, Vale RD et al (1998) Circularization of mRNA by eukaryotic translation initiation factors. Mol Cell 2: 135–140
4. Radford HE, Meijer HA, De Moor CH (2008) Translational control by cytoplasmic polyadenylation in Xenopus oocytes. Biochim Biophys Acta 1779:217–229
5. Kwak JE, Drier E, Barbee SA et al (2008) GLD2 poly(A) polymerase is required for long-term memory. Proc Natl Acad Sci U S A 105:14644–14649
6. LaCava J, Houseley J, Saveanu C et al (2005) RNA degradation by the exosome is promoted by a nuclear polyadenylation complex. Cell 121:713–724
7. Condon C (2007) Maturation and degradation of RNA in bacteria. Curr Opin Microbiol 10:271–278
8. Portnoy V, Evguenieva-Hackenberg E, Klein F et al (2005) RNA polyadenylation in Archaea: not observed in Haloferax while the exosome polynucleotidylates RNA in Sulfolobus. EMBO Rep 6:1188–1193
9. Lange H, Sement FM, Canaday J et al (2009) Polyadenylation-assisted RNA degradation processes in plants. Trends Plant Sci 14:497–504
10. Slomovic S, Schuster G (2011) Exonucleases and endonucleases involved in polyadenylation-assisted RNA decay. Wiley Interdiscip Rev RNA 2:106–123
11. Schmidt M-J, Norbury CJ (2010) Polyadenylation and beyond: emerging roles for noncanonical poly(A) polymerases. Wiley Interdiscip Rev RNA 1:142–151
12. Trippe R, Guschina E, Hossbach M et al (2006) Identification, cloning, and functional analysis of the human U6 snRNA-specific terminal uridylyl transferase. RNA 12: 1494–1504
13. Van Wolfswinkel JC, Claycomb JM, Batista PJ et al (2009) CDE-1 affects chromosome segregation through uridylation of CSR-1-bound siRNAs. Cell 139:135–148
14. Zhao Y, Yu Y, Zhai J et al (2012) The Arabidopsis nucleotidyl transferase HESO1 uridylates unmethylated small RNAs to trigger their degradation. Curr Biol 22: 689–694
15. Ren G, Chen X, Yu B (2012) Uridylation of miRNAs by HEN1 SUPPRESSOR1 in Arabidopsis. Curr Biol 22:695–700
16. Mullen TE, Marzluff WF (2008) Degradation of histone mRNA requires oligouridylation followed by decapping and simultaneous degradation of the mRNA both 5' to 3' and 3' to 5'. Genes Dev 22:50–65
17. Rissland OS, Norbury CJ (2009) Decapping is preceded by 3' uridylation in a novel pathway of bulk mRNA turnover. Nat Struct Mol Biol 16:616–623
18. Morozov IY, Jones MG, Razak AA et al (2010) CUCU modification of mRNA promotes decapping and transcript degradation in Aspergillus nidulans. Mol Cell Biol 30:460–469
19. Malecki M, Viegas SC, Carneiro T et al (2013) The exoribonuclease Dis3L2 defines a novel eukaryotic RNA degradation pathway. EMBO J 32(13):1842–1854
20. Sement FM, Ferrier E, Zuber H et al (2013) Uridylation prevents 3'trimming of oligoadenylated mRNAs. Nucl Acids Res 41(14): 7115–7127

Chapter 5

Cytoplasmic Polyadenylation Assays

Olga Coll, Ana Villalba, and Fátima Gebauer

Abstract

Basic research in *Drosophila melanogaster* has benefited from a plethora of powerful genetics tools. Detailed biochemical analysis, however, has often been difficult due to the lack of in vitro systems that faithfully recapitulate the observations made in vivo. In the field of posttranscriptional regulation, the recent establishment of robust in vitro systems from embryo and ovary material has fueled the mechanistic understanding of a variety of processes. Here we describe protocols to obtain and use extracts from *Drosophila* embryos that are competent for cytoplasmic polyadenylation and translation of exogenously added transcripts.

Key words Cytoplasmic polyadenylation, Translation, *Drosophila* extract, PAT assay

1 Introduction

In vitro systems have played a fundamental role in deciphering the details of eukaryotic translation. Cell-free systems derived from yeast, plant (wheat germ), or mammalian cells (e.g., rabbit reticulocyte or HeLa lysates) have been typically used to study the process of protein synthesis and its modulation by cis-regulatory elements and trans-acting factors. More recently, extracts from *Drosophila* embryos and ovaries that recapitulate in vivo properties of translation, such as the stimulatory effect of the m^7GpppN cap structure and the poly(A) tail, have been described [1–4]. These extracts and their derivatives have been used to study translational control by RNA-binding proteins in addition to a variety of other purposes including regulation by miRNAs [5], mRNA deadenylation [6], internal ribosome entry site (IRES)-mediated translation [7], and RNA interference [8]. The preparation of *Drosophila* embryo extracts competent for translation [3, 9], deadenylation [10], and RNA interference [11, 12] has been described in detail elsewhere.

Cytoplasmic polyadenylation is a widespread mechanism of translational control that consists of the elongation of the poly(A) tail in the cytoplasm (reviewed in ref. 13, 14). Polyadenylation correlates with translational activation, while deadenylation has the

Joanna Rorbach and Agnieszka J. Bobrowicz (eds.), *Polyadenylation: Methods and Protocols*, Methods in Molecular Biology, vol. 1125, DOI 10.1007/978-1-62703-971-0_5, © Springer Science+Business Media New York 2014

opposite effect. The biochemistry of cytoplasmic polyadenylation has been elucidated using *Xenopus* oocytes that are stimulated to undergo maturation (i.e., meiosis) by the addition of progesterone. Most usually, injection of oocytes with exogenous transcripts, antibodies, or other materials has been the method of choice, although egg extracts that are competent for cytoplasmic polyadenylation have been occasionally used. Recently, we have optimized extracts from *Drosophila* early embryos that support cytoplasmic polyadenylation [15]. The preparation of these extracts follows a similar protocol as that used to obtain translation extracts and, thus, they are also competent for translation. These extracts are a powerful tool to study polyadenylation-mediated translational control, and their use has already allowed to detect several machineries for cytoplasmic polyadenylation [15].

Here we describe the preparation of extracts from 90 min *Drosophila* embryos that are competent for both cytoplasmic polyadenylation and translation, and the assays we use to measure the length of the poly(A) tail.

2 Materials

2.1 Extract Preparation

1. Agar-apple juice plates: 2.9 % agar, 30 % apple juice, 4.4 % sugar-beet syrup, 0.25 % Nipagin.
2. Yeast paste: 0.68 g/mL dry yeast in deionized water, 0.6 % propionic acid.
3. Bleach (stocks from the market typically contain 13 % sodium hypochlorite).
4. Sieves (Neolab; mesh sizes 710, 355, and 125 μm).
5. EW buffer: 0.7 % NaCl, 0.04 % Triton X-100.
6. DE buffer: 10 mM HEPES pH 7.4, 5 mM DTT.
7. DEI buffer: DE buffer supplemented with 1× complete protease inhibitors (Roche). Obtained by dissolving one tablet of protease inhibitors in 50 mL DE buffer.

All buffers must be freshly prepared in autoclaved milliQ water. The DE and DEI buffers must be kept on ice.

2.2 Preparation of the Substrate RNA

1. RNA polymerase (Promega).
2. Linearized DNA template.
3. Fresh α^{32}P-UTP (3,000 Ci/mmol; 10 μCi/μL).
4. RQ DNase (Promega).
5. NTP mix: 20 mM ATP, 20 mM CTP, 2 mM UTP.
6. 100 mM GTP.
7. m^{7}GpppG anti-reverse cap analog (ARCAP, Kedar).
8. Sephadex G-25 columns.

2.3 Cytoplasmic Polyadenylation Assay

All solutions should be prepared in autoclaved milliQ water.

1. 10 mg/mL creatine kinase (CK): prepare in 20 mM HEPES pH 7.4, 50 % glycerol and store at −20 °C.
2. 1 M creatine phosphate (CP): prepare in water, aliquot, and store at −80 °C. Aliquots should be discarded after thawing.
3. 1 M HEPES pH 7.4.
4. 10 mM $Mg(OAc)_2$.
5. 1 M KOAc.
6. 2 mM complete amino acid mix (Promega).
7. TRIzol (Invitrogen).

2.4 PAT Assay

1. Oligo(dT) primers: mix of phosphorylated oligo(dT) primers of sizes ranging from dT_{12} to dT_{16}.
2. Oligo(dT)-anchor primer: 5′-GCGAGCTCCGCGGCCGC GTTTTTTTTTTTT-3′.
3. Specific forward oligonucleotide mapping at ~100 nt of the 3′ end of the gene of interest.
4. AMV reverse transcriptase (Promega).
5. T4 DNA ligase (Fermentas).
6. Taq Supratherm (GeneCraft).
7. Turbo-DNase kit (Ambion).

2.5 Denaturing Agarose Gel Electrophoresis

1. Gel tray of at least 15 cm in length.
2. 10× MOPS buffer: 200 mM MOPS acid, 50 mM AcONa, 10 mM EDTA. Adjust to pH 7.0 with 10 N NaOH.
3. Running buffer: 1× MOPS, 1.6 % formaldehyde. For 1 L, mix 100 mL 10× MOPS and 44 mL 37 % formaldehyde.
4. Northern dye: 50 % glycerol, 1 mM EDTA, 0.25 % xylene cyanol, 0.25 % bromophenol blue.

3 Methods

3.1 Preparation of Extracts

A procedure for large-scale preparation of polyadenylation/translation extracts is described in this section. The procedure can be scaled down. Extracts are prepared from synchronized embryos at 90 min of development, a time when cytoplasmic polyadenylation of several endogenous transcripts is maximal [16] (*see* **Note 1**).

Adult flies are maintained in plastic cylinders with a diameter of 40 cm, at a density of ~300 mL flies per cylinder, a constant temperature of 25 °C, and a humidity of 60–65 %. Fertilized eggs layed by 2–3-day-old flies are collected on agar-apple juice plates spread with yeast paste. In order to collect synchronized embryos, plates are changed every hour for a total of three times before the

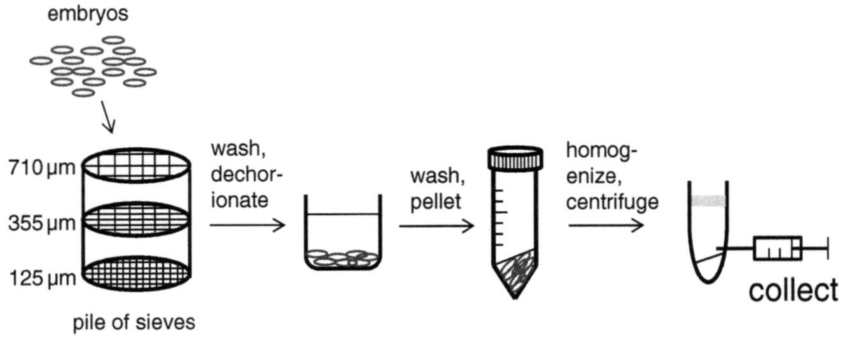

Fig. 1 Schematic representation of the procedure for preparation of cytoplasmic polyadenylation extracts. Embryos are collected in a pile of sieves, washed with EW buffer, dechorionated, washed again with DE buffer, and homogenized in DEI buffer. After centrifugation, the middle (cytoplasmic) phase is collected using a syringe and snap-frozen

final embryo collection, which is taken at 90 min after egg laying. This ensures that flies lay the partially developed embryos they have kept inside in the initial plates and that the last (fourth) plate contains a uniform population of 0–90 min embryos to be used for extract preparation. A schematic representation of the preparation procedure is shown in Fig. 1.

1. Rinse the embryos off the trays with distilled water and collect them using a pile of three sieves with decreasing cutoff sizes. The upper sieve retains adult flies, the middle sieve retains body parts, and the lower sieve retains the embryos. Use a brush to remove all the embryos from the tray.
2. Wash the embryos extensively (5–10 min) with distilled water to remove all debris coming from the trays (e.g., yeast food and agar fragments). Embryos look like sand on the last sieve.
3. Using a spatula, transfer the embryos to a beaker with 500 mL of EW buffer and allow them to settle. Remove as much EW buffer as possible and add again EW buffer to wash the embryos extensively. During the last wash, remove the floating embryos by suction with a pipette.
4. Fill the cylinder with 60 mL of EW buffer and provide vigorous agitation with a magnetic stirrer. Dechorionate embryos by adding 18 mL bleach (3 % sodium hypochlorite final) and incubate 3 min at room temperature with agitation.
5. Quickly transfer the embryos to a sieve and wash by flushing a strong stream of distilled water for about 5 min. It is important to remove well the bleach since it could be detrimental for subsequent assays. If dechorionation is efficient, the embryos become slightly sticky. Important: from this point of the protocol onward, every step should be performed on ice.

6. Transfer the embryos from the sieve to a 50 mL Falcon tube using a spatula. Add 30 mL of DE buffer and spin for 2 min at 1,000 × *g* and 4 °C. Remove the buffer and repeat this procedure adding equal volumes of DE and DEI buffers. After centrifugation, remove the buffer by decanting it.
7. Transfer the embryos to a dounce homogenizer using a spatula and add one volume (with respect to the embryo volume) of DEI buffer. Mechanically homogenize the embryos using a Potter-Elvehjem homogenizer or similar. Homogenization is complete after 20 strokes.
8. Spin the homogenate in a tabletop ultracentrifuge (Beckmann) at 24,000 rpm (40,000 × *g*) in a TLS-55 rotor at 4 °C for 20 min. If the homogenate volume is larger than 15 mL, use the SW41.1 rotor at 18,000 rpm (40,000 × *g*) for 30 min (or any other rotor that allows larger volumes).
9. After centrifugation, the sedimentation of the homogenate results in three layers: an upper fatty layer with a solid consistency, the soluble cytoplasmic interphase, and a pellet containing nuclei and cell debris. It is important to collect only the cytoplasmic interphase without any contamination from the other two layers. This can be done by puncturing the tube with a needle at the bottom of the phase and aspirating with a syringe.
10. Add glycerol to 10 % final, aliquot, flash freeze in liquid nitrogen, and store at −80 °C. For storage longer than 3 months, keep in liquid nitrogen. Keep a separate small aliquot to test the activity of the extract (*see* **Note 2**) and its protein concentration.

3.2 Preparation of Substrate RNA

Substrate RNAs are synthesized in vitro in the presence or absence of radioactive α^{32}P-UTP. In the case of trace-labeled RNAs, their poly(A) status can be directly visualized by gel electrophoresis after the polyadenylation reaction. In the case of cold RNAs, the poly(A) tails are visualized by PAT assay. RNAs should be capped at their 5′ end to minimize degradation in the extract. Either m^{7}GpppG- or ApppG-capped RNAs can be used with no differential effects on polyadenylation. In this section, we provide a protocol to synthesize trace-labeled RNAs containing a m^{7}GpppG cap. To prepare cold RNA, simply omit the α^{32}P-UTP from the reaction.

1. Assemble the transcription reaction as follows:

   4.4 μL 5× transcription buffer (provided with the RNA polymerase).

   2.5 μL 100 mM DTT.

   1.3 μL NTP mix (20 mM ATP, 20 mM CTP, 2 mM UTP).

   1.5 μL α^{32}P-UTP.

   4.4 μL 40 mM m^{7}GpppG cap analog.

1.0 μL RNasin (optional).

1.5 μL RNA polymerase (*see* **Note 3**).

5.0 μL linearized plasmid template (1 μg/μL).

2. Incubate 5 min at 37 °C (*see* **Note 4**).
3. Add 0.4 μL of 100 mM GTP, and incubate 1 h at 37 °C.
4. Add 1 μL RQ DNase, and incubate 15 min at 37 °C.
5. Add 60 μL of H_2O and treat with phenol and phenol/chloroform.
6. Pass the sample through a Sephadex G-25 column to eliminate the non-incorporated nucleotides.
7. Precipitate the RNA: add 1/10 vol of 3 M AcONa and 2.5 vol absolute ethanol, and incubate at 20 °C for 30 min.
8. Recover the RNA by centrifugation, resuspend in H_2O, and check its quality and concentration by agarose gel electrophoresis.

3.3 The Polyadenylation Assay

1. Assemble the polyadenylation reaction on ice. Mix 0.01 pmol of the substrate RNA (*see* **Note 5**) in a final volume of 12.5 μL containing 16.8 μM creatine phosphate, 80 ng/μL creatine kinase, 60 μM amino acids, 24 mM HEPES pH 7.4, 0.4 mM $Mg(OAc)_2$, 48 mM KOAc, and 40 % embryo extract.
2. Incubate at 25 °C for at least 90 min (*see* **Note 6**).
3. Extract the RNA using TRIzol, following the recommendations of the manufacturer.
4. Resuspend the RNA pellet in 7 μL water.

3.4 Measuring the Length of the Poly(A) Tail: PAT Assay

The poly(A) test (PAT) assay is a PCR-based method used to assess the poly(A) tail length of endogenous and exogenous mRNAs independently on the transcript size. Our protocol is based on that described by Salles and Strickland [17], although alternative protocols that include an initial RNA ligation step have been described [18]. A scheme of the PAT reaction and a typical result are shown in Fig. 2.

1. Treat the RNA sample with Turbo DNase following the recommendations of the manufacturer and resuspend it in 5–10 μL of water. Measure concentration.
2. Mix 700–2,500 ng RNA with 0.5 μL of 5 ng/μL oligo(dT) primers in a final volume of 5 μL (*see* **Note 7**). A duplicate reaction should be assembled for the –RT control (see below).
3. Incubate 5 min at 65 °C, spin briefly, and transfer to 42 °C.
4. Add 6.5 μL of a pre-warmed master mix containing:

2.0 μL 5× RT buffer (supplied with the AMV reverse transcriptase).

1.0 μL 100 mM DTT.

0.2 μL 25 mM dNTPs.

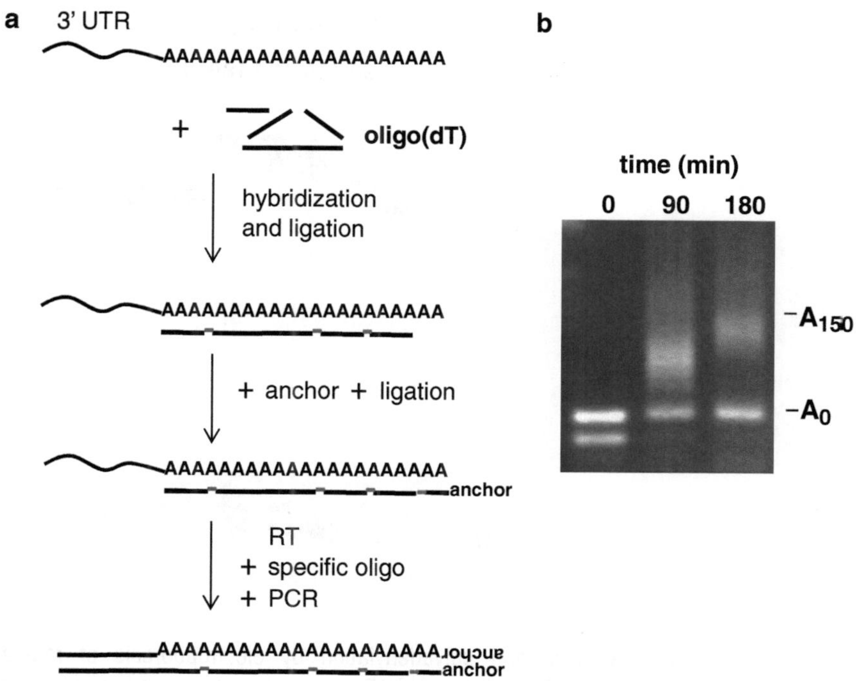

Fig. 2 The PAT assay. (**a**) Schematic representation of the PAT assay (see text for details). (**b**) PAT analysis of *Toll* mRNA. A capped, deadenylated 1.2 Kb piece of RNA corresponding to the 3′ UTR of *Toll* mRNA was added to a polyadenylation reaction. Samples were taken along the reaction; the RNA was extracted, analyzed by PAT, and visualized in a 2 % agarose gel containing ethidium bromide

0.5 μL 10 mM ATP.

0.5 μL 30 U/μL T4 DNA ligase (*see* **Note 8**).

2.0 μL H_2O.

5. Incubate 30 min at 42 °C and briefly spin samples.
6. Add 0.5 μL of oligo(dT)-anchor primer (200 ng/μL), and incubate 2 h to overnight at 12 °C (*see* **Note 9**).
7. Add 0.75 μL AMV reverse transcriptase, and incubate for 1 h at 42 °C. Do not add reverse transcriptase in the –RT control tube. Instead, add 0.75 μL H_2O.
8. Add 15 μL H_2O and incubate 30 min at 70 °C (*see* **Note 10**). Spin samples.
9. Assemble the PCR reaction:

   2.5 μL cDNA from **step 8**.

   2.5 μL 10× PCR buffer (provided with the Taq Supratherm enzyme).

   2.5 μL 2 μM oligo(dT)-anchor primer.

   2.5 μL 2 μM specific forward oligo (*see* **Note 11**).

   0.75 μL 50 mM $MgCl_2$.

   0.5 μL 10 mM dNTPs.

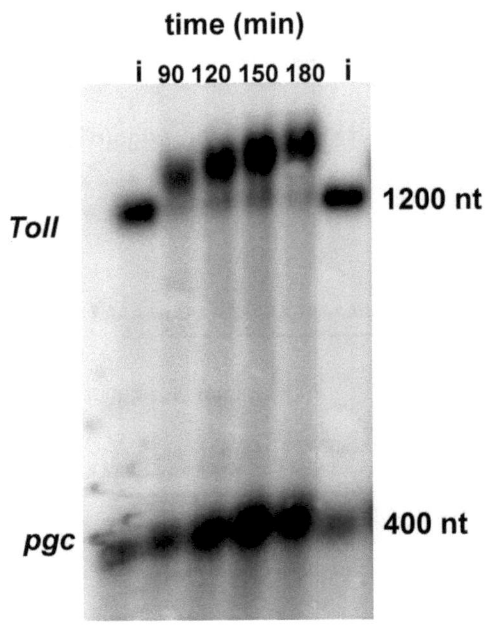

Fig. 3 Visualizing polyadenylation by electrophoresis of radioactive RNA. Radioactive *Toll* and *pgc* 3′ UTRs were added together to a cytoplasmic polyadenylation reaction. After the reaction, the RNA was extracted and directly loaded on a 1 % denaturing agarose gel. *i* input. The result shows that, while *Toll* 3′ UTR is polyadenylated efficiently, *pgc* 3′ UTR is not

0.25 μL Taq Supratherm (5 U/μL).

13.5 μL H_2O.

10. Amplify using the following conditions: 1 min at 94 °C, 1 min at 59 °C, and 1 min at 72 °C, during 25–30 cycles (*see* **Note 12**). Amplified products are visualized by 2 % agarose gel electrophoresis (*see* **Notes 13** and **14**).

3.5 Measuring the Length of the Poly(A) Tail: Electrophoresis of Radioactive RNA

This method is suitable to assess the length of the poly(A) tail of RNA substrates up to 2 Kb in length that are radioactively labeled and exogenously added to the polyadenylation reaction. For RNAs <500 nt, polyadenylation products are separated by denaturing polyacrylamide gel electrophoresis. For RNAs of 500–800, 800–1,500, and 1,500–2,000 nt, 1.5, 1, and 0.8 % denaturing agarose gel electrophoresis is used, respectively. In the following, instructions to run a denaturing 1 % agarose gel are given. A typical result is shown in Fig. 3.

1. Sterilize gel tray, combs, and gel chamber by incubating 10 min in (1:2) formaldehyde (you can reuse this solution many times). Rinse with autoclaved water.
2. For a 400 mL gel, mix 4 g of agarose with 300 mL sterile water and dissolve by heating in the microwave. Add 40 mL 10×

MOPS and 17.6 mL 37 % formaldehyde. Adjust to 400 mL with sterile water, pour into the gel tray, and let solidify.

3. Assemble the gel into the chamber and cover with running buffer.
4. Prepare the RNA sample mix:

   7.0 μL RNA.

   12.5 μL deionized formamide.

   0.8 μL 37 % formaldehyde.

   2 μL 10× MOPS.
5. Incubate samples 5 min at 65 °C and chill on ice.
6. Add 2 μL of Northern dye and 1 μL of 0.1 μg/μL ethidium bromide or other dye suitable to stain RNA (e.g., GelRed). Load samples.
7. Run at 100 V for 8–10 h or at 50 V for 16 h (*see* **Note 15**).
8. Dry gel and expose in the phosphorimager.

4 Notes

1. Polyadenylation extracts can be prepared at other times of development, but regulation might differ depending on the timing of the extract. For example, faithful regulation of cytoplasmic polyadenylation of maternal *Toll* mRNA is only supported by 90 min embryo extracts. Thus, we recommend that extracts are prepared at the developmental time when regulation of the studied endogenous transcript occurs.
2. It is necessary to test the polyadenylation activity of the extract, as preparations are subjected to batch-to-batch variation.
3. Concentrated (20–50 U/μL) T7, T3, or SP6 polymerase is used, depending on the promoter of the template.
4. Initial incubation without GTP is to promote efficient incorporation of the cap analog at the 5′ end of the RNA.
5. Polyadenylation is saturable in the system. We typically use 0.01 pmol of substrate RNA in a reaction of 12.5 μL, although amounts up to 0.1 pmol can also be used. RNAs to be compared must be synthesized in parallel and assayed in the same batch of extract.
6. Cytoplasmic polyadenylation is a slow, distributive process. Under the conditions described in this manuscript, polyadenylation of the substrate is usually visible at 40 min after the initiation of the reaction and increases steadily up to 3 h. The polyadenylation kinetics may change for different RNAs, so we typically assemble reactions in triplicate volumes and take time points at 60, 90, and 180 min.

7. The oligo(dT) primers need to be phosphorylated to support subsequent ligation.
8. The amount of ligase is in excess to ensure that significant activity remains for ligation of the oligo(dT)-anchor primer in a subsequent step.
9. The oligo(dT) mix used in the previous steps is expected to hybridize along the poly(A) tail length, leaving a few adenines uncovered at the 3′ end. The oligo(dT)-anchor primer will cover this gap. Low temperatures allow annealing of the oligo(dT) anchor.
10. This step allows heat inactivation of the ligase and the reverse transcriptase to prevent further modifications during subsequent manipulation of the PAT cDNAs.
11. We recommend primers that range between 20 and 25 nt in length, with a Tm around 65 °C and without significant complementarity to the oligo(dT)-anchor primer. Forward gene-specific primers mapping at 100 nt from the 3′ end of the transcript are optimal.
12. Optimally, the hybridization temperature during the PCR reaction should be 5 °C below the Tm of the oligonucleotides used in the amplification. During the reaction, we take samples at 26, 28, and 30 cycles to increase the chances of visualizing the amplified products under non-saturating conditions.
13. To evaluate the size of the non-adenylated transcript, an additional RT-PCR reaction can be assembled using a reverse oligo hybridizing specifically to the last nucleotides of the 3′ UTR and eliminating the oligo(dT) and oligo(dT) anchor.
14. To confirm the specificity of the PAT-amplified fragments, samples can be digested with a restriction endonuclease that cleaves at a known location from the 5′ end of the amplified products. This digestion should yield a 5′ fragment of constant size and a heterogeneous 3′ fragment containing the poly(A).
15. The length of the run depends on the size of the RNA. One should run the gel long enough to detect 100–150 nt differences. The separation can be monitored with appropriate RNA ladders or by visualization of the rRNAs from the extract. Note that the 28S rRNA from *Drosophila* is cleaved into two species (28Sa and 28Sb) and, thus, three RNAs of sizes ranging between 1.77 and 2.2 Kb will be distinguished if the gel is run long enough.

Acknowledgments

A. V. is supported by an FPI fellowship from the Spanish Ministry of Economy and Competitiveness. Work in our lab is supported by grants BFU2009-08243 and Consolider CSD2009-00080 from the same ministry.

References

1. Gebauer F, Corona DF, Preiss T et al (1999) Translational control of dosage compensation in Drosophila by Sex-lethal: cooperative silencing via the 5′ and 3′ UTRs of msl-2 mRNA is independent of the poly(A) tail. EMBO J 18:6146–6154
2. Castagnetti S, Hentze MW, Ephrussi A et al (2000) Control of oskar mRNA translation by Bruno in a novel cell-free system from Drosophila ovaries. Development 127:1063–1068
3. Lie YS, Macdonald PM (2000) In vitro translation extracts prepared from Drosophila ovaries and embryos. Biochem Biophys Res Commun 270:473–481
4. Andrews S, Snowflack DR, Clark IE et al (2011) Multiple mechanisms collaborate to repress nanos translation in the Drosophila ovary and embryo. RNA 17:967–977
5. Thermann R, Hentze MW (2007) Drosophila miR2 induces pseudo-polysomes and inhibits translation initiation. Nature 447:875–878
6. Jeske M, Meyer S, Temme C et al (2006) Rapid ATP-dependent deadenylation of nanos mRNA in a cell-free system from Drosophila embryos. J Biol Chem 281:25124–25133
7. Hernandez G, Vazquez-Pianzola P, Sierra JM et al (2004) Internal ribosome entry site drives cap-independent translation of reaper and heat shock protein 70 mRNAs in Drosophila embryos. RNA 10:1783–1797
8. Tuschl T, Zamore PD, Lehmann R et al (1999) Targeted mRNA degradation by double-stranded RNA in vitro. Genes Dev 13:3191–3197
9. Gebauer F, Hentze MW (2007) Studying translational control in Drosophila cell-free systems. Methods Enzymol 429:23–33
10. Jeske M, Wahle E (2008) Cell-free deadenylation assays with Drosophila embryo extracts. Methods Enzymol 448:107–118
11. Haley B, Tang G, Zamore PD (2003) In vitro analysis of RNA interference in Drosophila melanogaster. Methods 30:330–336
12. Lipardi C, Baek HJ, Wei Q et al (2005) Analysis of short interfering RNA function in RNA interference by using Drosophila embryo extracts and schneider cells. Methods Enzymol 392:351–371
13. Villalba A, Coll O, Gebauer F (2011) Cytoplasmic polyadenylation and translational control. Curr Opin Genet Dev 21:452–457
14. Fernandez-Miranda G, Mendez R (2012) The CPEB-family of proteins, translational control in senescence and cancer. Age Res Rev 11(4): 460–472. doi:10.1016/j.arr.2012.03.004
15. Coll O, Villalba A, Bussotti G et al (2010) A novel, noncanonical mechanism of cytoplasmic polyadenylation operates in Drosophila embryogenesis. Genes Dev 24:129–134
16. Salles FJ, Lieberfarb ME, Wreden C et al (1994) Coordinate initiation of Drosophila development by regulated polyadenylation of maternal messenger RNAs. Science 266: 1996–1999
17. Salles FJ, Strickland S (1995) Rapid and sensitive analysis of mRNA polyadenylation states by PCR. PCR Meth Appl 4:317–321
18. Charlesworth A, Cox LL, MacNicol AM (2004) Cytoplasmic polyadenylation element (CPE)- and CPE-binding protein (CPEB)-independent mechanisms regulate early class maternal mRNA translational activation in Xenopus oocytes. J Biol Chem 279: 17650–17659

Chapter 6

In Vitro Analysis of Transcriptional Activators and Polyadenylation

Takashi Nagaike and James L. Manley

Abstract

In vitro assays have provided a valuable tool to study the mechanism of 3′ processing of eukaryotic mRNA precursors and have contributed a great deal to the identification of factors that carry out and regulate 3′ processing. Previously, we have shown that transcriptional activators directly enhance polyadenylation by utilizing in vitro transcription-coupled polyadenylation with the prototypical transcription activator GAL4-VP16. In this chapter, we describe a detailed protocol for this assay, which will be useful in examining potential roles for other transcription-related factors in 3′ processing and other questions related to the coupling of transcription and mRNA polyadenylation.

Key words Transcription activators, Promoter, Nuclear extract, Oligo (dT) selection, Poly(A) site, siRNA

1 Introduction

3′ processing of nearly all mRNA precursors (pre-mRNA) in eukaryotic cells consists of an endonucleolytic cleavage followed by polyadenylation [1]. This process involves various polyadenylation factors such as cleavage-polyadenylation specificity factor (CPSF), cleavage stimulatory factor (CstF), cleavage factor I and II (CFI and CFII), and poly(A) polymerase. When 3′ processing takes place, these factors form a macromolecular complex on the pre-mRNA through recognition of the poly(A) site.

Considerable evidence now indicates that 3′ processing of pre-mRNA is often coupled to transcription [2–5]. This coupling is believed to ensure accurate, efficient, and rapid processing of nascent pre-mRNAs. An important factor in the coupling of transcription and 3′ processing is the carboxy-terminal domain (CTD) of the RNA polymerase II's (RNAP II) largest subunit. The CTD has a significant role in enhancing the efficiency of 3′ processing in both in vivo and in vitro assays [6, 7].

Joanna Rorbach and Agnieszka J. Bobrowicz (eds.), *Polyadenylation: Methods and Protocols*, Methods in Molecular Biology, vol. 1125, DOI 10.1007/978-1-62703-971-0_6, © Springer Science+Business Media New York 2014

In addition to the CTD, the general transcription factor TFIID, which directs specific transcription initiation from core promoter regions, also links transcription to polyadenylation. TFIID interacts with and integrates CPSF subunits into pre-initiation complexes. After transcription initiates, CPSF appears to be transferred from TFIID to elongating RNA pol II, likely via the CTD [8].

Transcription elongation also contributes to transcription-coupled 3′ processing. Positive transcription elongation factor b (P-TEFb) is a kinase that phosphorylates ser2 residues in the CTD heptapeptide repeats (YSPTSPS). Ser2 phosphorylation promotes the recruitment and/or stable interaction of 3′ processing factors to the site of 3′ processing on many genes [9, 10]. Another example is the PAF1 complex (PAF1c) which functions as an elongation factor in several ways. For example, it acts synergistically with another elongation factor TFIIS to enhance elongation [11]. We showed previously that PAF1c associates with 3′ processing factors and functions in transcription-coupled polyadenylation [12]. Depletion of PAF1c from cell extracts in vitro inhibited polyadenylation, but not transcription or splicing, and from cells in vivo reduced expression and extended transcription of a target gene [12]. The interaction between 3′ processing factors and PAF1c is also observed in yeast, and poly(A) tail length [13] and poly(A) site selection [14] were found to be altered in PAF mutant cells. These findings all support a role for PAF1c in helping to couple transcription to 3′ processing.

Other factors extend the coupling of transcription and polyadenylation. For example, overexpression of a mutant form of HSF1, the transcription activator required for the expression of heat-responsive genes, inhibits 3′ processing of Hsp70 mRNA in heat-shocked cells [15]. This suggests that sequence-specific transcription factors (TFs) can also function in transcription-coupled polyadenylation. In support of this, transient cotransfection assays showed that TFs can indeed increase not only transcription but also 3′-end processing [16].

Previously, by using in vitro transcription-coupled polyadenylation assays, we provided direct evidence that transcription activators enhance polyadenylation and that this requires PAF1c [17]. We showed that GAL4-VP16 can stimulate transcription-coupled (but not uncoupled) polyadenylation in vitro with HeLa nuclear extract (NE) using naked DNA templates and that depletion of PAF1c markedly reduced the VP16-induced polyadenylation, but not transcription. Moreover, the VP16 activation domain was shown to interact directly with and to recruit PAF1c to DNA templates [17].

This in vitro transcription-coupled polyadenylation system was also utilized to show that Ssu72, a phosphatase that dephosphorylates the CTD, is required for transcription-coupled (but not uncoupled) polyadenylation [18]. Thus, in vitro transcription-coupled polyadenylation with GAL4-VP16 has given us new insights into 3′ processing in eukaryotes. In this chapter, the detailed protocol of this in vitro assay will be described.

2 Materials

2.1 Preparation of Linear DNA Templates

1. pG5E4SVL [17].
2. Forward primer (5′-CAACTGTTGGGAAGGGCGATC-3′).
3. Reverse primer (5′-CTACAGCGTGAGCTATGAGAAAGCG-3′).
4. Taq DNA polymerase (Invitrogen).
5. Phenol:chloroform:isoamyl alcohol (25:24:1).
6. Ethanol.
7. 3 M sodium acetate (pH 5.2).
8. Agarose.
9. QIA quick Gel Extraction Kit (QIAGEN).

2.2 In Vitro Transcription-Coupled Polyadenylation

1. Linear DNA templates containing GAL4-binding sites upstream of E4 core promoter and SVL poly(A) site downstream.
2. HeLa nuclear extract [19] (*see* **Note 1**).
3. His-tagged GAL4-VP16 [17].
4. 20× NTP mixture containing 10 mM ATP, GTP, CTP, and 0.3 mM UTP.
5. [α-^{32}P]UTP (400 mCi/mmol).
6. 100 mM $MgCl_2$.
7. Creatine phosphate (di-tris, Sigma Aldrich): 0.5 M solution in water (*see* **Note 2**).
8. PEG 8000 (Sigma Aldrich): 40 % solution in water (*see* **Note 3**).
9. Proteinase K.
10. Proteinase K reaction buffer containing 10 mM Tris–HCl (pH 8.0), 1 % SDS, and 10 mM EDTA.
11. Phenol:chloroform:isoamyl alcohol (25:24:1).
12. Ethanol.
13. 3 M sodium acetate (pH 5.2).
14. 20 mg/mL glycogen solution (Roche).

2.3 Oligo (dT) Selection of Polyadenylated RNA

1. Oligo(dT) magnet beads (BioMag Oligo (dT)20, Polysciences).
2. Oligo(dT) selection binding buffer containing 10 mM Tris–HCl (pH 8.0) and 500 mM NaCl.
3. Oligo(dT) selection wash buffer containing 10 mM Tris–HCl (pH 8.0) and 150 mM NaCl.
4. Magnetic stand.
5. Ethanol.
6. 3 M sodium acetate (pH 5.2).
7. 20 mg/mL glycogen solution (Roche).

2.4 Analysis of Polyadenylated RNA

1. Gel loading buffer containing 95 % formamide, 20 mM EDTA, 0.025 % bromophenol blue, and 0.025 % xylene cyanol.
2. 40 % Acrylamide/Bis-acrylamide solution (19:1).
3. 1× TBE buffer containing 89 mM Tris–HCL, 89 mM boric acid, and 2 mM EDTA.
4. 5 % denaturing gel solution containing 5 % Acrylamide/Bis-acrylamide, 1× TBE, and 7 M urea.
5. 20 % ammonium persulfate solution.
6. *N*,*N*,*N*′,*N*′-tetramethyl-ethylenediamine.
7. 3MM filter paper (Whatman).
8. Gel drier (Bio-Rad).
9. Phosphor imager.

2.5 Preparation of Mini-Scale Nuclear Extract from siRNA-Treated Cells

1. Disposable cell scraper.
2. PBS buffer.
3. Buffer A containing 10 mM HEPES (pH 8.0), 1.5 mM $MgCl_2$, 10 mM KCl, and 1 mM DTT (*see* **Note 4**).
4. Buffer C containing 20 mM HEPES (pH 8.0), 1.5 mM $MgCl_2$, 25 % glycerol, 420 mM NaCl, 0.2 mM EDTA, 1 mM DTT, and 0.5 mM PMSF (*see* **Note 4**).
5. Buffer D containing 20 mM HEPES (pH 8.0), 20 % glycerol, 100 mM KCl, 0.2 mM EDTA, 1 mM DTT, and 0.5 mM PMSF (*see* **Note 4**).
6. 1 mL syringe.
7. Needles (25-gauge).
8. Small stir bar (*see* **Note 5**).
9. Slide-A-Lyzer MINI Dialysis Device (Pierce).

3 Methods

3.1 Preparation of Linear DNA Templates

1. Linear DNA templates containing GAL4-binding sites, adenovirus E4 core promoter, and SVL poly(A) site were prepared by PCR as follows. For 100 μL reaction, add 1 μL (100 ng) of pG5E4SVL (Fig. 1), 1 μL (100 pmol) of forward primer, 1 μL (100 pmol) of reverse primer, 10 μL of 10× PCR buffer minus Mg, 2 μL of 10 mM dNTP mixture, 3 μL of 50 mM $MgCl_2$, 0.5 μL of Taq DNA polymerase, and 81.5 μL of milli-Q water (*see* **Note 6**).
2. After PCR reaction, add 100 μL of phenol:chloroform:isoamyl alcohol and mix thoroughly.
3. Recover and transfer the aqueous phase to a new tube.

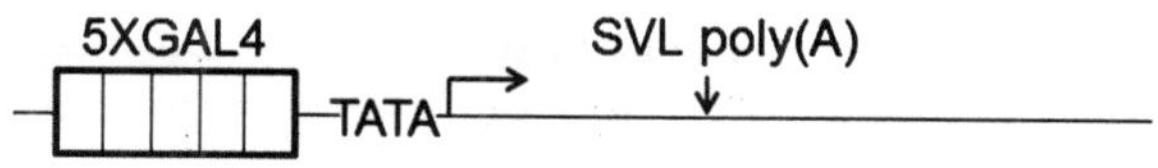

Fig. 1 DNA templates used in transcription-coupled polyadenylation in vitro. The schematic structure of pG5E4SVL. Five tandem repeats of GAL4-binding sites are located upstream of E4 core promoter and SVL poly(A) site downstream. Transcription run-off site is located about 600 base downstream of SVL poly(A) site

4. Add 10 μL of 3 M sodium acetate and 250 μL of 100 % ethanol. Mix thoroughly.
5. Place on ice for 20 min.
6. Spin down for 15 min at maximum speed in a microcentrifuge at 4 °C.
7. Wash the pellet with 500 μL of 70 % ethanol and air-dry the pellet.
8. Dissolve the pellet in 10 μL of 10 mM Tris–HCl (pH 8.0).
9. The PCR products were then gel-purified on 1 % agarose gel and using QIA quick Gel Extraction Kit.
10. The PCR products were again ethanol precipitated as described above (**steps 4–7**).
11. Dissolve the pellet in 10 mM Tris–HCl (pH 8.0) and adjust its concentration (500 ng/μL).

3.2 In Vitro Transcription-Coupled Polyadenylation Reaction

1. Transcription-coupled polyadenylation was carried out at 30 °C for 1 h in reaction mixtures (20 μL) as follows. Mix 3.2 μL of milli-Q water, 0.8 μL of 100 mM $MgCl_2$, 1 μL of 20× NTP mixture solution, 1 μL of [α-32P]UTP (10 μCi), 0.8 μL of 0.5 M creatine phosphate (di-tris), 1 μL of the DNA templates (500 ng), 1.2 μL of 40 % PEG, 1 μL of 100 ng/μL GAL4-VP16, and 10 μL NE (*see* **Note 7**).
2. The reaction was stopped by adding 30 μL of proteinase K reaction buffer and 1 μL of 10 mg/mL proteinase K.
3. Incubate for 15 min at 37 °C.
4. Add 50 μL of phenol:chloroform:isoamyl alcohol (25:24:1) and mix thoroughly.
5. Recover and transfer the aqueous phase to a new tube.
6. Add 1 μL of glycogen, 5 μL of 3 M sodium acetate, and 125 μL of 100 % ethanol. Mix thoroughly.
7. Place on ice for 20 min.
8. Spin down for 15 min at maximum speed in a microcentrifuge at 4 °C.
9. Wash the pellet with 200 μL of 80 % ethanol and air-dry the pellet.
10. Dissolve the pellet in 50 μL of 10 mM Tris–HCl (pH 8.0).

3.3 Oligo (dT) Selection of Polyadenylated RNA

1. Dispense 10 μL of BioMag Oligo dT(20) into a new tube. Using a magnetic stand, pull the magnetic particles to the side of the tube for 30 s.
2. Remove the supernatant.
3. Wash the BioMag Oligo dT(20) once with 50 μL of the binding buffer.
4. Magnetically separate with the magnet stand for 30 s and remove the supernatant.
5. Resuspend in 50 μL of the binding buffer.
6. Incubate the RNA sample at 55 °C for 5 min to disrupt secondary structures.
7. Add 5 μL of 5 M NaCl to the RNA sample.
8. Combine the RNA sample and the BioMag Oligo dT(20). Mix gently and hybridize at room temperature for 10 min.
9. Magnetically separate with the magnet stand for 30 s.
10. Take and keep supernatant as nonpolyadenylated RNA.
11. Wash the particles twice with 50 μL of the wash buffer.
12. Elute the bound polyadenylated RNA with 50 μL of milli-Q water at 55 °C for 5 min.
13. Magnetically separate with the magnet stand for 30 s and transfer the supernatant (polyadenylated RNA) to a new tube.

3.4 Analysis and Detection of Polyadenylated RNA

1. Add 1 μL of glycogen to polyadenylated RNA fraction.
2. Add 5 μL of 3 M sodium acetate and 125 μL of 100 % ethanol. Mix thoroughly.
3. Place on ice for 20 min.
4. Spin down for 15 min at maximum speed in a microcentrifuge at 4 °C.
5. Wash the pellet with 200 μL of 80 % ethanol and air-dry the pellet.
6. Dissolve the pellet in 5 μL of loading dye solution (poly(A)+ sample).
7. Likewise, ethanol precipitate nonpolyadenylated RNA and dissolve in 10 μL of loading dye solution (poly(A)– sample).
8. Load 100 % of poly(A)+ sample and 5–20 % of poly(A)– sample on 5 % denaturing gel (*see* **Note 8**).
9. Run the gel at 500 V until bromophenol blue reaches the bottom of the gel.
10. Separate the gel glasses and carefully transfer the gel to a piece of filter paper.
11. Wrap the gel with plastic film and dry on a gel dryer (Bio-Rad) at 80 °C for 30 min.
12. Radioactivity was detected using phosphorimager (Fig. 2) (*see* **Note 9**).

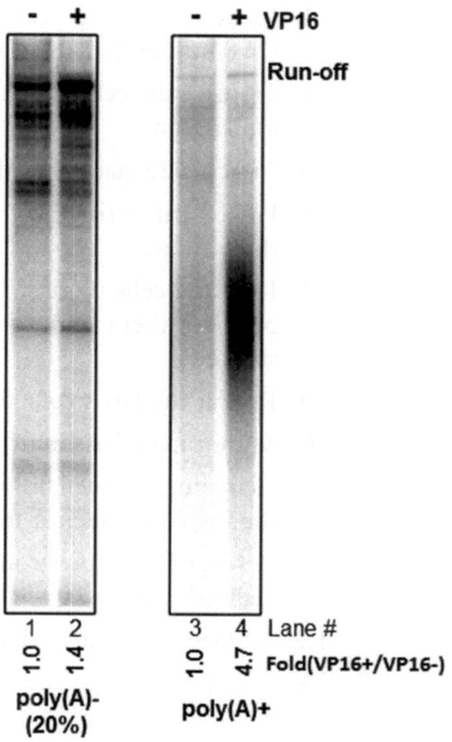

Fig. 2 Transcription-coupled polyadenylation assay with or without GAL4-VP16. After reactions in NE, transcripts were isolated, separated by oligo (dT) selection into poly(A)– and poly(A)+ fraction, and analyzed on 5 % denaturing gel containing 7 M urea. In the presence of GAL4-VP16, transcription was only modestly increased (compare lanes *1* and *2*), while polyadenylation was clearly induced by GAL4-VP16 (compare lanes *3* and *4*). Reprinted with permission from ref. 17 (Copyright 2011, Elsevier)

3.5 Preparation of Small-Scale NE from siRNA-Treated Cells

To analyze the function of protein of your interest in polyadenylation, it has been useful to immuno-deplete the protein from NE and test activity of the depleted extract for polyadenylation. Recently, however, siRNA technology has provided us with an alternate, often more efficient, and often better (avoids problem of co-depletion) method to prepare small-scale NE from cells where a target protein is specifically depleted by siRNA-mediated knockdown. The following protocol is based on the method described by Lee et al. [20] with some modifications. One plate (15 cm) of cells at 80 % confluence yields about 100 μL of NE:

1. Harvest siRNA-treated cells by cell scraper and transfer to 15 mL tube (*see* **Note 10**).
2. Wash the cells by resuspending with 10 mL of PBS buffer.
3. Centrifuge cells for 5 min at 100×*g*. Remove supernatant carefully.

4. Packed cells are resuspended in one packed cell volume (150 μL per 2×10^7 cells) of buffer A.
5. Transfer the cells to a 1.5 mL tube and leave it to swell on ice for 15 min.
6. Attach a 25-gauge needle to a 1 mL syringe.
7. Fill a 1 mL syringe with buffer A and expel all of the air from the syringe.
8. Lyse the cells by drawing the cells from the tube slowly and pushing them through the 25-guage needle with a rapid stroke.
9. Repeat the fast stroke four more times.
10. Monitor cell lysis under microscope.
11. Centrifuge the cell homogenate for 5 min at $1{,}500 \times g$ to pellet the insoluble nuclear fraction. Carefully remove supernatant.
12. Resuspend the pellet in two-thirds of one packed cell volume (100 μL per 2×10^7 cells) of Buffer C.
13. Add sterile small stir bar into the tube.
14. Stir the homogenate on the magnetic stirrer for 30 min at 4 °C.
15. Centrifuge for 5 min at maximum speed in a microcentrifuge at 4 °C.
16. Take supernatant (NE) carefully.
17. Dialyze the NE for 2 h against Buffer D using Slide-A-Lyzer MINI Dialysis Device.
18. The dialyzed NE (about 100 μL from 2×10^7 cells) is now ready for transcription-coupled polyadenylation (Fig. 3) (*see* **Note 11**), or it is flash-frozen in liquid nitrogen and stored at −80 °C.

4 Notes

1. Obtaining high quality of NE is the key to success of the assay. To achieve this, prepare NE from HeLa cells growing very well.
2. Creatine phosphate is added to enhance the efficiency of 3′ processing.
3. Concentrating reagent such as PEG was used to enhance both transcription and polyadenylation. PVA can be used as an alternate.
4. Buffers A, C, and D are stored at 4 °C without adding DTT and PMSF. DTT and PMSF should be added immediately before use.

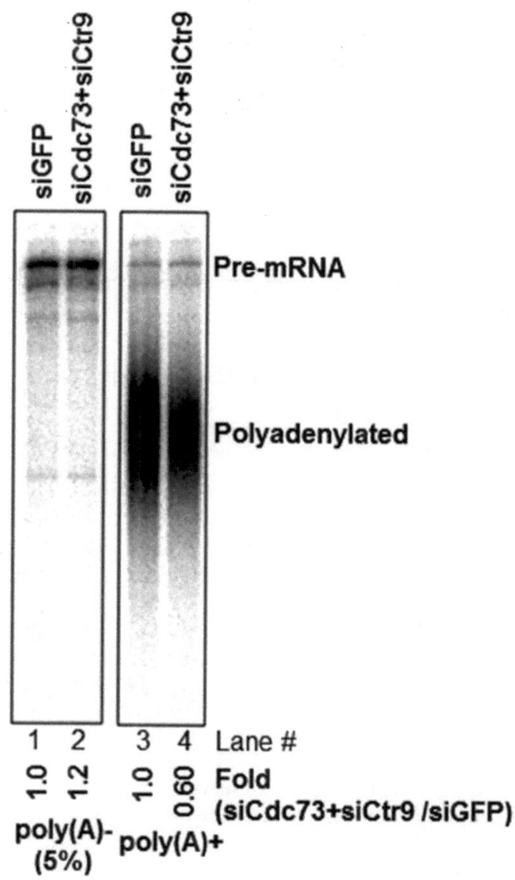

Fig. 3 Transcription-coupled polyadenylation with small-scale NE prepared from siRNA-treated cells. Small-scale NE was prepared from HEK293T cells where PAF1c subunit genes (Cdc73 and Ctr9) were knocked down. siRNA targeting GFP gene was used as a control siRNA. After reactions with respective NE, transcripts were analyzed as in Fig. 2. PAF1c depletion reduced polyadenylation markedly, while pre-mRNA accumulation was slightly increased. Reprinted with permission from ref. 17 (Copyright 2011, Elsevier)

5. Small stir bar is autoclaved before use.
6. DNA templates can be linearized by restriction enzymes (XhoI which cuts upstream of GAL4-binding sites and ApaLI which cuts about 600 base downstream of poly(A) site).
7. We recommend preparing a master mix for multiple reactions to enable accurate pipetting.
8. This is because the radioactivity of poly(A)– samples is much stronger than that of poly(A)+ sample.
9. To detect radioactivity and gain a clear image, overnight exposure is recommended.

10. When we prepared PAF1c-depleted NE by siRNA knockdown, HEK293 cells were used due to their high transfection efficiency (Fig. 3). Since HEK293T cells are very easy to detach from culture plate, we scrape the cells instead of trypsinization.
11. Check the knockdown efficiency by western blotting.

Acknowledgement

Work from the authors' lab is supported by grants from the NIH (J.L.M).

References

1. Colgan DF, Manley JL (1997) Mechanism and regulation of mRNA polyadenylation. Genes Dev 11:2755–2766
2. Hirose Y, Manley JL (2000) RNA polymerase II and the integration of nuclear events. Genes Dev 14:1415–1429
3. Bentley D (2002) The mRNA assembly line: transcription and processing machines in the same factory. Curr Opin Cell Biol 14:336–342
4. Maniatis T, Reed R (2002) An extensive network of coupling among gene expression machines. Nature 416:499–506
5. Proudfoot NJ, Furger A, Dye MJ (2002) Integrating mRNA processing with transcription. Cell 108:501–512
6. McCracken S, Fong N, Yankulov K et al (1997) The C-terminal domain of RNA polymerase II couples mRNA processing to transcription. Nature 385:357–361
7. Hirose Y, Manley JL (1998) RNA polymerase II is an essential mRNA polyadenylation factor. Nature 395:93–96
8. Dantonel JC, Murthy KG, Manley JL et al (1997) Transcription factor TFIID recruits factor CPSF for formation of 3′ end of mRNA. Nature 389:399–402
9. Ahn SH, Kim M, Buratowski S (2004) Phosphorylation of serine 2 within the RNA polymerase II C-terminal domain couples transcription and 3′ end processing. Mol Cell 13:67–76
10. Ni Z, Schwartz BE, Werner J et al (2004) Coordination of transcription, RNA processing, and surveillance by P-TEFb kinase on heat shock genes. Mol Cell 13:55–65
11. Kim J, Guermah M, Roeder RG (2010) The human PAF1 complex acts in chromatin transcription elongation both independently and cooperatively with SII/TFIIS. Cell 140:491–503
12. Rozenblatt-Rosen O, Nagaike T, Francis JM et al (2009) The tumor suppressor Cdc73 functionally associates with CPSF and CstF 3′ mRNA processing factors. Proc Natl Acad Sci U S A 106:755–760
13. Mueller CL, Porter SE, Hoffman MG et al (2004) The Paf1 complex has functions independent of actively transcribing RNA polymerase II. Mol Cell 14:447–456
14. Penheiter KL, Washburn TM, Porter SE et al (2005) A posttranscriptional role for the yeast Paf1-RNA polymerase II complex is revealed by identification of primary targets. Mol Cell 20:213–223
15. Xing H, Mayhew CN, Cullen KE et al (2004) HSF1 modulation of Hsp70 mRNA polyadenylation via interaction with symplekin. J Biol Chem 279:10551–10555
16. Rosonina E, Bakowski MA, McCracken S et al (2003) Transcriptional activators control splicing and 3′-end cleavage levels. J Biol Chem 278:43034–43040
17. Nagaike T, Logan C, Hotta I et al (2011) Transcriptional activators enhance polyadenylation of mRNA precursors. Mol Cell 41: 409–418
18. Xiang K, Nagaike T, Xiang S et al (2010) Crystal structure of the human symplekin-Ssu72-CTD phosphopeptide complex. Nature 467:729–733
19. Dignam JD, Lebovitz RM, Roeder RG (1983) Accurate transcription initiation by RNA polymerase II in a soluble extract from isolated mammalian nuclei. Nucleic Acids Res 11: 1475–1489
20. Lee KA, Green MR (1990) Small-scale preparation of extracts from radiolabeled cells efficient in pre-mRNA splicing. Methods Enzymol 181:20–30

Chapter 7

Fluorescence Detection of Polyadenylation Reaction Through the Coordination of Adenosine$_2$–Coralyne–Adenosine$_2$

Jia-Hui Lin and Wei-Lung Tseng

Abstract

A simple, sensitive, and selective-detection system for polyadenylation reaction was developed based on the fact that coralyne induces conformational change of polyadenosine [poly(A)] oligonucleotide through adenosine$_2$–coralyne–adenosine$_2$ coordination. Double-strand chelating dye SYBR Green I fluoresces weakly in the presence of random coil of poly(A). Coralyne enables poly(A) to change its conformation from a random coil to a folded structure, which leads to an increased fluorescence of SYBR Green I. When the 3′ end of RNA is polyadenylated with poly(A) polymerase, coralyne-induced fluorescence enhancement of the poly(A)-SYBR Green I complex is well suited for monitoring polyadenylation reaction.

Key words Coralyne, SYBR Green I, Fluorescence, Polyadenylation

1 Introduction

Small molecules and metal ions that drive the formation of non-Watson–Crick base pairs are of considerable interest in the design of novel sensor, DNA machine, DNA nanostructure, and new therapeutic agents [1–4]. In contrast to DNA aptamers, G-quadruplexes, DNAzyme, i-motifs, thymine(T)-containing DNA, and cytosine(C)-containing DNA [5–10], very little research has been devoted to understanding the binding of polyadenosine [poly(A)] to small molecules. In eukaryotic cells, invariably all mRNAs have a poly(A)-tail at their 3′-end, which plays a significant role in the initiation of translation, contributes to the stability and maturation of mRNA, and promotes an mRNA's translational efficiency [11, 12]. Poly(A) polymerase not only catalyzes the addition of poly(A) to the 3′-end of mRNA but also initiates the endonucleolytic cleavage of pre-mRNA. Neo-poly(A) polymerase has been demonstrated to be overexpressed in human tumor [13, 14]. Consequently, understanding of the binding of small

Joanna Rorbach and Agnieszka J. Bobrowicz (eds.), *Polyadenylation: Methods and Protocols*, Methods in Molecular Biology, vol. 1125, DOI 10.1007/978-1-62703-971-0_7,

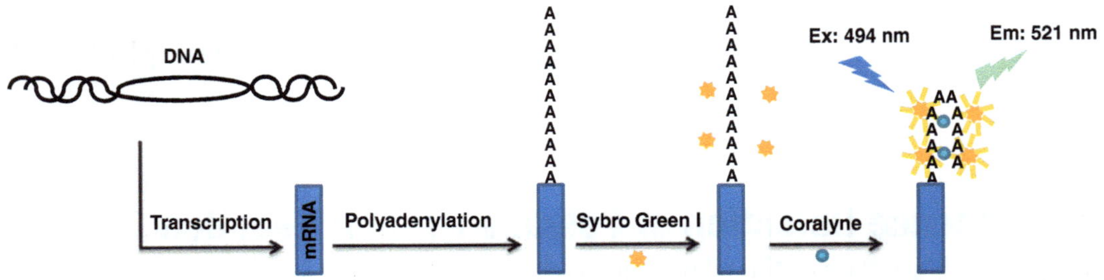

Fig. 1 Coralyne-induced fluorescence enhancement of the poly(A)-SYBR Green I complex was used for monitoring polyadenylation

molecules to poly(A)-tail of mRNA is prerequisite for screening of potential therapeutic agents.

Coralyne, a synthetic planar alkaloid, was identified to possess strong binding affinity to polydeoxyadenosine [poly(dA)] with an apparent association constant of 1.8×10^6 at pH 7 [15] and with a stoichiometry of one coralyne per four adenine bases [16]. The presence of coralyne triggers a self-structure formation of poly(A) or poly(dA). Asymmetric hydrogen bonds between Watson–Crick and Hoogsteen edges induce the formation of A·A base pairs. In contrast to other protoberberine alkaloids, such as berberine, palmatine, and sanguinarine, low-toxic coralyne has relatively high antitumor activity [17]. Based on the coordination of A_2–coralyne–A_2, current methods for determining the interaction of coralyne with poly(A) or poly(dA) include circular dichroism spectroscopy [18], UV–Vis melting experiments [19], gel electrophoresis [20], isothermal titration calorimetry [15], nanoparticle-based colorimetric assay [21], and fluorescence labeling [22].

In this chapter, we describe a sample preparation method for the monitoring of polyadenylation reaction through the coordination of A_2–coralyne–A_2 and the labeling of double-strand-chelating dye SYBR Green I (Fig. 1).

2 Materials

1. Solutions for the labeling of double-strand DNA: 10,000× SYBR Green I. Store at −20 °C, protected from light (*see* **Note 1**).
2. Reagents for the transcription of DNA to RNA: AmpliScribe T7-flash transcription kit (Epicentre Biotechnologies, Wisconsin, USA).
3. Reagents for the addition of poly(A)-tails to the 3′ end of any RNA: poly(A) polymerase tailing kit (Epicentre Biotechnologies, Wisconsin, USA).
4. Fluorescence spectrophotometer (F-7000, Hitachi, Tokyo, Japan).

3 Methods

3.1 Preparation of SYBR Green I

1. Dilute 10,000× of SG to 250× in dimethyl sulfoxide (*see* **Note 2**).
2. Dilute the resulting solution to 6.25× in ultrapure water in order to make a stock solution.

3.2 Preparation of RNA

1. Prepare a solution containing of AmpliScribe T7-flash reaction buffer (2 μL, 10×), adenosine triphosphate (1.8 μL, 100 mM), cytosine triphosphate (1.8 μL, 100 mM), guanosine triphosphate (1.8 μL, 100 mM), uridine triphosphate (1.8 μL, 100 mM), dithiothreitol (2 μL, 100 mM), RiboGuard RNase inhibitor (0.5 μL), and AmpliScribe T7-flash enzyme (2 μL).
2. Pipette 2 μL of 0.5 μg/μL control DNA template to the resulting solution (*see* **Notes 3** and **4**).
3. Adjust total reaction volume to 20 μL with RNase-free water.
4. Incubate at 37 °C for 1 h.
5. Treat the produced RNA with DNase I (1 μL, 1 MBU/μL), then incubate it at 37 °C for 15 min (*see* **Note 5**).

3.3 Sample Preparation in the Monitoring of Polyadenylation

1. Prepare a solution containing poly(A) polymerase reaction buffer (10 μL, 10×), adenosine triphosphate (10 μL, 10 mM), RiboGuard RNase inhibitor (2.5 μL), and the produced RNA (20 μL, 3 μg/μL) (*see* **Note 6**).
2. Pipette 5 μL of 4 units/μL poly(A) polymerase to the resulting solution.
3. Adjust total reaction volume to 100 μL with RNase-free water.
4. Incubate at 37 °C for different reaction times.
5. Add 5 μL of the mixture to 500 μL of 100 mM HEPES solution (pH 7.0) containing 200 mM NaCl, 1 μM coralyne, and 160 nM SYBR Green I at 5-min intervals over a period of 1 h (*see* **Notes 7** and **8**).
6. The fluorescence spectra of the resulting solutions are recorded when the excitation wavelength is set to 494 nm.

4 Notes

1. The wavelengths of maximum excitation and emission of SYBR Green I are centered at 494 nm and 521 nm, respectively. The fluorescence of SYBR Green I is a weak fluorescence in the presence of single-stranded DNA while the binding of SYBR Green I to double-stranded DNA provides more than 11-fold enhancement in its fluorescence.
2. The concentration of a solution of 10,000× SG is calculated to be 10 mg/mL, which corresponds to a molar concentration of 19.6 mM [23].

3. The control DNA template contains 4.2-kb linearized plasmid and 1.4-kb lambda DNA insert, which produces a 1,380-b runoff transcript.
4. 1 μg of control DNA template produces 160–180 μg of a ~1.4-kb RNA in a standard 20 μL AmpliScribe T7-flash reaction.
5. 1 Molecular Biology Unit (MBU) of DNase I digests 1 μg of pUC19 DNA to oligodeoxynucleotides in 10 min.
6. 10× poly(A) polymerase reaction buffer consists of 0.5 M Tris–HCl (pH 8.0), 2.5 M NaCl, and 0.1 M $MgCl_2$.
7. Coralyne is capable of trigging a self-structure formation of poly(A)-tails at the 3′ end of the produced RNA.
8. The fluorescence of SYBR Green I greatly increases upon interaction with the complex of poly(A)-tails and coralyne through intercalation and minor groove binding.

Acknowledgments

The author would like to thank the National Science Council (NSC 100-2628-M-110-001-MY4) and National Sun Yat-sen University-Kaohsiung Medical University Joint Research Center for the financial support of this work.

References

1. Huang CC, Chang HT (2008) Aptamer-based fluorescence sensor for rapid detection of potassium ions in urine. Chem Commun 1461–1463
2. Li D, Wieckowska A, Willner I (2008) Optical analysis of Hg^{2+} ions by oligonucleotide-gold-nanoparticle hybrids and DNA-based machines. Angew Chem Int Ed 47:3927–3931
3. Mitra D, Di Cesare N, Sleiman HF (2004) Self-assembly of cyclic metal-DNA nanostructures using ruthenium tris(bipyridine)-branched oligonucleotides. Angew Chem Int Ed 43:5804–5808
4. Zhao C, Qu K, Xu C et al (2011) Triplex inducer-directed self-assembly of single-walled carbon nanotubes: a triplex DNA-based approach for controlled manipulation of nanostructures. Nucleic Acids Res 39:3939–3948
5. Stojanovic MN, Landry DW (2002) Aptamer-based colorimetric probe for cocaine. J Am Chem Soc 124:9678–9679
6. Smirnov I, Shafer RH (2000) Lead is unusually effective in sequence-specific folding of DNA. J Mol Biol 296:1–5
7. Li J, Lu Y (2000) A highly sensitive and selective catalytic DNA biosensor for lead ions. J Am Chem Soc 122:10466–10467
8. Choi J, Kim S, Tachikawa T et al (2011) pH-induced intramolecular folding dynamics of i-motif DNA. J Am Chem Soc 133:16146–16153
9. Ono A, Togashi H (2004) Highly selective oligonucleotide-based sensor for mercury(II) in aqueous solutions. Angew Chem Int Ed 43:4300–4302
10. Ono A, Cao S, Togashi H et al (2008) Specific interactions between silver(I) ions and cytosine-cytosine pairs in DNA duplexes. Chem Commun 4825–4827
11. Wickens M, Anderson P, Jackson RJ (1997) Life and death in the cytoplasm: messages from the 3′ end. Curr Opin Genet Dev 7:220–232
12. Dower K, Kuperwasser N, Merrikh H et al (2004) A synthetic A tail rescues yeast nuclear accumulation of a ribozyme-terminated transcript. RNA 10:1888–1899

13. Topalian SL, Kaneko S, Gonzales MI et al (2001) Identification and functional characterization of neo-poly(A) polymerase, an RNA processing enzyme overexpressed in human tumors. Mol Cell Biol 21:5614–5623
14. Topalian SL, Gonzales MI, Ward Y et al (2002) Revelation of a cryptic major histocompatibility complex class II-restricted tumor epitope in a novel RNA-processing enzyme. Cancer Res 62:5505–5509
15. Xing F, Song G, Ren J et al (2005) Molecular recognition of nucleic acids: coralyne binds strongly to poly(A). FEBS Lett 579:5035–5039
16. Polak M, Hud NV (2002) Complete disproportionation of duplex poly(dT)*poly(dA) into triplex poly(dT)*poly(dA)*poly(dT) and poly(dA) by coralyne. Nucleic Acids Res 30:983–992
17. Gatto B, Sanders MM, Yu C et al (1996) Identification of topoisomerase I as the cytotoxic target of the protoberberine alkaloid coralyne. Cancer Res 56:2795–2800
18. Jain SS, Polak M, Hud NV (2003) Controlling nucleic acid secondary structure by intercalation: effects of DNA strand length on coralyne-driven duplex disproportionation. Nucleic Acids Res 31:4608–4615
19. Cetinkol OP, Hud NV (2009) Molecular recognition of poly(A) by small ligands: an alternative method of analysis reveals nanomolar, cooperative and shape-selective binding. Nucleic Acids Res 37:611–621
20. Persil O, Santai CT, Jain SS et al (2004) Assembly of an antiparallel homo-adenine DNA duplex by small-molecule binding. J Am Chem Soc 126:8644–8645
21. Song G, Chen C, Qu X et al (2008) Small-molecule-directed assembly: a gold nanoparticle-based strategy for screening of homo-adenine DNA duplex binders. Adv Mater 20:706–710
22. Lin YH, Tseng WL (2011) Fluorescence detection of coralyne and polyadenylation reaction using an oligonucleotide-based fluorogenic probe. Chem Commun 47:11134–11136
23. Zipper H, Brunner H, Bernhagen J et al (2004) Investigations on DNA intercalation and surface binding by SYBR Green I, its structure determination and methodological implications. Nucleic Acids Res 32:e103

Chapter 8

Small Molecule Recognition of Poly(A)

Özgül Persil Çetinkol

Abstract

The binding of small molecules to non-canonical nucleic acid structures has been a major focus of rational drug design. Among the non-canonical nucleic acid structures, targeting poly(A) using small molecules has attracted a special interest due to the cellular functions of poly(A) tails. Here, the methods for determining the binding of a small molecule to poly(A) using UV-visible(UV–Vis) and Circular Dichroism (CD) Spectroscopy are described. Experiments used in determining the melting temperature, binding stoichiometry and dissociation constant of poly(A)-small molecule systems are depicted.

Key words Poly(A), Small molecule binding, Intercalation, UV–Vis spectroscopy, Circular dichroism, Thermal denaturation, Job plot analysis, Dissociation constant, Assembly

1 Introduction

Rational drug design is known as the process of designing compounds that will bind selectively and tightly to cellular targets [1–3]. Due to the role of nucleic acids in cellular functions, targeting specific nucleic acid sequences with small molecules is an active area of rational drug design research [4–11]. Daunorubicin, dactinomycin and quarfloxin are examples of small molecules that are used today as therapeutic agents [12–15].

Poly(A) tails are specific nucleic acid sequences, found at the 3′end of all eukaryotic mRNAs. Targeting Poly(A) by small molecules has attracted special interest in the last decade due to the vital role of poly(A) tails in vivo [16–22]. The Poly(A) tails are about 250 nucleotides long, and they are necessary for the stability, maturation and the transport of mRNAs to cytoplasm as well as the success of eukaryotic translation. They are added to the 3′end of all eukaryotic mRNAs after transcription by poly(A) polymerase (PAP). Overexpression of a human PAP, neo-PAP, in some human cancer cells has suggested that poly(A) tails can act as possible malignancy specific targets [17, 23–26].

Joanna Rorbach and Agnieszka J. Bobrowicz (eds.), *Polyadenylation: Methods and Protocols*, Methods in Molecular Biology, vol. 1125, DOI 10.1007/978-1-62703-971-0_8, © Springer Science+Business Media New York 2014

Noncovalent binding of small molecules to nucleic acids can be classified into two major categories; groove binding and intercalation [27, 28]. In groove binding, the small molecule fits nicely into the groove of the nucleic acid structure, therefore preventing the functioning of other molecules such as enzymes by preventing their binding to the groove. Groove binding causes almost no perturbation in the structure of the nucleic acid. On the other hand, intercalation occurs by insertion of planar aromatic compounds between adjacent nucleic acid base pairs. The insertion of these heterocyles between the base pairs is driven by electrostatic interactions, dipole–dipole interactions, dispersive interactions and π–π stacking. Intercalators generally carry at least one positive charge. The nucleic acid structure unwinds and elongates, often causing changes in sugar puckering, as the result of intercalation [27–31].

Most of the small molecules reported so far as poly(A) binders are intercalators (Fig. 1) and the formation of a self structure is observed upon binding of these molecules to poly(A) [16, 19–21, 26]. Here, the commonly used protocols in determining the binding of small molecules to poly(A) by measuring the changes in UV–Vis absorbance and/or ellipticity are reported. In addition, protocols for determining melting temperature, stoichiometry and

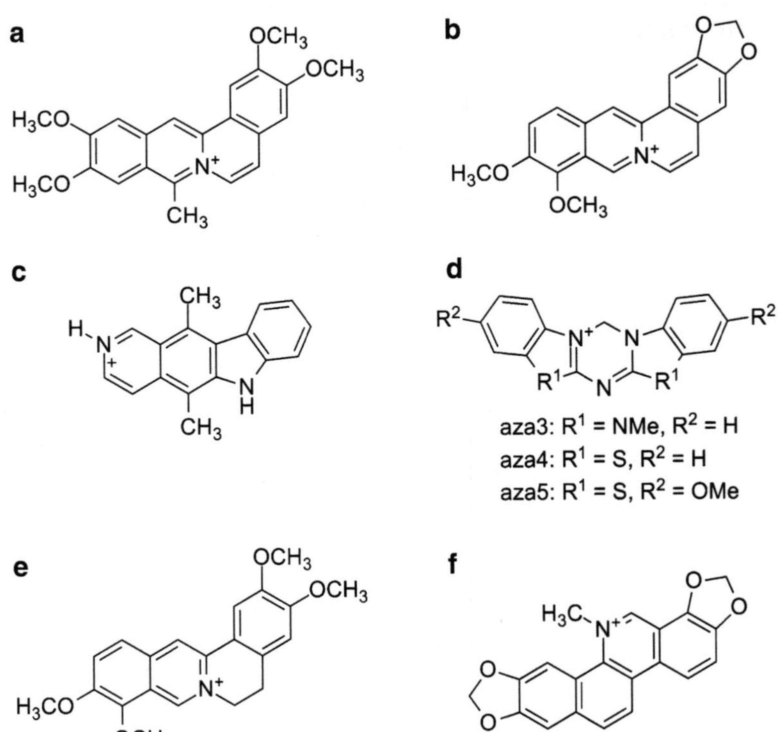

Fig. 1 Chemical structures of (**a**) coralyne, (**b**) berberine, (**c**) ellipticine, (**d**) azacyanines, (**e**) palmatine and (**f**) sanguinarine

the dissociation constant of poly(A)-small molecule systems are reported using poly(A)-coralyne system as the model system. However, the procedures described here should be easily applicable to all other poly(A)-small molecule systems where there is a change in UV-Vis and/or CD spectrum of poly(A) and/or the small molecule upon binding (*see* **Note 1**).

1.1 UV–Vis Absorbance Spectroscopy

Nucleic acids have strong absorbance in the region of 245–270 nm with the maximum absorbance at around 260 nm due to the $\pi \rightarrow \pi^*$ transitions of the ring systems of nucleobases [27, 30]. For poly(A), the maximum absorbance is at 258 nm. All the small molecules that bind to poly(A) with an appreciable binding constant are intercalators with one positive charge (Fig. 1). They generally have distinct absorption peaks between 270 and 500 nm. When an intercalator binds to poly(A), a red shift as well as hypochromicity is observed in the UV–Vis absorbance peaks of the intercalator due to the π-π stacking of the intercalator with adenine bases [19–21, 26, 32]. Changes in the UV–Vis absorbance peaks of the intercalator can be used in order to monitor small molecule binding to poly(A). Binding stoichiometry, binding constant and thermal denaturation of the system can be obtained by simply monitoring these UV–Vis absorbance peaks.

1.2 Circular Dichroism (CD) Spectroscopy

CD is one of the extensively used techniques in small molecule–nucleic acids research. It is based on the differential absorption of circularly polarized (left and right) light by chiral molecules. The CD is usually reported as ellipticity (θ), which is defined as the angle of polarization. Ellipticity is generally measured in millidegrees (mdeg). Nucleic acid solutions give rise to particular CD signals mainly due to their asymmetric carbon atoms in sugars and their secondary structure. For instance, CD is used extensively to differentiate between different nucleic acid secondary structures, such as B-form or A-form DNA [27, 33–36]. Achiral molecules such as coralyne are optically inactive and do not give rise to particular CD signals. However, the interactions of an achiral molecule with nucleic acids can give rise to "induced CD bands." Induced CD band formation is due to the coupling of the electronic transitions of the small molecule with nucleobases [36]. Observation of an induced CD band is a very fast and an easy way of detecting interactions between nucleic acids and small molecules. It is convenient to monitor induced CD bands in order to gain information about the poly(A)-small molecules system since they are only observed if the small molecule is binding to poly(A). The effect of experimental conditions on the system, such as temperature, salt or pH can be easily determined by monitoring the induced CD bands. The binding stoichiometry, denaturation temperature as well as the equilibrium constant of the system can also be obtained by monitoring induced CD bands [16, 27, 37, 38].

UV–Vis and CD spectroscopy are two of the most commonly used techniques in assessing poly(A)-small molecule systems and are complementary to all the other techniques used in examining nucleic acid-small molecule systems. For instance, fluorescence spectroscopy can be used in order to gather similar information about poly(A)-small molecule systems [19, 21]. Competition dialysis is another great technique for initial screening of small molecules that are poly(A) binders [39, 40]. DSC and ITC can be used to elucidate thermodynamic profile of the systems, whereas NMR and XRD can be used to elucidate structural information. There are many other techniques that can be used to investigate poly(A)-small molecule systems [33, 41–43]. The choice of method and instrumentation will always depend on availability, sensitivity and ease as well as the required information. However, one should always keep in mind that each technique is monitoring a different aspect of the interactions and the results could always be slightly different even when performed under the same experimental conditions.

2 Materials

Prepare all solutions using ultrapure water (sensitivity of 18 MΩ cm at 25 °C) and analytical grade reagents. Prepare all stock and buffer solutions in autoclaved/sterilized bottles. All reagents should be stored at 4 °C and protected from light (*see* **Note 2**) unless indicated otherwise. Follow the proper procedures when handling and disposing all reagents and materials.

1. Dilution Buffer Solution: 1× BPE, pH 7.0 buffer with 20 mM NaCl:

   Prepare the following stock solutions:

   0.5 M Na_2HPO_4: Weigh 7.1 g of Na_2HPO_4 and transfer to 0.1 L volumetric flask. Add about 80 mL of H_2O to dissolve the salt and then bring solution up to its final volume.

   0.5 M NaH_2PO_4: Weigh 6.9 g of $NaHPO_4 \cdot H_2O$ and transfer to 0.1 L volumetric flask. Add about 80 mL of H_2O to dissolve the salt and then bring solution up to its final volume.

   0.5 M Disodium Ethylene Diamine Tetra-Acetate (EDTA), pH 8.0: Weigh 18.61 g of $Na_2EDTA \cdot 2H_2O$ and transfer to 0.1 L bottle. Add about 80 mL of H_2O, and adjust the pH of the solution to pH 8.0 using NaOH pellets (about 2 g) or 10 M NaOH while stirring on a magnetic stirrer to dissolve the salt. Once the pH is 8.0, bring up to volume of 0.1 L (*see* **Note 3**).

   0.5 M NaCl: Weigh 2.922 g of NaCl and transfer to 0.1 L volumetric flask. Add about 400 mL of H_2O to dissolve the salt and then bring up to volume of 0.1 L.

Prepare Dilution Buffer Solution:

10× BPE buffer with 200 mM NaCl: Mix 12 mL 0.5 M Na_2HPO_4, 4 mL 0.5 M NaH_2PO_4, 2 mL 0.5 M EDTA, 40 mL 0.5 M NaCl and 42 mL H_2O.

1× BPE buffer, pH 7.0 with 20 mM NaCl: Dilute "10× BPE buffer with 200 mM NaCl" 1:10.

2. Coralyne Chloride Stock Solution: Weigh about 2–2.5 mg coralyne chloride monohydrate (Sigma-Aldrich, St. Louis, MO, USA) and dissolve in 10 mL H_2O. Use vortex and mild sonication to dissolve all the salt. Determine the concentration of the stock solution using UV–Vis spectrophotometer. The extinction coefficient for coralyne is $\varepsilon_{420} = 14{,}500\ M^{-1}\ cm^{-1}$. Wrap the solution bottle with aluminum foil and store at 4 °C (*see* **Note 4**).
3. Ellipticine Stock Solution: Weigh about 5 mg ellipticine (EMD Millipore Corporation, Billerica, MA, USA) and dissolve in 10 mL DMSO. Use vortex and mild sonication to dissolve all the solid. Determine the concentration of the stock solution using UV–Vis spectrophotometer. The extinction coefficient for ellipticine is $\varepsilon_{295} = 60{,}000\ M^{-1}\ cm^{-1}$. Wrap the solution bottle with aluminum foil and store at 4 °C (*see* **Note 4**).
4. Berberine Chloride Stock Solution: Weigh about 2.5–3.0 mg berberine chloride (Sigma-Aldrich, St. Louis, MO, USA) and dissolve in 10 mL H_2O. Use vortex and mild sonication to dissolve all the salt. Determine the concentration of the stock solution using UV–Vis spectrophotometer. The extinction coefficient for berberine is $\varepsilon_{344} = 22{,}500\ M^{-1}\ cm^{-1}$. Wrap the solution bottle with aluminum foil and store at 4 °C (*see* **Note 4**).
5. Polyadenylic acid (polyA) Solution: Dissolve the contents of poly(A) (Sigma-Aldrich, St. Louis, MO, USA) bottle in about 2 mL H_2O. Transfer about 500 μL of the poly(A) solution to an ependorf tube and determine the concentration spectrophotometrically using the extinction coefficient $\varepsilon_{258} = 9{,}800\ M^{-1}\ cm^{-1}$ nucleotide^{-1}. Store the poly(A) solution in the ependorf at 4 °C for immediate use. Store the poly(A) solution in the bottle at –20 °C for later use.
6. N_2 gas.
7. 1 and 10 cm path length high quality quartz cells with cut off wavelength of 170 nm (*see* **Note 5**).
8. A UV–Vis spectrophotometer equipped with a temperature control unit. HP-8453 UV–Vis diode array spectrophotometer equipped with an Agilent 89090A Peltier temperature control unit was used in the experiments described here (*see* **Note 6**).

9. A CD spectrophotometer equipped with a temperature control unit. JASCO-715 or JASCO-815 equipped with a peltier temperature control unit was used in the experiments described here.
10. Software program such as IGOR (Wavemetrics, Portland, OR) for analyzing data. The program of choice for analyzing the UV–Vis or CD spectra can be different based on the availability and the familiarity with the program.

3 Methods

Carry out all procedures at room temperature unless otherwise specified.

3.1 Determining the Concentration of Stock Solutions by UV–Vis Spectroscopy

1. Switch on the UV lamps and allow the instrument to stabilize for 30 minutes before making any measurements.
2. While waiting for the instrument to warm up, prepare a diluted sample (start with 1:200 dilution) of poly(A) in a new eppendorf tube from your stock solution in H_2O.
3. Once the instrument is warmed up, fill a 1 cm path length cell with H_2O to determine the baseline spectrum between 190 and 1,100 nm. This should be flat.
4. Empty and dry the cuvette and fill it with your diluted poly(A) sample solution. Take UV–Vis spectrum between 190 and 1,100 nm.
5. Calculate the concentration of diluted poly(A) sample using Beer-Lambert law, $A = \varepsilon cl$. For poly(A) use A_{258} with $\varepsilon_{258} = 9{,}800\ M^{-1}\ cm^{-1}$ nucleotide^{-1}. Calculate the concentration of your poly(A) stock solution by multiplying the concentration of your diluted sample with your dilution factor (*see* **Note 7**).
6. Determine the concentration of coralyne and berberine stock solutions using the very same procedure described in **step 4**. For coralyne use A_{420} with $\varepsilon_{420} = 14{,}500\ M^{-1}\ cm^{-1}$ and for berberine use A_{344} with $\varepsilon_{344} = 22{,}500\ M^{-1}\ cm^{-1}$. For ellipticine, use DMSO rather than H_2O (**steps 3–5**) as the baseline spectrum and A_{295} with $\varepsilon_{295} = 60{,}000\ M^{-1}\ cm^{-1}$(*see* **Note 7**).

3.2 Determining the Binding of a Small Molecule to Poly(A) by UV–Vis Spectroscopy

1. Switch on the UV lamps and allow the instrument to stabilize for 30 minutes before making any measurements.
2. While waiting for the instrument to warm up, prepare 850 μL of the following samples (*see* **Note 8**):
   (a) First sample: 15 μM small molecule with 60 μM nucleotide base poly(A) in “1× BPE with 20 mM NaCl” e.g.:

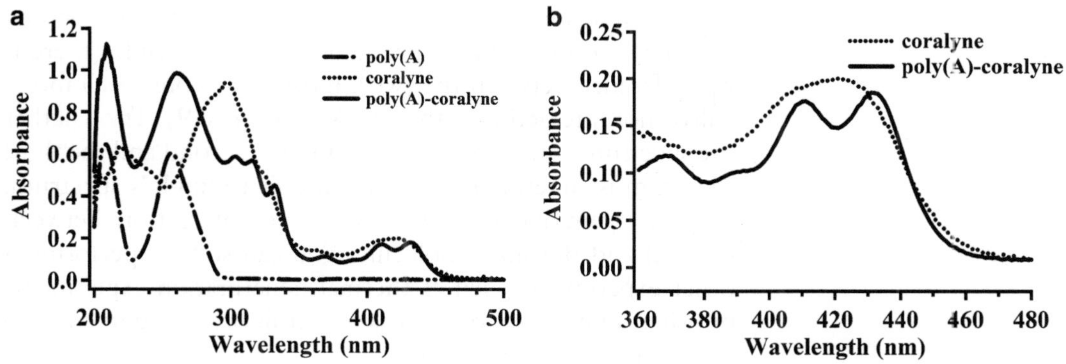

Fig. 2 (**a**) UV–Vis spectra of poly(A), coralyne and poly(A)–coralyne complex in 1× BPE buffer with 20 mM NaCl, pH 7.0. Samples were 60 μM poly(A) (in units of nucleotide), 15 μM coralyne and 60 μM poly(A) (in units of nucleotide) with 15 μM coralyne respectively. (**b**) The longer wavelength region of the UV–Vis spectra of coralyne and poly(A)–coralyne samples reported in Fig. 2a

15 μM coralyne with 60 μM nucleotide base poly(A) in "1× BPE with 20 mM NaCl."

(b) Second sample: 15 μM small molecule in "1× BPE with 20 mM NaCl" e.g.: 15 μM coralyne in "1× BPE with 20 mM NaCl."

(c) Third sample: 60 μM nucleotide base poly(A) in "1× BPE with 20 mM NaCl."

(d) 1× BPE with 20 mM NaCl.

3. Once the instrument is warmed up, fill a 1 cm path length cell with "1× BPE buffer with 20 mM NaCl" to determine the baseline spectrum between 190 and 1,100 nm. This should be flat.
4. Empty and dry the cuvette and fill it with 800 μL of your first sample. Take UV–Vis spectrum between 190 and 1,100 nm.
5. Repeat **step 4** using your second and third samples.
6. Using the program supplied by the instrument, or another program such as IGOR (Wavemetrics, Portland, OR), put all three spectra on the same graph to compare with each other (Fig. 2a).
7. Nucleic acids have strong absorbance in the region of 245–270 nm with the maximum absorbance at around 260 nm due to the $\pi \rightarrow \pi^*$ transitions of the ring systems of nucleobases[27, 30]. For poly(A) the maximum absorbance is at 258 nm. All the small molecules reported so far as binding to poly(A) with an appreciable binding constant are intercalators with one positive charge [5, 16, 19, 21, 44–46]. And the intercalators generally have distinct absorption peaks between 270 and 500 nm.

8. If UV–Vis absorbance peaks of the small molecule in the first and the second sample's spectra are superimposable (there is no difference between the two), most likely your small molecule is not interacting with poly(A) (*see* **Note 9**). Try whether the spectrum you obtained by adding the second and the third spectrum is superimposable with the first sample's spectrum. If it is, again most likely there is no interaction between poly(A) and the small molecule. You can still try performing other experiments such as thermal denaturation experiments or using other methods such as CD or fluorescence to confirm that there is no interaction.
9. If UV–Vis absorbance peaks of the small molecule in the first and the second sample's spectra are not superimposable, the small molecule of interest is interacting with poly(A).
10. When a small molecule binds to a nucleic acid structure via intercalation, generally a red shift (UV-absorbance maximum shifts to a higher wavelength) and hypochromicity in the UV–Vis absorbance peaks of the small molecule is observed. These changes are associated with the π-π stacking of the intercalator with nucleobases. A red shift of around 10 nm is observed in the absorption peaks between 400 and 450 nm upon binding of coralyne to poly(A) (Fig. 2b). On the other hand, observing a blue shift in the UV–Vis absorbance peaks of the small molecule is attributed to groove binding [27, 30, 44].

3.3 Determining the Binding of a Small Molecule to Poly(A) by Circular Dichroism (CD)

1. Turn on the N_2 and purge the instrument for 10–15 minutes before turning on the CD instrument (*see* **Note 10**).
2. Turn on the CD instrument and Peltier temperature controller.
3. From the computer, start the "Spectra Manager" program. Choose "Spectrum Measurement" as your choice of experiment. The instrument will run diagnostic tests and then open the shutter and the lamp respectively. Once the lamp is open, allow the instrument to stabilize for at least 15 minutes before making any measurements.
4. Initiliaze Peltier from the "Spectrum Measurement" program (SpectrumMeasurement/Measurement/Accessory/Temperature/JASCO Peltier Type (Single, RTE On)) (*see* **Note 10**). Press start button on the Peltier instrument.
5. From the "Accessory/Jasco Peltier" set temperature to 5 °C.
6. While waiting for the instrument to warm up, prepare 3,500 μL of the following samples (*see* **Note 8**):
   (a) First sample: 15 μM small molecule with 60 μM nucleotide base poly(A) in "1× BPE with 20 mM NaCl" e.g.: 15 μM coralyne with 60 μM nucleotide base poly(A) in "1× BPE with 20 mM NaCl."

(b) Second sample: 15 μM small molecule in "1× BPE with 20 mM NaCl" e.g.: 15 μM coralyne in "1× BPE with 20 mM NaCl."

(c) Third sample: 60 μM nucleotide base poly(A) in "1× BPE with 20 mM NaCl."

(d) 1×BPE with 20 mM NaCl.

7. Once the instrument is stabilized, fill the cuvette with "1× BPE buffer with 20 mM NaCl" and place the cuvette into the sample holder to determine the baseline spectrum. Equilibrate the sample at 5 °C for a couple of minutes. You can watch out the temperature of the sample from the software.

8. To start collecting CD spectra, click on the "Measurement" menu and select "Parameter" tab. Enter the following parameters into the program (*see* **Note 10**):

   Sensitivity: Standard (100 mdeg).

   Start Wavelength: 500 nm.

   End Wavelength: 190 nm.

   Data Pitch: 1 nm.

   Scanning Mode: Continuous.

   Response: 1 s.

   Scanning Speed: 100 nm/min.

   Bandwidth: 2 nm.

   Accumulation: 3.

9. Switch to "Data File" tab to save your spectra. Select "autosave" and specify a file name. Select a folder location. For your records, you can enter sample specific information under the "Option" tab.

10. Click "OK" to apply the parameters and close the dialog box.

11. Click "Start" to begin data acquisition.

12. Once the spectrum is acquired, it will appear on the new "Spectra Analysis" window. You can also save your spectrum at this point. You could save it either ".jws" or ".txt" from "File/Save As" menu. Saving the spectrum as ".txt" file makes it easier to transfer data to other analysis programs such as IGOR (Wavemetrics, Portland, OR) for further processing.

13. Take your cuvette out, empty and clean it. Refill with your first sample. Let the sample equilibrate at 5 °C for a couple of minutes.

14. Click on the "Measurement" menu and select "Baseline Correction" tab. Open the baseline spectrum file you just collected and click "OK".

15. Switch to "Data File" tab, and specify a new file name. Click "OK" and close the dialog box.

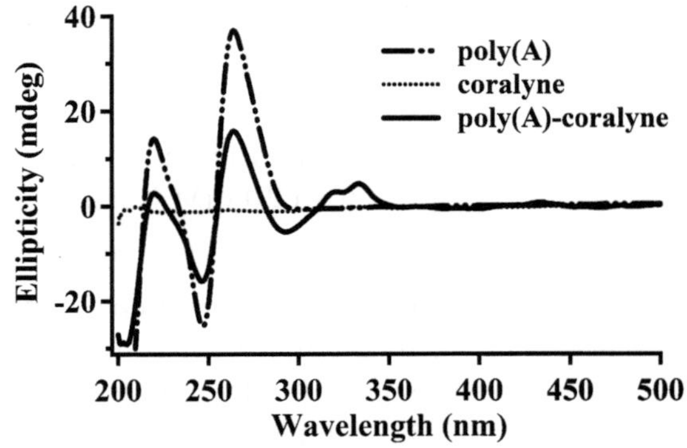

Fig. 3 CD spectra of poly(A), coralyne and poly(A)–coralyne complex in 1× BPE buffer with 20 mM NaCl, pH 7.0. Samples were 60 μM poly(A) (in units of nucleotide), 15 μM coralyne and 60 μM poly(A) (in units of nucleotide) with 15 μM coralyne respectively

16. Click "Start" to begin data acquisition. Again you could save your spectrum as ".txt" file.
17. Acquire a spectrum for your second and third sample by repeating **steps 13–16**.
18. Transfer your spectra to an analysis program such as IGOR (Wavemetrics, Portland, OR) and plot all three spectra (ellipticity vs. wavelength) on the same graph in order to compare with each other (Fig. 3).
19. Once you plot the spectra for your first, second, and third sample on the graph, look whether there is a change in the CD bands of poly(A) and the small molecule or whether there is a formation of a new induced CD band.
20. If the spectra obtained by adding the second and the third spectrum is superimposable with the first spectrum, most likely there is no interaction between the small molecule and poly(A). You can still try performing other experiments such as thermal denaturation experiments or fluorescence experiments to confirm that there is no interaction.
21. If CD spectra of the first and the second sample are not superimposable, that means the small molecule of interest is interacting with poly(A).
22. In the case of poly(A) and coralyne, coralyne is an achiral molecule and has no intrinsic CD signal. Poly(A) has three major peaks, maximums at 220 and 264 nm and a minimum at 247 nm (Fig. 3). Upon binding of coralyne to poly(A), hypochromic effect (almost 50 %) is observed in the CD bands of poly(A). A new CD band, called an induced CD band, is observed between 310 and 360 nm where only coralyne

absorbs. The chiral environment of poly(A) induces chirality in coralyne structure upon binding of coralyne between adenine base pairs. The binding of coralyne gives rise to the formation of the induced CD band. The presence of the induced CD band confirms the binding of coralyne to the chiral environment of poly(A) [16, 27]. Monitoring the induced CD band formation is a convenient way of investigating poly(A)-small molecule interactions.

3.4 Determining the Binding Stoichometry by Job Plot (Method of Continuous Variation) Analysis

Job plot (Method of Continuous Variation) is a widely used method in determining the stoichiometry of a binding event. In this method, the total molar concentration of the two species (nucleic acid and the small molecule) is held constant, but their mole fraction is varied. A measurable parameter that is proportional to complex formation (such as absorbance) is plotted against the mole fractions of these two components [27, 38, 47]. UV–Vis absorbance, ellipticity or fluorescence are some of the widely used measurable parameters in Job plots [16, 19, 21, 26, 44, 45]. Here, the UV–Vis absorbance is measured and the total molar concentration of [poly(A)]/4 (in nucleotides base) + [coralyne] is held constant.

1. Switch on the UV lamps and allow the instrument to stabilize for 30 minutes before making any measurements.
2. While waiting for the instrument to warm up, prepare the following samples:
   (a) First sample: 850 μL 60 μM in nucleotide base poly(A) in "1× BPE with 20 mM NaCl."
   (b) Second sample: 2.0 mL 15 μM small molecule in "1× BPE with 20 mM NaCl" e.g.: 15 μM coralyne in 1× BPE with 20 mM NaCl.
3. Once the UV instrument is warmed up, fill a 1 cm path length cell with "1× BPE buffer with 20 mM NaCl" to determine the baseline spectrum between 190 and 1,100 nm. This should be flat.
4. Empty and dry the cuvette and fill it with 800 μL of your first sample (60 μM in nucleotide base poly(A)). Take UV–Vis spectrum between 190 and 1,100 nm.
5. Take 40 μL of the sample out from the cuvette, and add 40 μL of the second sample (15 μM small molecule in "1× BPE buffer with 20 mM NaCl") instead. Mix the sample by pipetting up and down carefully. Take UV–Vis spectrum again between 190 and 1,100 nm.
6. Repeat **step 5** with the appropriate amount of the sample taken out from the cuvette and replaced with second sample according to Table 1.

Table 1
The volume that needs to be taken out and replaced with second sample during the Job Plot experiment in order to achieve the desired mole fraction using 1 cm path length UV–Vis cuvette whose volume is 800 μL

Ratio (*R*)	[poly(A)]/4 (μM)	[Coralyne] (μM)	Volume exchanged (μL)
0.000	15.00	0.00	0.00
0.050	14.25	0.75	40.00
0.100	13.50	1.50	42.11
0.150	12.75	2.25	44.44
0.200	12.00	3.00	47.06
0.250	11.25	3.75	50.00
0.300	10.50	4.50	53.33
0.350	9.75	5.25	57.14
0.400	9.00	6.00	61.54
0.450	8.25	6.75	66.67
0.500	7.50	7.50	72.73
0.550	6.75	8.25	80.00
0.600	6.00	9.00	88.89
0.650	5.25	9.75	100.00
0.700	4.50	10.50	114.29
0.750	3.75	11.25	133.33
0.800	3.00	12.00	160.00
0.850	2.25	12.75	200.00
0.900	1.50	13.50	266.67
0.950	0.75	14.25	400.00

7. Transfer all spectra to an analysis program such as IGOR (Wavemetrics, Portland, OR).
8. Determine the small molecule absorbance peaks which are changing significantly upon binding to poly(A). You could use your spectra for first and second sample from Subheading 3.2.
9. In the poly(A)-coralyne system, free coralyne spectrum has an absorption maximum at 420 nm, with a shoulder at 400 nm. When it binds to poly(A), a red shift, around 10 nm, is observed. Bound coralyne spectrum has absorbance maximum at 435 nm with a more distinct shoulder peak at 412 nm (Fig. 2).

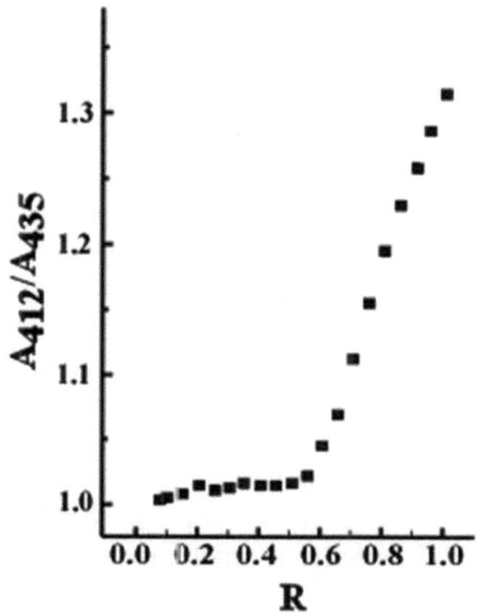

Fig. 4 Job Plot of poly(A) with coralyne in 1× BPE with 19 mM NaCl, pH 7.0. Figure is replotted from Fig. 4b, Xing et al. [26]

10. Determine A_{412}/A_{435} for each spectrum you collected.
11. Plot A_{412}/A_{435} vs. Ratio (R, Table 1) (Fig. 4).
12. You will observe an abrupt change in A_{412}/A_{435} at $R = 0.5$. Seeing one abrupt change means there is only one mode of binding. And $R = 0.5$ means one coralyne molecule is binding to four adenine bases. If you are seeing more than one abrupt point, it means there is more than one mode of binding. If the point where the abrupt change is occurring is not clear, try to do line fits on your data. For coralyne, you can fit data points with two lines with different slopes. The first line will fit to data points with R values up to 0.5 and the second line will fit to data points with R values between 0.5 and 1. The point where the two lines intersect will be the stoichiometric point, which is $R = 0.5$ for poly(A)-coralyne system (Fig. 4) (*see* **Note 11**) [26].

3.5 Determining the Secondary Structure Formation Using Thermal Denaturation Experiments

Thermal denaturation experiments are regularly used in detecting small molecule binding to nucleic acids [27, 30, 33, 41, 42]. Thermal denaturation refers to the dissociation of nucleic acid strands, e.g. duplex DNA, from each other with increasing temperature of the solution. In a typical denaturation experiment, a melting curve will be obtained by monitoring a change in an observable such as UV–Vis absorbance or ellipticity as a function of temperature. The melting temperature (Tm) which is defined as the temperature at which half of the nucleic acid strands are in the double-helical state and half are in the random coil state can easily be obtained from the normalized melting curves. Another way of obtaining Tm is taking the first derivative of the melting curve. The binding of small molecules to nucleic acids stabilizes the nucleic acid structure, therefore increases Tm of the nucleic acid structure. Comparing Tm of nucleic acid structure to the Tm of the nucleic acid structure in the presence of the small molecule is commonly used in the assessment of small molecule binding [27, 30, 41]. In addition, the amount of the increase in Tm due to small molecule binding can give information regarding the strength of binding.

Poly(A) is a single stranded nucleic acid and doesn't have a secondary structure at neutral pH. However, small molecule binding, such as coralyne binding, induces an anti-parallel double helical secondary structure formation in poly(A) structure. The secondary structure formation can be easily confirmed via the existence of Tm by performing thermal denaturation experiments of a poly(A)-small molecule solution [16, 26]. Here, the thermal denaturation experiment of poly(A)-coralyne solution in which induced CD bands are monitored is described. Since induced CD bands are directly associated with binding, Tm's obtained by monitoring these bands is a direct assessment of the denaturation of poly(A)-small molecule structure. If no induced CD band formation is observed upon binding of the small molecule to poly(A), the small molecule or poly(A) peaks can be monitored to obtain Tm. However, one should pay attention to analysis of data. Proper control experiments (small molecule denaturation by itself and poly(A) denaturation by itself) should be performed in parallel to poly(A)-small molecule denaturation.

1. Open the CD instrument as described in Subheading 3.3, **steps 1–5**.
2. While waiting for the instrument to warm up, prepare 3,100 μL of 15 μM small molecule with 60 μM in nucleotide base poly(A) in "1× BPE with 20 mM NaCl" sample, e.g.: 15 μM coralyne with 60 μM nucleotide base poly(A) in "1× BPE with 20 mM NaCl."
3. Once the instrument is stabilized, fill the cuvette with "1× BPE buffer with 20 mM NaCl" and place the cuvette into the sample holder to determine the baseline spectrum. Equilibrate the sample at 5 °C for a couple of minutes. You can watch out the temperature of the sample from the software.
4. Collect the baseline spectrum as described in Subheading 3.3.
5. Take your cuvette out, empty and clean it. Refill the cuvette with your sample. Let the sample equilibrate for a couple of minutes and collect spectrum as described in Subheading 3.3. It is always a good habit to run "Spectrum Measurement" before running thermal denaturation experiments to check that the sample and the instrumentation are OK.

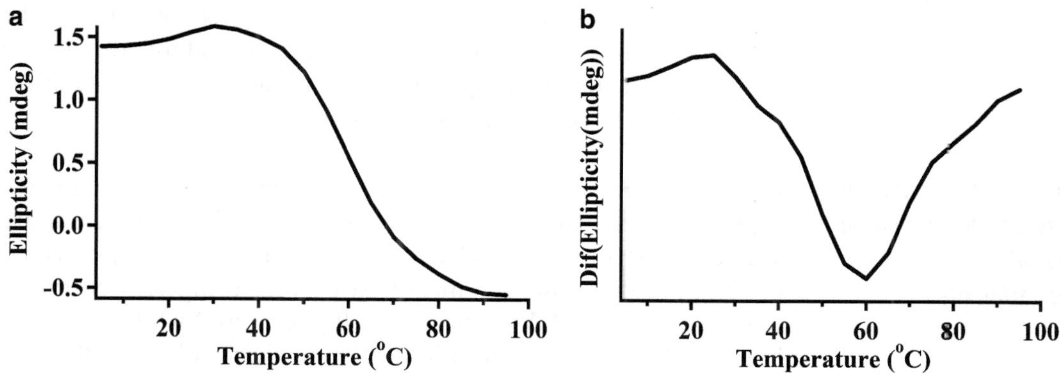

Fig. 5 (**a**) Melting curve for poly(A)-coralyne complex obtained by monitoring change in ellipticity at 320 nm at various temperatures. (**b**) Derivative of CD melting curve for poly(A)-coralyne complex. Sample is 60 μM poly(A) (in units of nucleotide) with 15 μM coralyne in 1× BPE with 20 mM NaCl, pH 7.0

6. If the CD spectrum looks fine, close the window and go back to "Spectra Manager" program. Choose "Temperature/ Wavelength Scan" as your choice of experiment (*see* **Note 12**). Open the parameters page and choose "Auto" as the "Spectrum Mode." Choose "Use range of temperature." Set 5 °C as your starting temperature and 95 °C as your final temperature. Choose 5 °C as your data pitch, and click "Append" to append all the temperatures (*see* **Note 13**). Click on the "Wavelength" and recheck your parameters for the spectrum measurement. Click "OK" to return.
7. Click "OK" to apply the parameters and close the dialog box.
8. Click "Start" to begin data acquisition.
9. Transfer all spectra to an analysis program such as IGOR (Wavemetrics, Portland, OR).
10. Plot ellipticity vs. wavelength for all temperatures on the same graph. Examine the spectra for different wavelengths, and try to find a wavelength which is a representation of binding. Choosing a wavelength at induced CD band region is a good choice, since these bands are direct representation of binding. If you would like to use a wavelength in poly(A) region, make sure you run the proper control experiments, which are the thermal denaturation of poly(A) in buffer and thermal denaturation of small molecule in buffer. Also always make sure that your buffer solution do not absorb in the region you are interested.
11. For poly(A)-coralyne system, induced CD band formation is observed between 310 and 360 nm. Thermal denaturation profiles can be created using the ellipticity changes in this range of wavelengths. In this case, ellipticity change at 320 nm provides a clear melting curve. Plot ellipticity at 320 nm vs. temperature to create the melting curve (Fig. 5a).

12. Take the first derivative of the melting curve. Define the minimum as Tm of the system (Fig. 5b). Here, the Tm of poly(A)-coralyne complex is 60 °C. If melting curves are noisy, the curves can be smoothened a little.

3.6 Determining the Dissociation Constant by Dilution Experiments

Determination of the binding affinity is one of the fundamental experiments in drug design. The drug candidate molecules are sought to be tight and selective binders. In classical approaches equilibrium constants are measured by measuring an observable (e.g. fluorescence, UV absorbance, ellipticity etc.) while titrating a ligand into nucleic acid solution. High nucleic acid to small molecule concentrations are used, so that the binding sites are excessive, and only the strongest binding event occurs. In these experiments, the binding of a small molecule to a pre-structured nucleic acid, such as duplex DNA, is generally treated as a single event during the calculation of equilibrium constants [48].

Compared to the classical approaches, dilution experiments are proven to be a better technique in estimating equilibrium constants in poly(A)-small molecule systems especially if there is a self-structure formation upon small molecule binding [16]. In dilution experiments, a concentrated poly(A)-small molecule complex, in which the poly(A) self structure is intact, is diluted with the buffer solution until below the point at which the small molecule dissociates from the poly(A) self-structure and the poly(A) self-structure is lost [16]. The binding of the small molecule drives the formation of the poly(A) self-structure and this poly(A) self-structure is only stable upon binding of the small molecule. The binding of the small molecule and the secondary structure formation due to the assembly of poly(A) strands by the small molecule are coupled events. Therefore, the binding of a molecule to poly(A) is different from the binding of a molecule to a pre-structured nucleic acid. Thus, the classical approaches for analyzing small molecules–nucleic acid interactions can over- or under-estimate the equilibrium constants of poly(A)-small molecule systems. Here, we demonstrate the determination of binding constants for coralyne, ellipticine and berberine to poly(A) by dilution studies.

1. Switch on the UV lamps and allow the instrument to stabilize for 30 minutes before making any measurements.
2. While waiting for the instrument to warm up, prepare 850 μL of dilution sample which is 200 μM poly(A) nucleotide and 50 μM small molecule in "1× BPE buffer with 20 mM NaCl." Prepare also 200 mL of buffer solution (1× BPE buffer with 20 mM NaCl) by diluting "10× BPE buffer with 200 mM NaCl" (*see* **Note 8**).
3. Once the instrument is warmed up, fill a 1 cm path length cell with "1× BPE buffer with 20 mM NaCl" to determine the baseline spectrum between 190 and 1,100 nm. This should be flat.

4. Empty and dry the cuvette and fill it with 800 μL of your dilution sample. Take UV–Vis spectrum between 190 and 1,100 nm.
5. Take 80 μL of your dilution sample out from the cuvette, and add 80 μL of buffer solution (1× BPE buffer with 20 mM NaCl) instead. Mix the sample by pipetting up and down carefully. Take UV–Vis spectrum again between 190 and 1,100 nm.
6. Repeat **step 5** with the appropriate amount of dilution sample replaced with the buffer solution each time (Table 2) until the signal to noise ratio of the absorbance peaks of the small molecule is high (*see* **Note 14**). Monitor the absorbance of the peaks between 380 and 480 nm for coralyne, 290 and 350 nm for ellipticine and 305 and 390 nm for berberine (Fig. 6) (*see* **Notes 1** and **15**).
7. Once the signal to noise ratio is high with the 1 cm path length UV–Vis cuvette, switch the UV–Vis spectrometer set up to use 10 cm path length cell. If possible, set the integration time to 10 s rather than 1 s to get a better resolution.
8. Fill 10 cm path length cell with buffer solution (1× BPE buffer with 20 mM NaCl) to determine the baseline spectrum between 190 and 1,100 nm.
9. Empty and dry the cuvette and fill it with 30 mL of the dilution sample solution which is 40 μM nucleotide and 10 μM small molecule in "1× BPE buffer with 20 mM NaCl." Take UV–Vis spectrum between 190 and 1,100 nm (*see* **Note 15**).
10. Take 3,000 μL of the dilution sample out from the cuvette, and add 3,000 μL buffer solution (1× BPE buffer with 20 mM NaCl) instead. Mix the sample by pipetting up and down carefully. Take UV–Vis spectrum again between 190 and 1,100 nm.
11. Repeat **step 10** with the appropriate amount of dilution sample replaced with the buffer solution each time (Table 3) (*see* **Note 14**) until the signal to noise ratio of the absorbance peaks that you are monitoring is high. Monitor the absorbance of the peaks between 380 and 480 nm for coralyne, 290 and 350 nm for ellipticine and 305 and 390 nm for berberine (Fig. 6).
12. Prepare 850 μL of a control sample which is 50 μM small molecule in "1× BPE buffer with 20 mM NaCl."
13. Using 1 cm path length cuvette, take UV–Vis spectrum of this sample between 190 and 1,100 nm.
14. Prepare another control sample which is 30 mL and 10 μM small molecule in "1× BPE buffer with 20 mM NaCl."
15. Using 10 cm path length cuvette, take UV–Vis spectrum of this sample between 190 and 1,100 nm.

Table 2
The volume that needs to be taken out and replaced with the buffer solution during the dilution experiment in order to achieve the desired concentration using 1 cm path length UV–Vis cuvette

Concentration of polyA (μM in nucleotide)	Total sample volume (μL), V_t	Volume exchanged with buffer (μL), V_e
200	800	0.0
180	800	80.0
160	800	88.9
140	800	100.0
120	800	114.3
110	800	66.7
100	800	72.7
80	800	160.0
60	800	200.0
55	800	66.7
50	800	72.7
40	800	160.0
30	800	200.0
27.5	800	66.7
25	800	72.7
20	800	160.0
17.5	800	100.0
15	800	114.3
12.5	800	133.3
10	800	160.0
9	800	80.0
8	800	88.9
7	800	100.0
6	800	114.3
5	800	133.3
4	800	160.0
3	800	200.0
2.5	800	133.3
2	800	160.0
1.75	800	100.0

(continued)

Table 2
(continued)

Concentration of polyA (μM in nucleotide)	Total sample volume (μL), V_t	Volume exchanged with buffer (μL), V_e
1.5	800	114.3
1.25	800	133.3
1	800	160.0
0.75	800	200.0
0.5	800	266.7
0.25	800	400.0
0.125	800	400.0
0.1	800	160.0

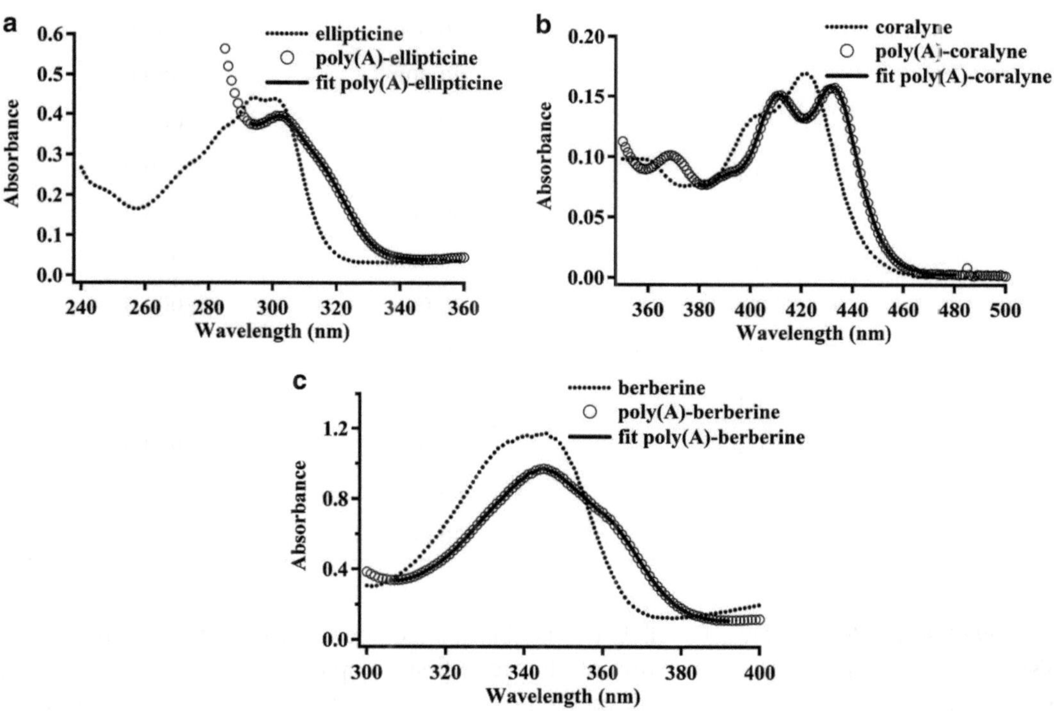

Fig. 6 (**a**) Longer wavelength region of the UV–Vis spectra of unbound (free) ellipticine (*dotted line*) and poly(A)-ellipticine (*circles*) complex. Poly(A)-ellipticine complex is 180 μM poly(A), in units of nucleotide, and 45 μM ellipticine. The *straight line* represents the best fit composite spectrum from the two state model, which corresponds to 99 % bound and 1 % unbound ellipticine. (**b**) Longer wavelength region of the UV–Vis spectra of unbound (free) coralyne (*dotted line*) and poly(A)-coralyne (*circles*) complex. Poly(A)-coralyne complex is 4 μM poly(A), in units of nucleotide, and 1 μM coralyne. The *straight line* represents the best fit composite spectrum from the two state model, which corresponds to 96 % bound and 4 % unbound coralyne. (**c**) Longer wavelength region of the UV–Vis spectra of unbound (free) berberine (*dotted line*) and poly(A)-berberine (*circles*) complex. Poly(A)-berberine complex is 180 μM poly(A), in units of nucleotide, and 45 μM berberine. The *straight line* represents the best fit composite spectrum from the two state model, which corresponds to 89 % bound and 11 % unbound coralyne. Samples were in 1× BPE buffer with 20 mM NaCl, pH 7.0. Spectra collected at 22 °C

Table 3
The volume that needs to be taken out and replaced with the buffer solution during the dilution experiment in order to achieve the desired concentration using 10 cm path length UV–Vis cuvette

Concentration of polyA (μM in nucleotide)	Total sample volume (μL), V_t	Volume exchanged with buffer (μL), V_e
10	30,000	0.0
9	30,000	3,000.0
8	30,000	3,333.3
7	30,000	3,750.0
6	30,000	4,285.7
5	30,000	5,000.0
4	30,000	6,000.0
3	30,000	7,500.0
2.5	30,000	5,000.0
2	30,000	6,000.0
1.75	30,000	3,750.0
1.5	30,000	4,285.7
1.25	30,000	5,000.0
1	30,000	6,000.0
0.75	30,000	7,500.0
0.5	30,000	10,000.0
0.25	30,000	15,000.0
0.125	30,000	15,000.0
0.1	30,000	6,000.0

3.7 Determining the Dissociation Constant by Dilution Experiments: Spectra Analysis

Transfer all UV–Vis absorbance spectra to Igor program. Assign the spectrum of the very first sample (200 μM nucleotide and 50 μM small molecule) as the "bound small molecule" spectrum and the control sample (small molecule sample in buffer solution) as the "unbound small molecule" spectrum. Using least-squares fits, fit each of the dilution series spectrum by a weighted sum of the two absorption spectra. For coralyne, fit only the region between 380 and 480 nm, for ellipticine between 290 and 350 nm and for berberine between 305 and 390 nm (Fig. 6) (*see* **Note 1**).

$y = f_0(f_1 \times \text{bound small molecule } (x) + f_2 \times \text{unbound small molecule } (x)) + f_3$

Here f_0, f_1, f_2 and f_3 are all variables where $0 < f_n < 1$

Calculate % complex formed $= [f_1/(f_1 + f_2)] \times 100$

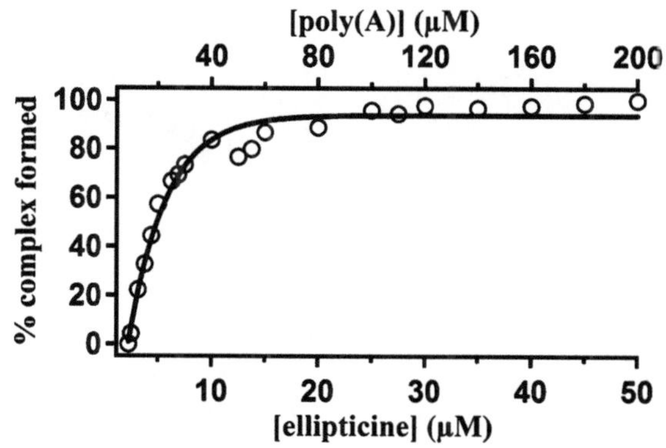

Fig. 7 Plot of percentage of poly(A)-ellipticine complex formed as a function of ellipticine and poly(A) concentration. % of poly(A)-ellipticine complex formed is obtained using least square fits of dilution data with a two state model. The *line* represents the single exponential fit of the % poly(A)-ellipticine complex formed data. Samples were in 1× BPE buffer with 20 mM NaCl, pH 7.0. Spectra collected at 22 °C. Figure is replotted from Persil Cetinkol et al. [16]

Plot % complex formed vs. the small molecule concentration (Fig. 7).

In the graph a plateau region will be observed at high small molecule concentrations. The formation of the plateau region is due to the formation of poly(A)-small molecule complexes. All the small molecules are bound to poly(A) at these concentration range. The plateau region will be followed by a step dissociation region, over a narrow range of small molecule concentrations. Dissociation constant, K_d, can be assigned as the concentration of the small molecule where complex formation is 50 %. To determine the point where complex formation is 50 %, % complex formed vs. concentration of the small molecule data can be fitted using a single exponential. Determination of the dissociation constant, K_d, for three different cases is described below.

3.7.1 Ellipticine Binding to Poly(A): Moderate Binding

In % complex formed vs. ellipticine concentration graph (Fig. 7), the plateau region is observed between ellipticine concentrations of 50 and 10 μM. Up to 25 μM ellipticine concentration, the percent of complex formed is greater than 95 %. At 10 μM ellipticine concentration still the percent of complex formed is greater than 80 %. Rapid dissociation of ellipticine is observed below 10 μM of ellipticine concentration. At 2.5 μM ellipticine concentration, the percent of complex formation is only about 4 %. Here, the binding constant, K_d, can be precisely assigned using a single exponential fit as the point where % complex formed = 50 %, which is 4.9 μM (Fig. 7).

3.7.2 Coralyne Binding to Poly(A): Tight Binding

When % complex formed vs. coralyne concentration is plotted (Fig. 8), a plateau region is observed between coralyne

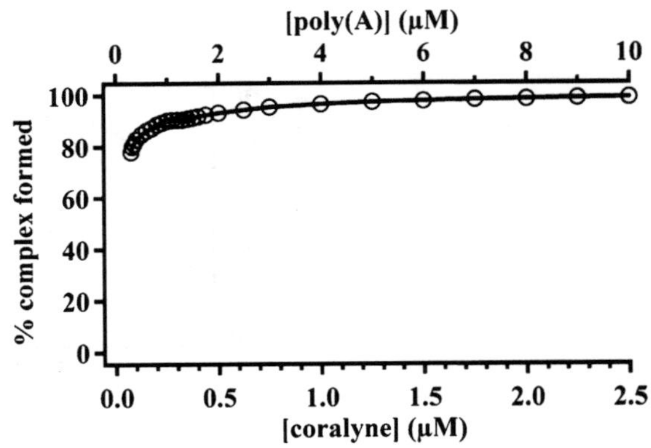

Fig. 8 Plot of percentage of poly(A)-coralyne complex formed as a function of coralyne and poly(A) concentration. Samples were in 1× BPE buffer with 20 mM NaCl, pH 7.0. Spectra collected at 22 °C. Figure is replotted from Persil Cetinkol et al. [16]

concentrations of 50 and 0.25 μM. All spectra were well-fit by the two-state model in this region. More than 95 % complex formation was observed between coralyne concentrations of 50 and 1.0 μM and more than 90 % complex formation was observed between coralyne concentrations of 1.0 and 0.25 μM. The amount of unbound coralyne to bound coralyne increases rapidly below 0.25 μM coralyne concentration. However, even in the case of a sample diluted to 0.1 μM coralyne concentration, still the complex formation was 80 %, meaning the dissociation constant, K_d, of coralyne to poly(A) is higher than 100 nM (K_a of $>10^7$ M^{-1}) (Fig. 8). The exact dissociation constant could not be obtained in this case, due to the high signal to noise ratio in the UV–Vis spectra of the samples whose coralyne concentrations are below 0.1 μM.

3.7.3 Berberine Binding to Poly(A): Weak Binding

When % complex formed vs. berberine concentration is plotted, the plateau region is not observed (Fig. 9b). Rapid dissociation of berberine is observed starting from the most concentrated sample, which is 200 μM poly(A) with 50 μM berberine. The absence of the plateau region establishes the fact that the UV–Vis spectrum of the most concentrated sample cannot be used as the representative of fully bound berberine system. The K_d cannot be assigned using a two state model making the assumption that the spectrum of the most concentrated sample represents 100 % bound berberine spectrum. One way to mitigate this problem is to start dilution series with a more concentrated sample, where the very first sample is the representative of fully bound berberine system. However, if the plateau region is not observed (e.g. the melting temperature of the poly(A)-small molecule complex might be close to room temperature as in poly(A)-berberine complex) even with the more concentrated samples, an alternative method based on the ratio of the

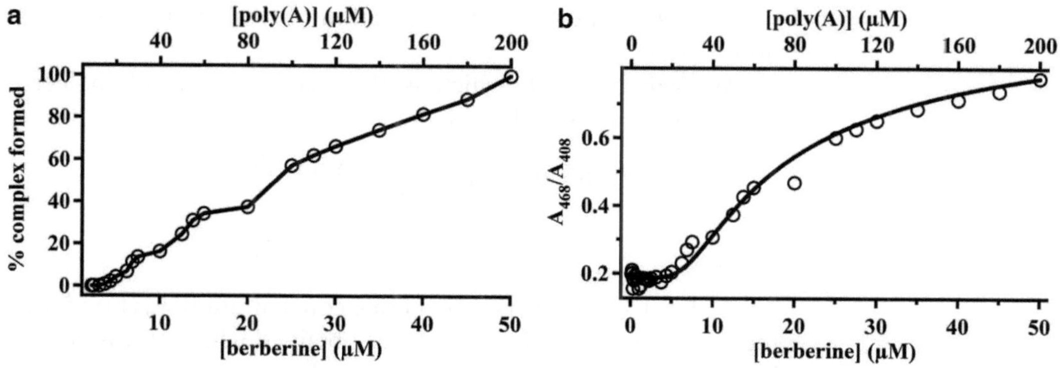

Fig. 9 (**a**) Plot of percentage of poly(A)-berberine complex formed as a function of berberine and poly(A) concentration. % of poly(A)-berberine complex formed is obtained using least square fits of dilution data with a two state model. (**b**) Plots of A_{468}/A_{408} as a function of berberine and poly(A) concentration. The *curve* represents the best fit obtained by CMC equation where six ligands were required for the stabilization of poly(A)-berberine complex. Samples were in 1× BPE buffer with 20 mM NaCl, pH 7.0. Spectra collected at 22 °C. Figure is replotted from Persil Cetinkol et al. [16]

berberine absorbance peaks can be used to mitigate this problem. Binding of berberine to poly(A) results in a red shift in the UV–Vis absorbance of berberine, which results in a different value of A_{468}/A_{408} for bound and unbound berberine (Fig. 6c). If berberine absorption at 468 nm versus 408 nm is plotted as a function of berberine concentration, a cooperative transition is observed between berberine concentrations of 0 and 25 μM (Fig. 9b). Similar cooperative transitions are observed in the formation of multimolecular assemblies such as micelle formation. Therefore, the poly(A)-berberine system can be modeled as the micelle formation [16, 49]. Certain number of surfactant molecules, q, are needed for the formation of micelles in solution. And the critical micelle concentration, C^* (CMC) is defined as the concentration at which half of these surfactant molecules are found in micelles. All the added surfactants will go into micelles if their concentration is higher than CMC. We can describe the equilibrium equation for the formation of a cooperative multimolecular system, C_q (fully assembled molecules), by Eq. 1](q, monomers; C_1, individual molecules) [16, 49]:

$$qC_1 \rightleftarrows C_q \tag{1}$$

$$\ln K_{eq} = \ln\left(\left(X_q\right)/\left(X_1^q\right)\right)$$

X defines the mole fraction.

C^* is the concentration at which half of the molecules is absorbed into assemblies.

$$C^* = q\left[C_q\right]/\left[C_t\right] = 1/2$$

Here C_t is the molar concentration of small molecules in solution; $C_t = [C_1] + q[C_q]$.

We can define f, which is the fraction of small molecules absorbed in the assembly, as

$$f = q[C_q]/[C_t], \text{ which is a function of } C_t/C^*.$$

$$C_t / C^* = f^{1/(q-1)} / \left(2(1-f)^{q/(q-1)}\right).$$

If the A_{468}/A_{408} versus berberine concentration was fitted using the equation above, the best fit was obtained where at least six berberine molecules were needed for stabilization of the poly(A)-berberine complex (Fig. 9b). The dissociation constant, where only half of the strands were assembled, can be assigned as 25 μM. Analysis also revealed that the most concentrated sample, 200 μM in poly(A) and 50 μM in berberine, was only 71.4 % assembled.

4 Notes

1. UV–Vis absorbance and CD spectra of each poly(A)-small molecule system is unique such that they depend on the structure of the small molecule. If there is a change in the UV–Vis absorbance or ellipticity, such as UV–Vis spectrum shifted to red, it means that small molecule is interacting with poly(A). During thermal denaturation, job plot or binding studies, monitor the peaks of the small molecule that are changing upon binding to poly(A). It is always easier to monitor the peaks specific to the small molecule where there is no interference from the nucleotide absorbance. The choice of the peaks chosen to monitor will depend on the system and the person who is doing the analysis. There might be more than one peak that might be changing upon poly(A) binding. In these cases, it is always a good habit trying to monitor different peaks, and determine which peak might represent the poly(A)-small molecule system the best (e.g. looking at the signal to noise ratio might be one way of determining).
2. You can wrap the solutions with aluminum foil in order to protect from direct light.
3. EDTA doesn't dissolve completely in H_2O until the pH is about 8.0.
4. We find that it is best to prepare the small molecule solutions fresh every 2–3 weeks. It is best to prepare solutions whose concentrations are lower than the dimerization constant of the small molecule to ensure the homogeneity of the solution. If you do not know the dimerization constant: (1) Prepare a

small molecule stock solution, (2) Centrifuge your stock solution for about 10 min at about 8,000 rpm (~7,000 × *g*), (3) Take the supernatant into a new tube without disturbing precipitates at the bottom, (4) Dilute the supernatant a little more with water, make sure there is no precipitate in it. We prefer to use small molecule stock solutions whose concentrations are around 500 μM. It is also good habit to recheck the concentration of your stock solution using UV–Vis spectrophotometer each time before you use it.

5. Make sure you have the proper cuvettes, with the right cut off values which do not scatter light. Use cuvettes suggested by the manufacturer of your instrument. In UV–Vis experiments cuvettes from Starna (Starna Cells, Inc, Atascadero, CA) and in CD experiments cuvettes supplied by the manufacturer of the instrument or from Hellma (Hellma USA, Planview, NY) were used in the experiments described here.
6. HP-8453 UV–Vis diode array spectrophotometer equipped with an Agilent 89090A Peltier temperature control unit was used in these experiments. Any other instrument can be used based on the availability. For thermal denaturation experiments, temperature control unit is essential. However, for other experiments temperature control unit is necessary just to ensure that the temperatures do not drift during the course of the experiment. You do not need one if you have a lab with temperature control and if you are interested in poly(A)-small molecule interactions at room temperature. For instance, the dilution experiments described here are all performed at room temperature.
7. When measuring the UV–Vis absorbance to back calculate the concentration of your stock solutions, try to stay in the most sensitive range of your instrumentation. With our instrument, we always use UV–Vis absorbance readings between 0.3 and 0.5. If the value of the UV–Vis absorbance of the diluted sample is out of your instrumentation's range, prepare a more dilute sample.
8. Different UV–Vis and CD cuvettes with different path lengths and volumes can be used based on the need and availability. The cuvettes used in the experiments described here were: 1 cm path length UV–Vis cuvette with 800 μL volume, 10 cm path length UV–Vis cuvette with 30 mL volume and 1 cm path length CD cuvette with 3,000 μL volume. For melting studies, use cuvettes with stoppers. Sample volumes should be adjusted according to the volume of the cuvettes used. The reason behind using UV–Vis cuvettes with different path lengths is to get decent UV–Vis spectra both for concentrated and dilute samples.

9. It is always easier to detect the changes in the absorption peaks where only the small molecule is absorbing, generally above 300 nm.
10. Operating the CD instrument without N_2 gas will damage the optics of the instrument. Make sure you have enough N_2 gas to run the whole experiment before you turn on the instrument. Make yourself familiar with N_2 regulators and decide on the required amount of N_2 gas flow based on the instrumentation manual. For our instrument, we generally let the N_2 gauger around 40 psi. CD is a delicate instrument, and a number of parameters need to be set before each experiment. These parameters depend on the sample as well as the instrumentation. Therefore, it is always best to work under the guidance of an experienced personnel and obey their instructions as well as the user's manual of the instrument when running a CD experiment.
11. According to the Job plots one coralyne is binding to four adenine bases. Poly(A)-coralyne system obeys the nearest neighbor exclusion principle, which states one ligand can bind to maximum of two base pairs (every other binding site will be occupied by the small molecule). In future experiments, the ratio of concentration of poly(A) (in nucleotide) to small molecule will be held constant at 4, e.g. poly(A) and small molecule concentrations will be 60 μM in nucleotide and 15 μM respectively.
12. In your CD Instrument's software, if you do not have the "Temperature/Wavelength Scan," you can use "Temperature Scan" program. "Temperature/Wavelength Scan" program collects data over a range of temperatures at the designated wavelength region. On the other hand, "Temperature Scan" program collects data only at a single wavelength over a range of temperatures. Therefore, you need to know which wavelength you would like to monitor beforehand if you are using "Temperature Scan" program. For small molecule binding, monitoring a wavelength in the induced CD band region can be a good start point.
13. If you want your data points to be finer, you should set data pitch to a lower value. However, setting the data pitch to a lower value will extend the duration of your experiment.
14. The amount of the sample solution replaced with buffer solution is calculated as:

$$C_f = \frac{n}{V_t} \rightarrow C_f = \frac{C_i\,(V_t - V_e)}{V_t} \rightarrow V_e = \frac{(C_i - C_f)^* V_t}{V_i}$$

C_i = Concentration of the poly(A) sample in the UV–Vis cuvette, C_f = Desired concentration of the poly(A) sample in the UV–Vis cuvette after the buffer is replaced, n = Number of moles of poly(A), V_t = Total sample volume, V_e = Sample volume exchanged with buffer solution.

15. As mentioned in **Note 8**, the reason behind switching UV–Vis cuvettes is to get decent UV–Vis spectra for less concentrated small molecule samples. The concentration at which you switched from using 1 cm path length cuvette to 10 cm path length cuvette will depend on the extinction coefficient of the small molecule. When signal to noise ratio of the absorbance peaks that you are monitoring is high, switch the cuvettes.

Acknowledgements

I am grateful to Prof. Dr. Nicholas V. Hud for his support and guidance. The methods described here were developed and executed in his laboratory. I also thank the members of Hud laboratory for their helpful discussions.

References

1. Dailey MM, Hait C, Holt PA et al (2009) Exp Mol Pathol 86:141–150
2. Madsen U, Krogsgaard-Larsen P, Liljefors T (eds) (2002) Textbook of drug design and discovery. Taylor & Francis, Washington, DC
3. Reddy MR, Parill AL (1999) Overview of rational drug design. In: Reddy MR, Parill AL (eds) Rational drug design: novel methodology and practical applications, vol 719. American Chemical Society, Washington, DC, pp 1–11
4. Cetinkol OP, Engelhart AE, Nanjunda RK et al (2008) Chembiochem 9:1889–1892
5. Islam MM, Kumar GS (2009) Biochim Biophys Acta 1790:829–839
6. Martinez R, Chacon-Garcia L (2005) Curr Med Chem 12:127–151
7. Mergny JL, Helene C (1998) Nat Med 4:1366–1367
8. Palchaudhuri R, Hergenrother PJ (2007) Curr Opin Biotechnol 18:497–503
9. Thomas JR, Hergenrother PJ (2008) Chem Rev 108:1171–1224
10. Thurston DE (1999) Br J Cancer 80:65–85
11. Balasubramanian S, Hurley LH, Neidle S (2011) Nat Rev Drug Discov 10:261–275
12. Papadopoulos K, Mita A, Ricart A et al (2007) Mol Cancer Ther 6:3462S–3463S
13. Quigley GJ, Wang AH, Ughetto G et al (1980) Proc Natl Acad Sci 77:7204–7208
14. Sobell HM (1985) Proc Natl Acad Sci U S A 82:5328–5331
15. Tan C, Tasaka H, Yu KP et al (1967) Cancer 20:333–353
16. Cetinkol OP, Hud NV (2009) Nucleic Acids Res 37:611–621
17. Di Giammartino DC, Nishida K, Manley JL (2011) Mol Cell 43:853–866
18. Dower K, Kuperwasser N, Merrikh H et al (2004) RNA 10:1888–1899
19. Giri P, Kumar GS (2010) Mini Rev Med Chem 10:568–577
20. Giri P, Kumar GS (2010) Mol Biosyst 6:81–88
21. Giri P, Kumar GS (2009) Curr Med Chem 16:965–987
22. Wickens M, Anderson P, Jackson RJ (1997) Curr Opin Genet Dev 7:220–232
23. Schmidt MJ, Norbury CJ (2010) Wiley Interdiscip Rev RNA 1:142–151
24. Topalian SL, Gonzales MI, Ward Y et al (2002) Cancer Res 62:5505–5509
25. Topalian SL, Kaneko S, Gonzales MI et al (2001) Mol Cell Biol 21:5614–5623
26. Xing FF, Song GT, Ren JS et al (2005) FEBS Lett 579:5035–5039
27. Bloomfield VA, Crothers DM, Tinoco I (2000) Nucleic acids: structures, properties, and functions. University Science, Sausalito, CA

28. Strekowski L, Wilson B (2007) Mutat Res Fund Mol Mech Mutagen 623:3–13
29. Neville DM Jr, Davies DR (1966) J Mol Biol 17:57–74
30. Parish JH (1985) Biochem Educ 13:92
31. Wartell RM, Larson JE, Wells RD (1974) J Biol Chem 249:6719–6731
32. Biver T (2012) Appl Spectrosc Rev 47:272–325
33. Jaumot J, Gargallo R (2012) Curr Pharm Des 18:1900–1916
34. Chang YM, Chen CKM, Hou MH (2012) Int J Mol Sci 13:3394–3413
35. Eriksson M, Norden B (2001) Drug Nucl Acid Interact 340:68–98
36. Berova N, Nakanishi K, Woody RW (eds) (2000) Circular dichoism. Wiley-VCH, New York, NY
37. Persil O, Santai CT, Jain SS et al (2004) J Am Chem Soc 126:8644–8645
38. Polak M, Hud NV (2002) Nucleic Acids Res 30:983–992
39. Chaires JB (2005) Structural selectivity of drug-nucleic acid interactions probed by competition dialysis. In: Waring MJ, Chaires JB (eds) Dna binders and related subjects (Topp Curr Chem), vol 253. Springer, Berlin, pp 33–53
40. Ragazzon PA, Garbett NC, Chaires JB (2007) Methods 42:173–182
41. Demeunynck M, Baily C, Wilson WD (eds) (2003) Small molecule DNA and RNA binders: from synthesis to nucleic acid complexes. Wiley-VCH Verlag GmbH & Co. KGaA, Weinhem
42. Fox KR (ed) (2010) Drug-dna interaction protocols. Humana Press, New York, NY
43. Wanunu M, Tor Y (eds) (2011) Methods for studying nucleic acid/drug interactions. CRC, Baco Raton, FL
44. Biver T, Ciatto C, Secco F et al (2006) Arch Biochem Biophys 452:93–101
45. Giri P, Hossain M, Kumar GS (2006) Int J Biol Macromol 39:210–221
46. Spiegel K, Magistrato A (2006) Org Biomol Chem 4:2507–2517
47. Bloomfield DA, Crothers DM, Tinono I Jr (1974) Physical chemistry of nucleic acids. Harper & Row, New York, NY
48. Qu X, Chaires JB et al (2000) Methods Enzymol 321:353–369
49. Kegel WK, van der Schoot P (2004) Biophys J 86:3905–3913

Chapter 9

The Stem-Loop Luciferase Assay for Polyadenylation (SLAP) Method for Determining CstF-64-Dependent Polyadenylation Activity

J. Andrew Hockert and Clinton C. MacDonald

Abstract

Polyadenylation is an essential cellular process in eukaryotic cells (Edmonds M and Abrams R, J Biol Chem 235, 1142–1149, 1960; Zhao J et al., Microbiol Mol Biol Rev 63, 405–445, 1999; Edmonds M, Progr Nucleic Acid Res Mol Biol 71, 285–389, 2002). For this reason, it has been difficult to examine the functions of specific polyadenylation proteins in vivo. Here, we describe a cell culture assay that allows structure-function experiments on CstF-64, a protein that binds to pre-mRNAs downstream of the cleavage site for accurate and efficient polyadenylation. We also demonstrate that the stem-loop luciferase assay for polyadenylation (SLAP) accurately reflects CstF-64-dependent polyadenylation. This assay could be easily adapted to the study of other important RNA-binding proteins in polyadenylation.

Key words Polyadenylation, CstF-64, Luciferase reporter gene, Dual-Luciferase Assay, RNA-binding protein, Epitope tagging

1 Introduction

Many stages in gene expression require the action of RNA-binding proteins. One such stage, polyadenylation, requires at least five separate protein complexes [4–6] requiring the action of many RNA-binding proteins [7, 8]. One of the best studied of these proteins is the 64,000 M_r subunit of the cleavage stimulation factor, CstF-64 [9, 10]. CstF-64 binds to the downstream G/U-rich sequence element in pre-mRNAs [11, 12] and is essential for polyadenylation of the pre-mRNA [6, 13, 14].

Because it is essential for normal cellular growth and development, examination of the structure-function relationships of the CstF-64 protein by disrupting its activities in polyadenylation in vivo has been difficult [13, 15, 16]. However, in vitro studies have allowed descriptions of the linear domains of CstF-64: the N-terminal RNA-binding domain (RBD), the hinge region, the

Joanna Rorbach and Agnieszka J. Bobrowicz (eds.), *Polyadenylation: Methods and Protocols*, Methods in Molecular Biology, vol. 1125, DOI 10.1007/978-1-62703-971-0_9, © Springer Science+Business Media New York 2014

MEAR(A/G) region, and the highly conserved C-terminal domain [12, 17–23]. But these proposed functions of putative domains had not been demonstrated in vivo due to the absence of a straightforward assay system.

Here, we present the stem-loop luciferase assay for polyadenylation (SLAP) that allows structure-function studies of CstF-64 in vivo. The SLAP procedure involves the downstream G/U-rich sequence element of a luciferase reporter gene to be replaced with two copies of the MS2 bacteriophage stem-loop element [24]. This modification makes luciferase expression dependent on the co-expression of a CstF-64-MS2 coat protein RNA-binding domain [25] fusion protein. These constructs are expressed in HeLa cells and the varying levels of luciferase expression correlate with changes in mRNA polyadenylation. The SLAP system has allowed us to verify not only that SLAP is a valid method of measuring polyadenylation in vivo but also implicated that the RBD, hinge, and CTD domains of CstF-64 were critical for polyadenylation [26].

2 Materials

All DNA, buffers, and other solutions were prepared using double deionized water. All reagents were stored as indicated.

2.1 Transfection Components (Store at −20 °C)

1. SL-Luc luciferase reporter: Generated by removing the SV40 polyadenylation cassette from pRL-SV40 (Promega, Madison, Wisconsin) using NotI and ClaI restriction enzymes. A SV40 polyadenylation cassette with MS2 stem loops replacing the G/U-rich downstream polyadenylation element was cloned in by PCR from p3Z SVL-MS2 (Maciolek and McNally (Medical College of Wisconsin)) also using NotI and ClaI primers (primer sequences available upon request). Additionally, two other SL-Luc constructs were made: one that contains neither the G/U-rich downstream element nor the MS2 stem loops (SL-LucΔ_{SL}) and one in which the AAUAAA polyadenylation signal was mutated by site-directed mutagenesis to AGGAGA (SL-Luc$_{AGGAGA}$). All SL-Luc constructs are shown in Fig. 1b.
2. MS2-CstF-64 fusion plasmids were obtained from Maciolek and McNally (Medical College of Wisconsin, ref. 24) and then cloned into p3XFLAG 7.1 (Sigma-Aldrich, St Louis, Missouri) using NotI, BglII, and KpnI restriction enzymes. Various primer combinations were used to generate different CstF-64 deletion constructs (primers available upon request). CstF-64 constructs that were made are as follows and diagramed in Fig. 1c: MS2-CstF-64, MS2-CstF-64Δ_{RBD}, MS2-CstF-64ΔHinge, MS2-CstF-64$\Delta_{P/G}$, MS2-CstF-64Δ_{MEARA}, MS2-CstF-64ΔCTD, and MS2-CstF-64$_{Y85A\text{-}L86S}$ (point mutations demonstrated to inactivate the MS2 coat protein RNA-binding domain [27]).

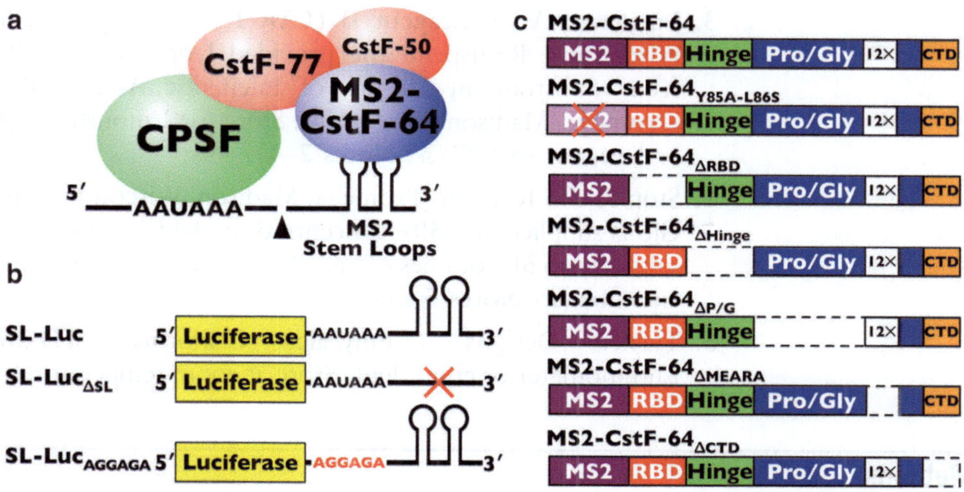

Fig. 1 The SLAP method and components. (**a**) Simplified cartoon of components in polyadenylation. Shown are the pre-mRNA (*line*) with the MS2 stem-loop RNA elements, the cleavage and polyadenylation specificity factor (CPSF, *green ball*), the subunits of the cleavage stimulation factor (CstF-77, CstF-50, *red balls*), and the MS2-CstF-64 (*blue ball*). (**b**) Stem-loop luciferase (SL-Luc) constructs used in SLAP. (**c**) MS2-CstF-64 and deletion constructs

3. pGL3-Control (firefly luciferase plasmid, Promega, Madison, Wisconsin).
4. p3XFLAG 7.1 empty vector (Sigma-Aldrich, St. Louis, Missouri).

2.2 HeLa Cell Culture (Store at 4 °C Except Where Indicated)

1. Cell culture media: DMEM with 4.5 g/L-glucose, L-glutamine, and sodium pyruvate. DMEM was supplemented with 10 % Cosmic Calf Serum (Hyclone, Logan, Utah), 100 U/mL penicillin, and 100 μg streptomycin/mL.
2. Cell culture media without penicillin and streptomycin.
3. Serum-free cell culture media: DMEM with 4.5 g/L-glucose, L-glutamine, and sodium pyruvate.
4. Cell culture media for post-transfection: Media from #2 with 20 % Cosmic Calf Serum.
5. Opti-MEM I Reduced Serum Medium (Invitrogen).
6. Lipofectamine (Invitrogen).
7. HeLa cells (ATCC, Manassas, Virginia)—grow at 37 °C in 5 % CO_2 (see below) and store in liquid nitrogen.

2.3 Luciferase Assay Components (Store at −80 °C Except Where Indicated)

1. Passive Lysis Buffer from Dual-Luciferase® Reporter Assay System (Promega, Madison, Wisconsin). Prepare a 1× working concentration by adding 1 volume of 5× PLB to 4 volumes of distilled water (*see* **Note 1**).
2. Phosphate buffered saline (PBS; 1×): 8 g NaCl, 0.2 g KCl, 1.44 g Na_2HPO_4, 0.24 g KH_2PO_4 pH 7.4 in 1 L H_2O. Store at room temperature.

3. Luciferase Assay Reagent II (LAR II) (Promega, Madison, Wisconsin). Resuspend the provided lyophilized Luciferase Assay Substrate in 10 mL of Luciferase Assay Buffer II (Promega, Madison, Wisconsin). Store for 1 month at −20 °C or 1 year at −80 °C (*see* **Notes 2–4**).
4. Stop & Glo Reagent (Promega, Madison, Wisconsin). Stop & Glo is supplied at a 50× concentration. Add 1 volume of 50× substrate to 50 volumes of Stop & Glo buffer in a polypropylene tube (*see* **Notes 5** and **6**).
5. Turner Designs (Sunnyvale, California) TD-20/20 Luminometer or equivalent. Store at room temperature.

3 Methods

Carry out all procedures at room temperature unless otherwise specified.

3.1 HeLa Cell Transfection (All Procedures to Be Carried Out Under a Sterile Hood)

1. Approximately 24 h before transfection, plate 2.75×10^4 HeLa cells per well on 24-well plates. Cells are to be plated in 500 μL of media *without* antibiotics. 15 wells will need to be plated for each experimental data point (one triplicate for each experiment and five replicates of each experiment, *see* **Note 7**).
2. On the day of transfection, set up 1.5 mL tubes for each experiment (each triplicate) as follows: 100 μL of Opti-MEM with 48 ng of SL-Luc reporter, 16 ng of pGL3 control, 400 ng of MS2-CstF-64 fusion plasmid, and 536 ng of p3XFLAG 7.1 for a total of 1,000 ng of DNA (*see* **Notes 8** and **9**).
3. In a separate 1.5 mL tube for each experiment, add 100 μL of Opti-MEM with 8 μL of Lipofectamine Transfection Reagent (*see* **Note 10**).
4. Pipette the Opti-MEM/Lipofectamine mixture into the tube containing the DNA, mix gently with a pipetter, and incubate for 15 min at room temperature.
5. Add 600 μL of Opti-MEM to the combined DNA/Lipofectamine mix and mix gently (*see* **Note 11**).
6. Under a tissue culture hood, remove growth medium from the HeLa cells and replace it with 200 μL of serum-free DMEM in each well. Add 250 μL of DNA/Lipofectamine/Opti-MEM (from **step 5**) to each well (three times for each triplicate). Mix by rocking the plate gently.
7. Incubate the plate for 5 h at 37 °C.
8. Add 400 μL of DMEM with 20 % Cosmic Calf Serum to each well without removing the transfection media. Incubate at 37 °C for approximately 19 h.

9. Remove transfection media and replace with 500 μL per well of regular cell culture media with 10 % Cosmic Calf Serum and antibiotics (penicillin and streptomycin). Incubate plates at 37 °C for 24 h.

3.2 Cell Lysis

1. Label enough 1.5 mL tubes for each well of the transfection (*see* **Note 12**).
2. Prepare enough Passive Lysis Buffer by diluting 5×PLB with water (*see* **Note 1**).
3. Remove media from the cells and gently add approximately 500 μL of 1×PBS to each well to wash media from the cells.
4. Wash cells a second time with 1×PBS.
5. Remove PBS from cells and add 100 μL of Passive Lysis Buffer per well.
6. Place the plates onto an orbital shaker platform and agitate gently for 15 min at room temperature.
7. Transfer the lysate into the appropriate labeled 1.5 mL tube (from **step 1**).

3.3 Luciferase Assay

1. Prepare Luciferase Assay Reagent II as described above (*see* **Note 2**).
2. Prepare Stop & Glo Reagent as described above (*see* **Note 3**).
3. Set up enough 1.5 mL tubes for each sample in the experiment (15 for each data point, *see* **Note 13**).
4. To each tube (from **step 3**), add 100 μL of LAR II.
5. Program the luminometer to perform a 2-s premeasurement delay, followed by a 10-s measurement period for each assay.
6. Transfer 20 μL of cell lysate (from Subheading 3.2, **step 7**) into the luminometer tube containing LAR II. Mix by pipetting four or five times.
7. Place tube in the luminometer and initiate reading immediately after mixing.
8. If the luminometer is not connected to a printer or computer, record the firefly luciferase activity measurement (*see* **Note 14**).
9. Remove the tube and dispense 100 μL of Stop & Glo Reagent. Mix by pipetting, and replace the tube in the luminometer and initiate reading.
10. Record the Renilla luciferase activity measurement (*see* **Note 15**).
11. Discard the reaction tube and proceed to the next tube.
12. When all samples have been measured, analyze as follows (Table 1):
    - Divide each Renilla activity value by its corresponding firefly luciferase activity value. This produces the luciferase activity for each transfected well.

Table 1
Typical Luciferase Values

	Renilla[a]	Firefly	Renilla/firefly	Average Renilla/firefly	Standard deviation
SL-Luc	390.3	138.9	2.810	2.889	0.993
	994.9	253.8	3.920		
	556.2	287.0	1.938		
SL-Luc + MS2-CstF-64	8,966.0	211.5	42.392	41.066	1.152
	9,238.0	229.2	40.305		
	11,466.0	283.1	40.502		
SL-Luc$_{AGGAGA}$	131.2	318.6	0.412	0.427	0.016
	124.8	292.9	0.426		
	123.2	278.0	0.443		
SL-Luc$_{AGGAGA}$ + MS2-CstF-64	508.7	158.5	3.209	3.021	0.187
	646.1	227.9	2.835		
	589.4	195.2	3.019		

[a]Shown are typical luciferase values from three technical replicates using HeLa cells transfected as indicated using the Dual-Luciferase® Reporter Assay System and the TD-20/20 Luminometer. For simplicity, only technical replicates are shown. However, we routinely perform five independent measurements (independently transfected wells) with three technical replicates each

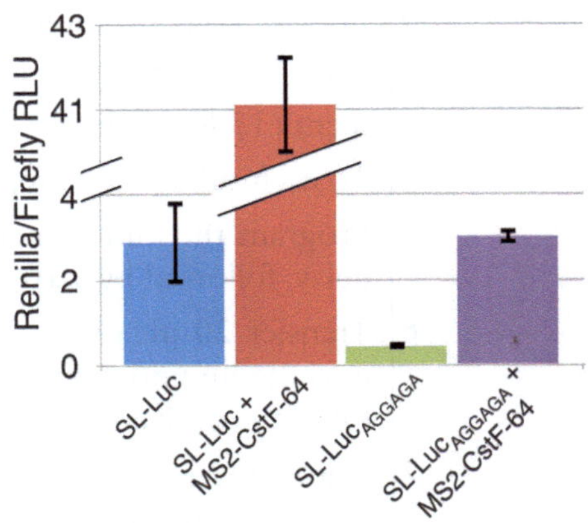

Fig. 2 Renilla/firefly ratios as calculated from the data in Table 1. Shown are relative luciferase units calculated from the Renilla/firefly ratios for SL-Luc, SL-Luc + MS2-CstF-64, SL-Luc$_{AGGAGA}$, and SL-Luc$_{AGGAGA}$ + MS2-CstF-64. Note that the Y-axis is broken to accommodate the higher values for SL-Luc + MS2-CstF-64

- Then average the triplicate luciferase values (three transfected well values).
- Then average the five triplicate groups that are the replicates for each MS2 or MS2-CstF-64 construct.
- Use the values for the replicates to determine the standard deviation for each construct.
- Graphs for these results can be produced using a program such as Microsoft Excel (Fig. 2).

13. Use the values for the replicates along with a statistics program such as InStat (Graphpad Software, Inc, San Diego, California) to determine the validity of the results using an ANOVA and a Tukey Post Test.

4 Notes

1. Prepare 1×PLB on the day of the cell lysis and make sure that enough is made for all samples plus 10 % for overage.
2. When planning SLAP experiments, it is best to plan a number of wells so that all of the LAR II is used. With each data point representing 15 transfected wells and each well using 100 μL of LAR II, only about 6 or 7 data points (constructs) are needed to use this amount of LAR II.
3. Prepare the LAR II reagent on the day of the luciferase assay (it should not be refrozen).
4. Prepare a new LAR II solution for each new experiment. However, because the LAR II reagent is expensive, if you need to use LAR II that has been frozen (left over from a previous assay), make sure to mix it with the new LAR II so that each sample gets the same LAR II for consistency.
5. Make Stop & Glo on the day of the luciferase assay. Fresh reagent always works best.
6. Make enough Stop & Glo for all samples plus 10 % to ensure that each sample gets enough of the reagent.
7. For organizational purposes it is easiest to plate cells by experiment (each one of the five replicates for each data point being tested). For example, if transfecting MS2 alone, MS2-CstF-64, ΔRBD, ΔHinge, ΔP/G, ΔMEARA, and ΔCTD wells can be arranged for each triplicate (three wells of MS2 alone, etc.) until all triplicates are done for each replicate, then do the same for the next replicate. This breaks the samples into different replicates for ease of organization.
8. It is easiest to make one large master mix tube for each data point (experimental construct) that you will be testing for. So prepare five times the amount of Opti-MEM and DNA plus 10 %. This provides each well with the same amount of Opti-MEM and DNA.
9. When making the master mix, it is most convenient to use a 5 mL (or 15 mL) tube to ensure that all components can be mixed.
10. When using a master mix, add 550 μL of Opti-MEM with 44 μL of Lipofectamine.
11. If making a master mix (*see* **Note 8**), add 3,300 μL of Opti-MEM to the master mix tubes.

12. Keep the same organization for the sample tubes as for the wells on the plates. Not all of the sample will be used in each luciferase assay, so samples can be retained and stored at –70 °C and used for additional luciferase assays.
13. Prepare enough tubes to do all samples. All luciferase samples from one transfection should be performed on the same day. This eliminates day-to-day variation between the samples and the LAR II and Stop & Glo Reagents that can occur.
14. Record the firefly luciferase activity as quickly as possible if not using a printer or make sure to take the same amount of time between samples. For greatest reproducibility, we find it most effective to use a stopwatch to time the addition of the reagents and luciferase readings.
15. Record the Renilla luciferase activity quickly to ensure timely transfer from one sample to the next. For greatest reproducibility, we find it most effective to use a stopwatch to time the addition of the reagents and luciferase readings.

References

1. Edmonds M, Abrams R (1960) Polynucleotide biosynthesis: formation of a sequence of adenylate units from adenosine triphosphate by an enzyme from thymus nuclei. J Biol Chem 235: 1142–1149
2. Zhao J, Hyman L, Moore C (1999) Formation of mRNA 3′ ends in eukaryotes: mechanism, regulation, and interrelationships with other steps in mRNA synthesis. Microbiol Mol Biol Rev 63:405–445
3. Edmonds M (2002) A history of poly A sequences: from formation to factors to function. Prog Nucleic Acid Res Mol Biol 71: 285–389
4. Christofori G, Keller W (1998) 3′ cleavage and polyadenylation of mRNA precursors in vitro requires a poly(A) polymerase, a cleavage factor, and a snRNP. Cell 54:875–889
5. Takagaki Y, Ryner LC, Manley JL (1989) Four factors are required for 3′-end cleavage of pre-mRNAs. Genes Dev 3:1711–1724
6. Gilmartin GM, Nevins JR (1989) An ordered pathway of assembly of components required for polyadenylation site recognition and processing. Genes Dev 3:2180–2190
7. Mandel CR, Bai Y, Tong L (2008) Protein factors in pre-mRNA 3′-end processing. Cell Mol Life Sci 65:1099–1122
8. Shi Y, Di Giammartino DC, Taylor D et al (2009) Molecular architecture of the human pre-mRNA 3′ processing complex. Mol Cell 33:365–376
9. Moore CL, Chen J, Whoriskey J (1988) Two proteins crosslinked to RNA containing the adenovirus L3 poly(A) site requires the AAUAAA sequence for binding. EMBO J 7:3159–3169
10. Wilusz J, Shenk T (1988) A 64 kd nuclear protein binds to RNA segments that include the AAUAAA polyadenylation motif. Cell 52: 221–228
11. Chou ZF, Chen F, Wilusz J (1994) Sequence and position requirements for uridylate-rich downstream elements of polyadenylation signals. Nucleic Acids Res 22:2525–2531
12. MacDonald CC, Wilusz J, Shenk T (1994) The 64-kilodalton subunit of the CstF polyadenylation factor binds to pre-mRNA downstream of the cleavage site and influences cleavage site location. Mol Cell Biol 14:6647–6654
13. Takagaki Y, Manley JL (1998) Levels of polyadenylation factor CstF-64 control IgM heavy chain mRNA accumulation and other events associated with B cell differentiation. Mol Cell 2:761–771
14. Proudfoot NJ, O'Sullivan J (2002) Polyadenylation: a tail of two complexes. Curr Biol 12:855–857
15. Minvielle-Sebastia L, Winsor B, Bonneaud N et al (1991) Mutations in the yeast RNA14 and RNA15 genes result in an abnormal mRNA decay rate; sequence analysis reveals an RNA-binding domain in the RNA15 protein. Mol Cell Biol 11:3075–3087

16. Dass B, Tardif S, Park JY et al (2007) Loss of polyadenylation protein τCstF-64 causes spermatogenic defects and male infertility. Proc Natl Acad Sci U S A 104:20374–20379
17. Pérez Cañadillas JM, Varani G (2003) Recognition of GU-rich polyadenylation regulatory elements by human CstF-64 protein. EMBO J 22:2821–2830
18. Takagaki Y, Manley JL (1997) RNA recognition by the human polyadenylation factor CstF. Mol Cell Biol 17:3907–3914
19. Takagaki Y, Manley JL (2000) Complex protein interactions within the human polyadenylation machinery identify a novel component. Mol Cell Biol 20:1515–1525
20. Bai Y, Auperin TC, Chou CY et al (2007) Crystal structure of murine CstF-77: dimeric association and implications for polyadenylation of mRNA precursors. Mol Cell 25: 863–875
21. Richardson JM, McMahon KW, MacDonald CC et al (1999) MEARA sequence repeat of human CstF-64 polyadenylation factor is helical in solution. A spectroscopic and calorimetric study. Biochemistry 38:1403–1407
22. Qu X, Pérez Cañadillas JM, Agarawal S et al (2007) The C-terminal domains of vertebrate CstF-64 and its yeast orthologue Rna15 form a new structure critical for mRNA 3′-end processing. J Biol Chem 282:2101–2115
23. Ruepp MD, Schweingruber C, Kleinschmidt N et al (2011) Interactions of CstF-64, CstF-77, and symplekin: implications on localisation and function. Mol Biol Cell 22:91–104
24. Maciolek NL, McNally MT (2008) Characterization of Rous sarcoma virus polyadenylation site use in vitro. Virology 374: 468–476
25. Lago H, Fonseca SA, Murrau JB et al (1998) Dissecting the key recognition features of the MS2 bacteriophage translational repression complex. Nucleic Acids Res 26:1337–1344
26. Hockert JA, Hsiang-Jui Y, MacDonald CC (2010) The hinge domain of the cleavage stimulation factor protein CstF-64 is essential for CstF-77 interaction, nuclear localization, and polyadenylation. J Biol Chem 285: 695–704
27. Valegard K, Murrary JB, Stonehouse NJ et al (1997) The three-dimensional structures of two complexes between recombinant MS2 capsids and RNA operator fragments reveal sequence-specific protein-RNA interactions. J Mol Biol 270:724–738

Chapter 10

Accurate Mapping of Cleavage and Polyadenylation Sites by 3′ Region Extraction and Deep Sequencing

Mainul Hoque, Wencheng Li, and Bin Tian

Abstract

Deep sequencing of RNA (RNA-seq) is becoming a standard method to study gene expression. While RNA-seq reads cover most regions of an mRNA sequence, they are often depleted in the 3′ end region, making them less amenable for mapping the cleavage and polyadenylation site (pA). A major problem in identification of pA is mispriming at internal A-rich regions and oligo(A) tails when an oligo(dT) primer is used for reverse transcription or sequencing. We recently developed a method named 3′ region extraction and deep sequencing (3′READS), which completely addresses mispriming issues and is straightforward to use. The method accurately maps pAs and allows quantitative analysis of alternative cleavage and polyadenylation (APA) isoforms and gene expression.

Key words Deep sequencing, Gene expression, Cleavage and polyadenylation, PolyA site, RNA-seq

1 Introduction

Almost all mRNAs in eukaryotes have a poly(A) tail, which plays a key role in mRNA metabolism [1]. Cleavage and polyadenylation is the process responsible for adding the poly(A) tail to nascent RNA and is coupled to termination of transcription [2]. The cleavage and polyadenylation site, or polyA site (pA), is defined by surrounding *cis*-elements [3, 4]. Over half of the human mRNA genes have multiple pAs, leading to mRNA isoforms that contain different coding sequences, 3′ untranslated regions (3′UTRs), or both [5]. Alternative cleavage and polyadenylation (APA) can play a significant role in mRNA metabolism by controlling the length of 3′UTR and its encoded *cis*-elements [6, 7]. Dynamic regulation of 3′UTRs by APA has been reported in different tissue types [8, 9], cell proliferation, differentiation and development [10, 11], cancer cell transformation [12, 13], and response to extracellular stimuli [14].

Joanna Rorbach and Agnieszka J. Bobrowicz (eds.), *Polyadenylation: Methods and Protocols*, Methods in Molecular Biology, vol. 1125, DOI 10.1007/978-1-62703-971-0_10,

Identification of pA is typically based on cDNA sequence corresponding to the poly(A) tail, which is generated by oligo(dT)-based reverse transcription [15, 16]. However, oligo(dT) can also prime at internal A-rich sequences, which produce a string of sequenced As that is indistinguishable from a real poly(A) tail [17]. This internal priming problem is usually addressed computationally by discarding putative pAs located in genomic A-rich regions, but the approach cannot guarantee full elimination of false positives caused by internal priming and can discard real pAs. In addition, oligo(A) tails, typically shorter than 15 nt in length [18], are synthesized by noncanonical poly(A) polymerases, such as those involved in exosome-mediated RNA decay [19]. They can potentially be targeted by oligo(dT) for priming, leading to false identification of pA.

We recently developed a method named 3′ region extraction and deep sequencing (3′READS) for accurate mapping of pA [20]. This method solves the mispriming issues that commonly plague pA mapping. At the core of this method, a special oligonucleotide (oligo), named chimeric uridine thymidine (CUT) oligo, was used to capture poly(A)-containing RNA fragments under a stringent condition, followed by removal of most of the poly(A) sequence by RNase H, ligation of RNA fragment with adapters, reverse transcription and PCR amplification, and deep sequencing. We also found that the 3′READS data is quantitative to examine expression of APA isoforms and, when multiple isoforms are combined, this method can be used for gene expression analysis. Here we present the detailed protocol of this method with some recent improvements, as illustrated in Fig. 1.

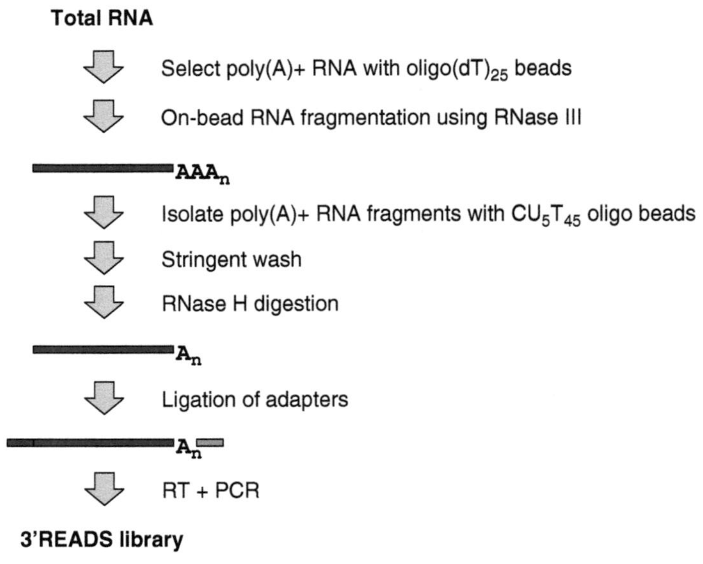

Fig. 1 Schematic of the 3′READS protocol

2 Materials

2.1 Reagents

1. Ultrapure water.
2. T4 RNA ligase 2 truncated (Bioo Scientific).
3. T4 RNA ligase 1 (New England Biolabs, NEB).
4. Superscript III Reverse Transcriptase (Life Technologies).
5. RNase III (NEB).
6. RNase H (NEB).
7. RNasin Ribonuclease Inhibitor (Promega).
8. Phusion DNA Polymerase (Finnzymes).
9. Glycogen (Ambion).
10. 10 mM ATP (NEB).
11. 25 mM dNTP Mix (Life Technologies).
12. 1 Kb Plus DNA Ladder (Life Technologies).
13. Oligo(dT)$_{25}$ magnetic beads (NEB).
14. Dynabeads® MyOne™ Streptavidin C1 (Life Technologies).
15. Spin-X® centrifuge tube filter (Corning).

2.2 Adapters, PCR Primers, and Sequencing Primers

1. 5′RNAadapter:5′-CCUUGGCACCCGAGAAUUCCNNNN.
2. 3′ DNA adapter: 5′-rApp/NNNNGATCGTCGGACTGTAG AACTCTGAAC/3ddC/.
3. Reverse transcription (RT) primer: 5′-GTTCAGAGTTCT ACAGTCCGACGATC.
4. PCR primer 1: 5′-AATGATACGGCGACCACCGAGATCT ACACGTTCAGAGTTCTACAGTCCGA.
5. PCR primer 2 (index region in bracket): 5′-CAAGCAGAAGA CGGCATACGAGAT[CGTGAT]GTGACTGGAGTT CCTTGGCACCCGAGAATTCCA.
6. Chimeric U_5T_{45} (CU_5T_{45}) oligo: 5’-Biotin-TTTTTTTTTTTTTTTTTTTTTTTTTTTTTTT TTTTTTTTTTTTTTUUUUU.
7. Sequencing primer for the insert: 5′-CTACACGTTCAGAG TTCTACAGTCCGACGATC.
8. Sequencing primer for the index region: 5′-GGAATTCTC GGGTGCCAAGGAACTCCAGTCAC.
9. quantitative real-time PCR (qPCR) primers: 5′-AATGATA CGGCGACCACCGAGAT and 5′-CAAGCAGAAGACGGC ATACGA.

2.3 Instruments

1. MagnaRack™ (Life Technologies).
2. X-tracta™ gel extraction tool (USA Scientific).
3. Agilent 2100 Bioanalyzer.

4. Thermocycler.
5. Real-time/qPCR machine.
6. NanoDrop spectrophotometer.
7. Vertical gel electrophoresis apparatus.
8. UV transilluminator.
9. Microcentrifuge.
10. Rotator.

2.4 Solutions/Buffers

1. Binding buffer: 10 mM Tris–HCl pH 7.5, 150 mM NaCl, 1 mM EDTA.
2. RNase III digestion buffer: 10 mM Tris–HCl pH 8.3, 60 mM NaCl, 10 mM $MgCl_2$, 1 mM DTT.
3. Coupling buffer: 10 mM Tris–HCl pH 7.5, 1 M NaCl, 1 mM EDTA.
4. Stringent washing buffer: 10 mM Tris–HCl pH 7.5, 1 mM NaCl, 1 mM EDTA, 10 % formamide.
5. Elution buffer: 1 mM EDTA, 1 mM NaCl.
6. RNase H digestion buffer: 50 mM Tris–HCl pH 8.3, 1 mM NaCl, 3 mM $MgCl_2$, 10 mM DTT.
7. TE: 10 mM Tris–HCl pH 8.0, 1 mM EDTA.

3 Methods

3.1 RNA Extraction and Quality Control (QC)

Total RNA is extracted from cells/tissues using the RNeasy kit (Qiagen) or Trizol (Invitrogen). A total of 25 μg of good quality RNA is required to prepare a 3′READS cDNA library (*see* **Note 1**). The RNA quality is determined on an Agilent Bioanalyzer using the RNA 6000 Pico kit. A sample with RNA integrity number (RIN) >7.0 is considered acceptable for further processing (Fig. 2a).

3.2 Poly(A)+ RNA Isolation

RNA samples are first heated at 65 °C for 5 min and then cooled down on ice for 2 min. This step disrupts RNA structures. The RNA samples are then mixed with oligo$(dT)_{25}$ magnetic beads in 200 μL binding buffer, followed by incubation at room temperature for 1 h on a rotator. The beads are washed three times with binding buffer (1 mL each time), followed by one final wash with 1 mL of RNase III digestion buffer.

3.3 On-Bead RNA Fragmentation

RNA bound to beads is fragmented by incubation with 2.5 U RNase III in a volume of 50 μL at 37 °C for 15 min. Remove unbound RNA by washing the beads three times with 1 mL of binding buffer. Collect the bound poly(A)+ RNA from beads using 200 μL of TE. Purify RNA with phenol-chloroform extraction and

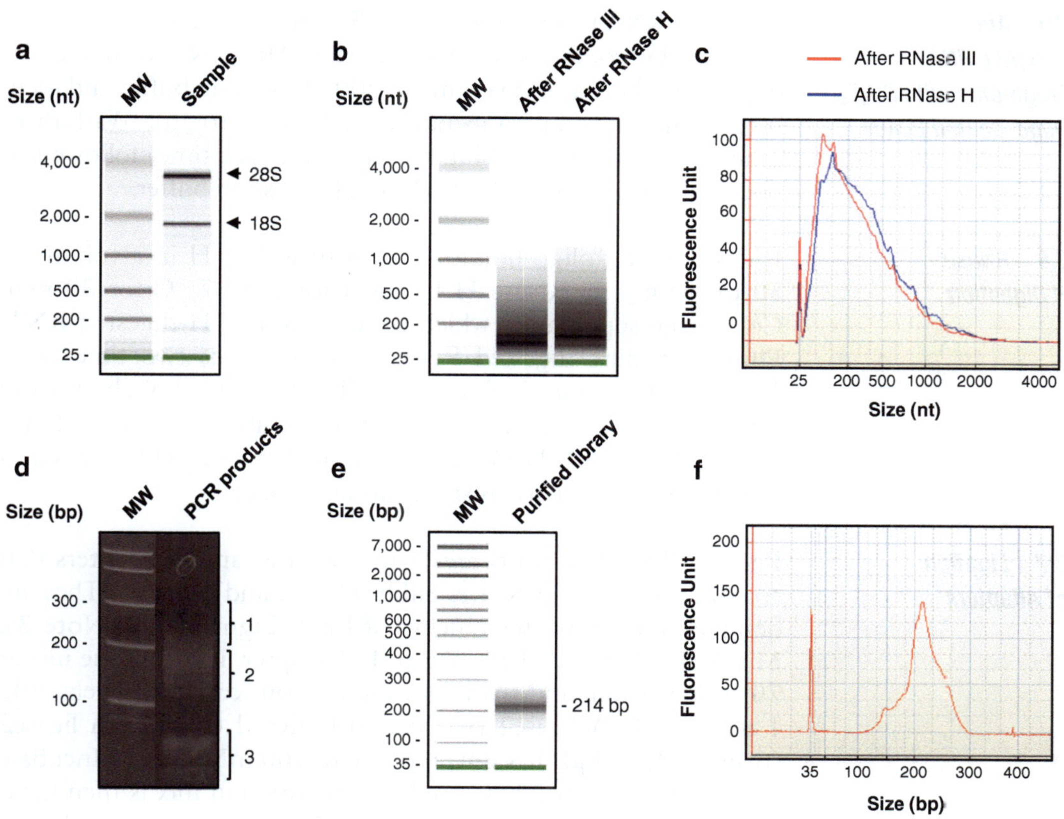

Fig. 2 Preparation of 3′READS cDNA library. (**a**) Quality control of input total RNA by Agilent Bioanalyzer. MW, molecular weight marker ladder. Sample is 1 ng of total RNA extracted from mouse testis. An RNA 6000 pico kit was used. The RNA integrity number (RIN) was 9.8. Bands corresponding to 28S and 18S rRNAs are indicated. (**b**) Fragmented RNAs. After RNase III, poly(A)+ RNA fragments eluted from oligo(dT)$_{25}$ beads after RNase III digestion; after RNase H, poly(A)+ RNA fragments eluted from CU$_5$T$_{45}$ oligo beads after RNase H digestion. As in (a), an RNA 6000 pico kit was used. (**c**) Size distribution of RNAs shown in (b). (**d**) Purification of cDNA library after PCR. Amplified cDNA libraries were resolved in an 8 % polyacrylamide gel. Library DNAs with sizes ranging from 200 to 300 bp (region *1*) were purified, corresponding to insert size of 70–170 bp. Region *2* contains library DNAs with shorter or no inserts, and region *3* contains PCR primers. (**e**) Qualitative and quantitative analysis of purified cDNA library in Agilent Bioanalyzer using a high sensitivity DNA kit. The peak is at 214 bp, corresponding to an insert size of 85 bp. (**f**) Size distribution of cDNAs shown in (**e**)

ethanol precipitation (with 6 μg/mL glycogen to facilitate precipitation). After washing with 70 % ethanol, RNA is dissolved in 10 μL H_2O.

3.4 Coupling Biotin-CU$_5$T$_{45}$ Oligo with Streptavidin Beads

Wash Dynabeads® MyOne™ Streptavidin C1 beads two times with 0.1 N NaOH and 50 mM NaCl (1 mL each time) and then two times with coupling buffer (1 mL each time). Conjugate 100 pM biotin-CU$_5$T$_{45}$ oligo with 0.2 mg of beads in 200 μL coupling buffer by incubation at room temperature for 30 min on a rotator. Wash three times with binding buffer (1 mL each time).

3.5 Annealing Poly(A)+ RNA Fragments with CU_5T_{45} Oligo-Coated Beads

Denature RNA from Subheading 3.3 by heating at 65 °C for 5 min followed by cooling on ice for 2 min. Mix RNA with CU_5T_{45} oligo-coated beads in a volume of 200 μL binding buffer and incubate the mix at room temperature for 1 h on a rotator. Wash three times with stringent washing buffer (1 mL each time) followed by one final wash with 1 mL of RNase H digestion buffer.

3.6 RNase H Digestion

Digest CU_5T_{45}-oligo-bound RNA with RNase H in a volume of 50 μL with 1 U RNase H by incubating at 37 °C for 30 min. Collect the supernatant which contains RNase H-digested RNA. Elute remaining digested RNA from the beads two times with 100 μL of elution buffer. Purify RNA by phenol-chloroform extraction and ethanol precipitation, as in Subheading 3.3. RNA is dissolved in 10 μL H_2O. Check the quality and yield of RNA in Bioanalyzer (optional), as shown in Fig. 2b, c.

3.7 Ligation of Adapters

Purified RNA fragments are ligated with 5′ and 3′ adapters that contain four random Ns (*see* Subheading 2 and **Note 2**). They are designed for reverse sequencing of RNA fragments (*see* **Note 3**). Mix 5 μL RNA with 1 μL of 1 μM 3′ adapter, incubate the mix at 70 °C for 2 min and then cool the mix on ice immediately. Add 1 μL 10× RNA ligase2 (truncated) buffer, 1.0 μL RNA ligase2 (truncated), 0.5 μL RNasin, and 0.8 μL 100 mM $MgCl_2$. Incubate the reaction mix at 22 °C for 1 h. The reaction mix is then incubated with 1 μL of 5 μM RT primer at 70 °C for 2 min, and then at 37 °C for 5 min, followed by cooling down to 4 °C. This step can block ligation of free 3' adapters with 5' adapters. To the same reaction tube, add 1 μL of 5 μM heat-inactivated 5′ adapter (70 °C for 2 min, then immediately transfer onto ice), 1 μL 10 mM ATP, and 1 μL T4 RNA ligase 1. Incubate the mix at 20 °C for 1 h.

3.8 Reverse Transcription

Transfer 5 μL of reaction mixture from Subheading 3.7 to a separate tube for RT using Superscript III Reverse Transcriptase. Since RT primer is already annealed to the ligated RNA at Subheading 3.7, there is no need to add RT primer again. The RT reaction mix contains (1) 5 μL of reaction mix from Subheading 3.7, (2) 2 μL 5× RT buffer, and (3) 0.5 μL 12.5 mM dNTP mix. Incubate the mix at 70 °C for 5 min, followed by immediate cooling on ice. Add 1 μL 100 mM DTT, 0.5 μL RNasin, and 1 μL Superscript III Reverse Transcriptase. Incubate the reaction mix at 50 °C for 60 min, then inactivate the enzyme by heating at 70 °C for 15 min, and then cool down the mix to 4 °C.

3.9 PCR Amplification of cDNA Library

Amplify RT product with 15 cycles of PCR in the same tube. The PCR reaction mix includes 10 μL 5× PCR buffer, 0.5 μL 25 mM dNTPs, 1 μL 25 μM forward primer, 1 μL 25 μM reverse primer, 0.5 μL of Phusion polymerase (2 U/μL), and 10 μL RT mix. The final reaction volume is adjusted to 50 μL with H_2O. PCR reaction starts with heating at 98 °C for 30 s, followed by 15 cycles of reactions, each with 98 °C for 10 s, 60 °C for 30 s, and 72 °C for 30 s. A final product extension cycle is performed at 72 °C for 10 min followed by cooling down to 4 °C.

3.10 Purification and Quantification of cDNA Library

Purify the PCR product from an 8 % polyacrylamide gel run in the TBE buffer (Tris/Borate/EDTA). When multiple samples are run in the same gel, put them in alternate lanes to avoid cross-contamination. Use a molecular weight DNA ladder with a good resolution for the 100–400 bp range for accurate estimation of cDNA size, such as the 1 Kb Plus DNA Ladder (Fig. 2d). X-tracta gel extractor is used to precisely cut the expected cDNA library from the gel. To extract DNA, the gel piece is placed in a 0.5 mL microcentrifuge tube housed in a 2 mL microcentrifuge tube. The inner tube has a hole at its bottom created by punching through it with a 21-gauge aspiration needle. The gel piece is shredded by passing through the hole from inner tube to outer tube at the maximum speed for 1 min. The shredded pieces are then resuspended in TE, followed by elution of DNA at room temperature overnight on a rotator. The suspension is transferred into a Spin-X centrifuge tube filter, followed by spinning at the maximum speed in a microcentrifuge for 2 min. Eluted DNA is purified by ethanol precipitation (with 6 μg/mL glycogen to facilitate precipitation). After washing with 70 % ethanol, DNA is dissolved in TE. Check DNA concentration by NanoDrop spectrophotometer. Take ~0.5 ng DNA to check the quality and quantity of DNA in Agilent Bioanalyzer with the high sensitivity DNA kit, as shown in Fig. 2e, f. The amount of library needs to be adjusted according to Illumina's suggestions to avoid over- or under-clustering (*see* **Note 4**). Primers used for sequencing the insert region and index region are shown in Subheading 2.

3.11 Mapping pAs in Genome

The reverse sequencing protocol of 3′READS generates reads corresponding to the antisense strand of transcript. As such, a read derived from a poly(A)+ fragment should start with four random bases (part of the 3′ adapter), followed by a number of Ts corresponding to the poly(A) sequence and then a specific sequence corresponding to the region before pA (illustrated in Fig. 3a). The 5′ adapter can also be sequenced if the number of cycles of sequencing exceeds the insert size of a cDNA. For the 3′READS data, we first remove 5′ adapter sequences from reads by searching the 3′-most 7 nt of the 5′ adapter in each read. If a match is found, the region from the first matched position to the 3′ end of read is removed. Reads with length <15 nt after this step are not used for further analysis. We then map reads against genome sequence using Bowtie 2 (version 2.1.0) [21] with the following setting: "--local -5 4." The parameter "-5 4" ensures trimming of the first four random bases from the 5′ end before alignment. The parameter "--local" is

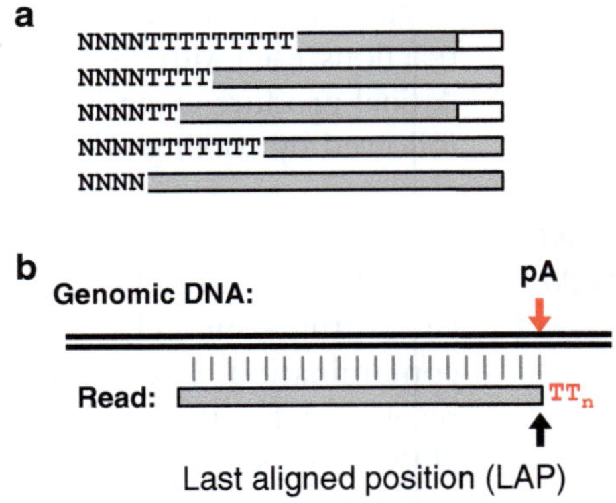

Fig. 3 Identification of pA using reads from 3′READS. (**a**) Schematic showing reads from 3′READS. Reads begin with four random nucleotides, followed by a number of Ts. *Grey box* represents insert sequence, *open box* represents 5′ adapter sequence. (**b**) Reads from 3′READS are aligned to genomic DNA. The last aligned position (LAP) and the putative pA are indicated by *arrows*

for the local alignment mode, which allows trimming of the alignment at one or both ends, a necessary step for aligning reads with a poly(T) sequence that is unalignable to genome sequence. Reads containing a poly(T) sequence have a "CIGAR" string containing "S," indicating soft clipping. For example, a sequence with eight unmatched Ts followed by 60 nt matched to the genome will get CIGAR = "8S60M." We only use reads with mapping quality score (MAPQ) ≥10, which corresponds to no more than 1 in 10 chances that the read truly originates elsewhere in the genome. In addition to using this intrinsic cutoff to filter unreliable alignments, we require mismatches to be ≤5 % of the read. For the remaining reads, those with ≥2 unaligned Ts at the 5′ end are called polyA site supporting (PASS) reads. While straightforward, this approach has one shortcoming: nearby pAs downstream of 3' splice sites cannot be mapped when the sequence from the 3′ splice site to the pA is too short for accurate alignment or its alignment has a lower score than the alignment for the rest of the sequence (*see* **Note 5** for discussion on this issue).

3.12 Clustering of pAs

Because multiple cleavage and polyadenylation can take place in a small window of sequence, the last aligned positions (LAPs) of PASS reads (Fig. 3b) are grouped to form pA clusters. We first progressively cluster together LAPs of PASS reads located within 24 nt from one another. If a cluster size is ≤24 nt, the LAP with the greatest number of PASS reads is used as the representative position for the pA cluster; if a cluster is >24 nt, we identify the LAP with the greatest number of PASS reads and re-cluster LAPs located >24 nt from the position. This process is repeated until all LAPs of PASS reads in the cluster are assigned to pAs.

3.13 Assigning pAs to Genes

pAs mapped to genome are further assigned to genes, according to gene models defined by RefSeq, Ensembl, and UCSC Known Gene databases. The 3′ ends of these gene models can be extended using strand-specific RNA-seq data to achieve better coverage of genic region [20]. To reduce false pAs, we require (1) the number of PASS reads for a pA is ≥5 % of all PASS reads for the gene and (2) a "real" pA has at least two distinct PASS reads as defined by the number of As and the four random Ns in the read. After this step, expression of each APA isoform can be examined, and all PASS reads for a gene can be summed up to represent the expression level of the gene. *See* **Note 6** for using PASS reads for quantitative analysis.

4 Notes

1. *Amount of input RNA*. The original 3′READS method [20] using zinc-based RNA fragmentation required ~50 μg of input RNA. We have found that RNA fragmentation by RNase III yields more poly(A)+ RNA fragments and allows the fragmentation to be performed on beads. As such, the modified method presented here uses RNase III, which requires ~25 μg of total RNA and involves on-bead RNA fragmentation.
2. *Random Ns used in 5′ and 3′ adapters.* The random four Ns in 5′ and 3′ adapters can help mitigate the bias of RNA ligases [22, 23]. In addition, the four Ns in the 3′ adapter can (1) facilitate cluster calling by the Illumina sequencing software, which uses the first four nucleotides to distinguish clusters, and (2) make it possible to quantify RNA fragments based on the distinct sequence of four Ns. The latter is important because some RNA fragments may be overrepresented in the result because of ligation bias and/or over-amplification by PCR.
3. *Reverse sequencing.* We previously designed adapters for sequencing the insert region from the 5′ end (forward sequencing) or from the 3′ end (reverse sequencing). The protocol we present here is based on reverse sequencing. The advantage of using this approach is that the terminal As in RNA fragments always get sequenced, making more reads usable for analysis. As such, it is not necessary to sequence the whole insert, which is desirable in forward sequencing, and it is possible to obtain more material for sequencing by selecting cDNAs with a wider size range.
4. *Adjusting amount of library.* In addition to Agilent Bioanalyzer, the amount of a library can be calculated by qPCR using standard libraries, either provided by Illumina or calibrated, custom-made libraries, as reference and PCR primers used for cluster generation (see Subheading 2). Since the same primers

are used for clustering on the flowcell, the result reflects the amount of "clusterable" DNA. Adjusting library amount is particularly important for sequencing multiple samples in one lane (multiplexing). We have found that the qPCR method is more accurate than Agilent Bioanalyzer in adjusting libraries.

5. *pAs near 3′ splice site.* Since Bowtie 2 does not identify exon-exon junctions, a read containing multiple exon sequences can only be partially mapped. Only the portion of read containing the exon sequence with the highest alignment score is in the result file; a nearby pA downstream of 3′ splice site is not identified if its distance to the 3′ splice site is shorter than the preceding exon sequence in the read. A solution to this problem is to map reads against a complete splice junction sequence database in addition to genome sequence [20].
6. *Quantification of pA isoform.* While the read number for each pA can be used to represent the expression level of the isoform(s) using the pA, the random four Ns in the adapter offers another approach for quantitative analysis of 3′READS data. In principle, the latter method can help address the issue of ligation bias and nonlinear PCR amplification.

Acknowledgments

We thank other members of the BT lab for helpful comments and suggestions. This work was funded by an NIH grant (GM084089) to BT.

References

1. Edmonds M (2002) A history of poly A sequences: from formation to factors to function. Prog Nucleic Acid Res Mol Biol 71:285–389
2. Colgan DF, Manley JL (1997) Mechanism and regulation of mRNA polyadenylation. Genes Dev 11:2755–2766
3. Proudfoot NJ (2011) Ending the message: poly(A) signals then and now. Genes Dev 25:1770–1782
4. Tian B, Graber JH (2012) Signals for pre-mRNA cleavage and polyadenylation. Wiley Interdiscip Rev RNA 3(3):385–396
5. Tian B, Hu J, Zhang H et al (2005) A large-scale analysis of mRNA polyadenylation of human and mouse genes. Nucleic Acids Res 33:201–212
6. di Giammartino DC, Nishida K, Manley JL (2011) Mechanisms and consequences of alternative polyadenylation. Mol Cell 43:853–866
7. Lutz CS, Moreira A (2011) Alternative mRNA polyadenylation in eukaryotes: an effective regulator of gene expression. WIREs RNA 2:23–31
8. Zhang H, Lee JY, Tian B (2005) Biased alternative polyadenylation in human tissues. Genome Biol 6:R100
9. Wang ET, Sandberg R, Luo S et al (2008) Alternative isoform regulation in human tissue transcriptomes. Nature 456:470–476
10. Ji Z, Lee JY, Pan Z et al (2009) Progressive lengthening of 3′ untranslated regions of mRNAs by alternative polyadenylation during mouse embryonic development. Proc Natl Acad Sci U S A 106:7028–7033
11. Sandberg R, Neilson JR, Sarma A et al (2008) Proliferating cells express mRNAs with shortened 3′ untranslated regions and fewer microRNA target sites. Science 320:1643–1647
12. Mayr C, Bartel DP (2009) Widespread shortening of 3′UTRs by alternative cleavage and

polyadenylation activates oncogenes in cancer cells. Cell 138:673–684
13. Singh P, Alley TL, Wright SM et al (2009) Global changes in processing of mRNA 3′ untranslated regions characterize clinically distinct cancer subtypes. Cancer Res 69: 9422–9430
14. Flavell SW, Kim TK, Gray JM et al (2008) Genome-wide analysis of MEF2 transcriptional program reveals synaptic target genes and neuronal activity-dependent polyadenylation site selection. Neuron 60:1022–1038
15. Lee JY, Yeh I, Park JY et al (2007) PolyA_DB 2: mRNA polyadenylation sites in vertebrate genes. Nucleic Acids Res 35:D165–D168
16. Brockman JM, Singh P, Liu D et al (2005) PACdb: polyA cleavage site and 3′-UTR database. Bioinformatics 21:3691–3693
17. Nam DK, Lee S, Zhou G et al (2002) Oligo(dT) primer generates a high frequency of truncated cDNAs through internal poly(A) priming during reverse transcription. Proc Natl Acad Sci U S A 99:6152–6156
18. Wlotzka W, Kudla G, Granneman S et al (2011) The nuclear RNA polymerase II surveillance system targets polymerase III transcripts. EMBO J 30:1790–1803
19. Schmidt MJ, Norbury CJ (2010) Polyadenylation and beyond: emerging roles for noncanonical poly(A) polymerases. Wiley Interdiscip Rev RNA 1:142–151
20. Hoque M, Ji Z, Zheng D et al (2013) Analysis of alternative cleavage and polyadenylation by 3′ region extraction and deep sequencing. Nat Methods 10:133–139
21. Langmead B, Salzberg SL (2012) Fast gapped-read alignment with Bowtie 2. Nat Methods 9:357–359
22. Jayaprakash AD, Jabado O, Brown BD et al (2011) Identification and remediation of biases in the activity of RNA ligases in small-RNA deep sequencing. Nucleic Acids Res 39:e141
23. Zhuang F, Fuchs RT, Sun Z et al (2012) Structural bias in T4 RNA ligase-mediated 3′-adapter ligation. Nucleic Acids Res 40:e54

Chapter 11

Polyadenylation Site Prediction Using PolyA-iEP Method

Ioannis Kavakiotis, George Tzanis, and Ioannis Vlahavas

Abstract

This chapter presents a method called PolyA-iEP that has been developed for the prediction of polyadenylation sites. More precisely, PolyA-iEP is a method that recognizes mRNA 3′ends which contain polyadenylation sites. It is a modular system which consists of two main components. The first exploits the advantages of emerging patterns and the second is a distance-based scoring method. The outputs of the two components are finally combined by a classifier. The final results reach very high scores of sensitivity and specificity.

Key words Data mining, Machine learning, Classification, Emerging patterns, Bioinformatics, Polyadenylation

1 Introduction

PolyA-iEP [1] is a method that addresses the problem of discriminating sequences that contain polyadenylation sites from the ones that do not. The discrimination of mRNA 3′ ends that contain polyadenylation sites from intronic or 5′ UTR sequences without polyadenylation site seem to be very difficult and the performance of the existing methods is moderate. PolyA-iEP has emerged from a previously published method called PolyA-EP [2], although the new one is more robust and sophisticated.

PolyA-iEP is a modular system which consists of two main components. The first component exploits the concept and the advantages of the *frequent itemsets.* The term frequent itemset has been proposed in the framework of association rule mining. *Association rule mining* is a popular field of data mining which has been proposed by Ragesh Agrawal [3]. It is a research field that aims to discover interesting relations between variables in large databases. This field was initially introduced in the framework of market basket analysis, but currently it is applied in many application areas including bioinformatics with outstanding results.

More precisely, the first component uses the concept of *interesting emerging patterns* [4]. *Emerging patterns* are defined as

Joanna Rorbach and Agnieszka J. Bobrowicz (eds.), *Polyadenylation: Methods and Protocols*, Methods in Molecular Biology, vol. 1125, DOI 10.1007/978-1-62703-971-0_11,

itemsets whose supports increase significantly from one dataset to another. A significant drawback of this method is that the number of emerging patterns that can occur may be huge. An approach to overcome this drawback is to use a measure of interestingness in order to reduce the number of the mined patterns to those that carry the most information. In our method we used as interestingness measure the chi-test and therefore the emerging patterns are called chi emerging patterns [5]. The formal definition of frequent patterns, emerging patterns, and chi emerging patterns is going to be presented in detail in the following section.

The second component is completely independent from the first. It is a distance-based scoring for the sequences. In order to calculate the distance, the method uses the Manhattan distance. The equation that calculates the Manhattan distance is also presented in the following section.

Every component calculates some scores. The first component calculates eight scores and the second one calculates five scores. The total of 13 scores is used as input to a classifier, which decides whether a sequence contains a polyadenylation site or not. In that step any classifier that handles real-valued attributes can be used. Some of the state-of-the-art machine learning algorithms that have been used are *support vector machines*, *neural networks*, and *classification trees*.

2 Materials

In this section we are going to present the definitions and the equations of every method that is used by PolyA-iEP.

2.1 Datasets

The datasets that our method can handle are divided in two major categories, namely, the positive and the negative datasets. The positive dataset are mRNA 3′ end sequences. The negative examples are a combination of 5′ UTR, coding, and intronic sequences. The datasets are provided to the method in text files which contain the sequences. Each sequence has length of 400 nucleotides. In the positive sequences, the polyadenylation site is found at the 301st position.

2.2 Frequent Patterns and Association Rules

Let $I = \{i_1, i_2, \ldots, i_N\}$ be a finite set of binary attributes called *items* and $D = \{t_1, t_2, \ldots, t_N\}$ be a finite multiset of transactions, which is called the *database*. Each *transaction* t_i contains a subset of items chosen from I and has a unique transaction ID. A set of items is referred to as an *itemset*. If an itemset contains k items, it is called a k-itemset. The number k is called *size* or *length* of the itemset. The itemset that does not contain any items is called an empty itemset. A transaction $T \in D$ is said to contain an itemset $X \subseteq I$, if $X \subseteq T$.

An *association rule* is an implication of the form $X \Rightarrow Y$ where $X \subset I$, $Y \subset I$, and $X \cap Y = \emptyset$. The itemset X is called *antecedent* or

Left-Hand-Side (LHS) of the rule, and the itemset Y is called *consequent* or *Right-Hand-Side* (RHS) of the rule.

There are many measures that have been proposed in order to evaluate a rule's interestingness. The most popular are *support* and *confidence*. They, respectively, reflect the usefulness and certainty of discovered rules. More specifically, support determines how often a rule is applicable to a given dataset, whereas confidence determines how frequently items in Y appear in transactions that contain X. The *support* of a rule $X \Rightarrow Y$ is equal to the support of the itemset $X \cup Y$ and is defined as the fraction of transactions in the database which contain the itemset. The support of an itemset X is calculated as presented in the following equation:

$$\text{support}_D(X) = \frac{|\{T \in D \mid X \subseteq T\}|}{|D|}$$

The *confidence* of the rule $X \Rightarrow Y$ is defined as the fraction of transactions in database that contains $X \cup Y$ over the number of transactions that contain only X. In other words, confidence is equal to the fraction of the support of $X \cup Y$ in D over the support of X in D. The equation that defines confidence is presented below:

$$\text{confidence}_D(X \Rightarrow Y) = \frac{\text{supp}_D(X \cup Y)}{\text{supp}_D(X)}$$

2.3 Emerging Patterns

Emerging patterns are itemsets whose supports increase significantly from one dataset to another. Given two datasets, D_1 and D_2, the *growth rate* of an itemset X from D_1 to D_2 is defined as (indices 1 and 2 are used instead of D_1 and D_2):

$$\text{gr}_{1\to2}(X) = \begin{cases} 0, & \text{if } \text{supp}_1(X) = 0 \text{ and } \text{supp}_2(X) = 0 \\ \infty, & \text{if } \text{supp}_1(X) = 0 \text{ and } \text{supp}_2(X) > 0 \\ \dfrac{\text{supp}_2(X)}{\text{supp}_1(X)}, & \text{otherwise} \end{cases}$$

Given a minimum growth rate threshold $\rho > 1$, an itemset X is said to be *ρ-emerging pattern*, or simply *emerging pattern*, from D_1 to D_2, if $\text{gr}_{1\to2}(X) \geq \rho$. D_1 is called *background dataset* and D_2 is called *target dataset*.

The *strength* of an emerging pattern X from D_1 to D_2 is defined as:

$$\text{strength}_{1\to2}(X) = \begin{cases} \text{supp}_2(X), & \text{if } \text{gr}_{1\to2}(X) = \infty \\ \text{supp}_2(X)\dfrac{\text{gr}_{1\to2}(X)}{\text{gr}_{1\to2}(X)+1}, & \text{otherwise} \end{cases}$$

2.4 Interesting Emerging Patterns

Given a background dataset D_1 and a target dataset D_2, an itemset X is called a *chi emerging pattern*, if all the following conditions are true:

1. $\text{supp}_2(X) \geq \sigma$, where σ is a minimum support threshold.
2. $\text{gr}_{1\rightarrow 2}(X) \geq \rho$, where ρ is a minimum growth rate threshold.
3. $\forall Y \subset X,\ \text{gr}_{1\rightarrow 2}(Y) < \text{gr}_{1\rightarrow 2}(X)$.
4. $|X| = 1 \vee |X| > 1 \wedge (\forall Y \subset X \wedge |Y| = |X| - 1 \wedge \text{chi}(X, Y) \geq \eta)$, where $\eta = 3.84$ is a minimum chi value threshold and $\text{chi}(X, Y)$ is computed using chi-squared test.

More information about the used method can be found in [1].

2.5 Manhattan Distance

The *Manhattan distance* between two items is the sum of the differences of their corresponding components

$$d(x,y) = \sum (x_i - y_i)$$

2.6 Machine Learning and Classification

Supervised learning is probably the most common type of machine learning problems. In general, it is the task in which a function is generated in order to map inputs to desired outputs. The function, which is called classifier if the output is a discrete value, is inferred through analyzing training data. The problem of inferring this function is called classification or prediction. The training data which are given to the machine learning algorithm are in most cases a pair of a vector and the category to which the example belongs. In the field of machine learning, the category which is assigned to each training data is called label. The classification process is presented in Fig. 1.

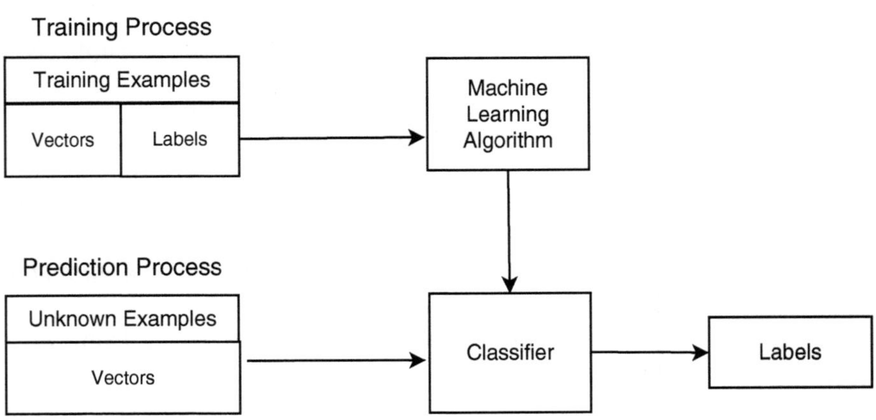

Fig. 1 Training and prediction processes

3 Methods

In this section we are going to present in detail every step of the PolyA-iEP method. Figure 2 presents the architecture of our method. The upper side presents the first component which is related to chi emerging patterns. The lowest side presents the distance-based scoring. It is clear that the results from the two components are used as inputs to a classification algorithm which will decide whether the sequence is a positive or negative example, i.e., contains or not a polyadenylation site.

3.1 Extraction of Elements

3.1.1 Input

Nucleotide sequences in the form that is presented in Subheading 2.1.

3.1.2 Processing

Previous studies have shown that the region near the polyadenylation site can be divided in four elements. These elements contain different nucleotide frequencies and so they must contain different patterns which our method is intended to mine. The elements, which are presented in Fig. 3, are FUE (far-upstream element), NUE (near-upstream element), CE (cleavage element), and NDE (near-downstream element).

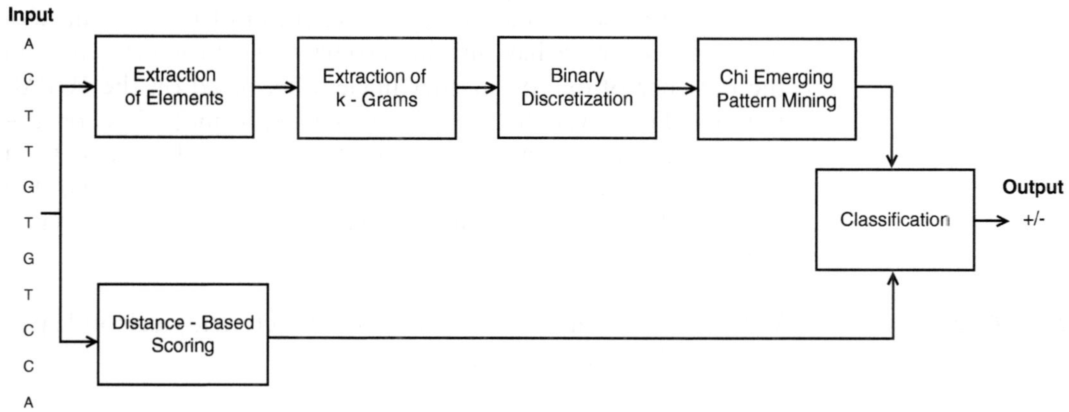

Fig. 2 PolyA-iEP architecture

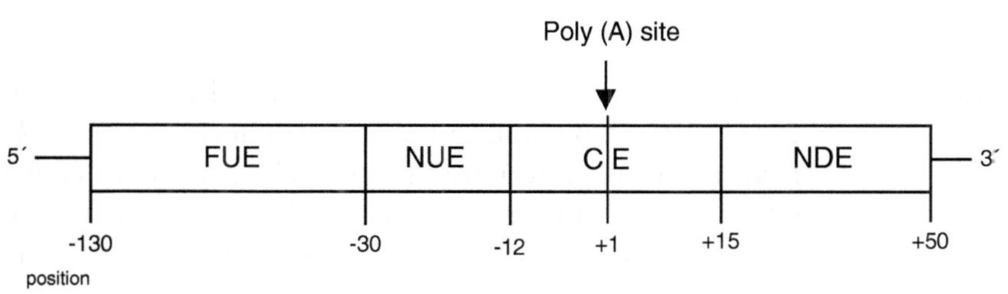

Fig. 3 The four sequence elements used in PolyA-iEP

The application is fully customizable. The user can choose the boundaries of the elements. The main reason behind the consideration of the elements is to search for extended patterns in the sequences, i.e., patterns in different elements that occur simultaneously. An example of a simple pattern could be {ATTA}. An example of an extended pattern would be {FUE_ATCT, NUE_AAA, CE_TT, NDE_AAAG}. This pattern can be interpreted as following: simultaneously appear the ATCT in the FUE element, the AAA in the NUE element, the TT in the CE element, and the AAAG in the downstream element. Extended patterns can definitely be more informative than the simple patterns, which can lead in better distinction of the positives and negative examples and the increasing of the overall accuracy of the method.

3.1.3 Output

Each sequence divided in the four elements.

3.2 Extraction of k-Grams

3.2.1 Input

Each element of the sequence has been produced in the previous step (*see* Subheading 3.1.3).

3.2.2 Processing

In this step each sequence is going to be presented by a number of vectors one for each element, i.e., FUE, NUE, CE, and NDE. These vectors will contain the frequencies of each valid nucleotide pattern. The user can specify the maximum length of the pattern. The patterns can contain every combination of the four nucleotides. Moreover, we have included some wildchars which represent the presence of one or another nucleotide, based on the IUPAC notions. These wildchars are R (A or G—puRine), Y (C or T—pYrimidine), M (A or C—aMino), K (G or T—Keto), S (C or G—Strong, 3H bonds), and W (A or T—Weak, 2H bonds). For instance, if CCT and CTT are valid patterns then also the valid pattern CYT occurs.

3.2.3 Output

Vectors which contain the frequencies of each valid nucleotide pattern found in the sequences.

3.3 Binary Discretization

3.3.1 Input

The vectors with the frequencies of each valid nucleotide pattern that where calculated in the previous step.

3.3.2 Processing

Information entropy was used as the discretization method in PolyA-iEP. All possible cut points are checked for each *k*-gram pattern among all pattern frequencies. The cut point that has the maximum information gain is finally selected. Given a set of training examples *S*, entropy (*E*) is defined by the following equation:

$$E(S) = -\sum_{i=1}^{c} p_i \log_2 (p_i)$$

where c is the number of classes and p_i is the proportion of examples in S that belong in class i. By definition, if p_i is zero, then the term $p_i \log_2(p_i)$ is set to zero.

Given an ordered set of candidate N cut points $T = \{t_1, \ldots, t_N\}$ for the values of an attribute A, that partition the set of examples in $N+1$ subsets $(S_1, \ldots, S_{N+1})$, the ***information gain*** (G) is defined by the following equation:

$$G(S;,A;,T) = E(S) - \sum_{i=1}^{N+1} \frac{|S_i|}{|S|} E(S_i)$$

where $S_i = \{s \in S_i | s[A] \in [t_i, t_{i+1})\}$.

As mentioned below PolyA-iEP uses binary discretization, and a single cut point which maximizes information gain is sought among all attribute values.

3.3.3 Output

The k-gram vectors that were extracted as described in Subheading 3.2 are transformed into a transaction of items. The items included in the transaction are those k-grams that have frequency greater than the corresponding cut point, which was previously calculated. In this step the data are transformed in a format that permits the extraction of emerging patterns.

3.4 Mining Interesting Emerging Patterns

The transactional data that have been produced in the previous step are used in this step for mining interesting emerging patterns.

3.4.1 Input

3.4.2 Processing

For mining the interesting emerging patterns FP-Growth, a frequent itemsets mining algorithm [6] has been modified accordingly. The modified algorithm receives, as input two datasets, the background and the target dataset and discovers all chi emerging patterns, based on the user-specified parameters (i.e., minimum support threshold and minimum growth rate threshold). For this reason, two sets of emerging patterns, E_+ and E_-, are generated for the positive and the negative class, respectively. As already mentioned a dataset that contains three types of negative sequences (5′ UTR, coding, and intronic) has been used in the proposed setup. These negative sequences express quite different nucleotide distributions. If all negatives were dealt as a whole only, then the effectiveness of classification would be moderate. So, PolyA-iEP mines four pairs of E_+/E_- sets of emerging patterns, one for discriminating positives from all negatives as a whole and three for discriminating positives from each type of negatives separately. An example of an "extended" interesting emerging pattern that can be mined by PolyA-iEP is {FUE_AGT, NUE_CT}: 0.25. This interesting emerging pattern associates the appearance of pattern

"AGT" in the far-upstream element, with pattern "CT" in the near-upstream element. The strength of this interesting emerging pattern is 0.25.

3.4.3 Output

The four E_+/E_- pairs of sets of emerging patterns.

3.5 Distance-Based Scoring

3.5.1 Input

Nucleotide sequences in the form that is presented in Subheading 2.1.

3.5.2 Processing

The distance-based scoring of PolyA-iEP is independent from the previous steps. This step includes the calculation of the frequencies of nucleotides at each position of a sequence and the construction of a nucleotide frequency matrix for each class, as shown in Table 2. For example, nucleotide A which has 0.14 frequency is position 1 of the sequences used to generate the matrix presented in Table 1. Then, for each position in the sequence, the rankings of the nucleotides are calculated according to their frequency at this particular position (Table 2). In our setup five nucleotide frequency ranking matrices are constructed, one for each of the following categories: positives, all negatives, 5′ UTR negatives, coding negatives, and intronic negatives.

3.5.3 Output

The five nucleotide frequency ranking matrices that are constructed, for each of the following categories, are the following: positives, all negatives, 5′ UTR negatives, coding negatives, and intronic negatives.

3.6 Classification

3.6.1 Input

Any unlabeled sequence that has been processed through Subheadings 3.1–3.3 and is represented in transactional format.

3.6.2 Processing

As mentioned before the two PolyA-iEP components are used in order to produce 13 scores. The first component produces eight scores and the second one five scores. These scores represent the

Table 1
An example of a nucleotide frequency matrix for sequences of length 5

	Position in sequence				
Nucleotide	**1**	**2**	**3**	**4**	**5**
A	0.14	0.06	0.30	0.11	0.16
C	0.21	0.21	0.18	0.29	0.28
G	0.35	0.36	0.26	0.40	0.28
T	0.30	0.37	0.26	0.20	0.28

Table 2
The nucleotide frequency ranking matrix that corresponds to Table 1 data

Nucleotide	Position in sequence				
	1	2	3	4	5
A	4	4	1	4	4
C	3	3	4	2	2
G	1	2	2.5	1	2
T	2	1	2.5	3	2

attributes of each sequence from the initial dataset. These scores are given as inputs to any classifier that can handle real-valued numeric attributes and decides whether the sequence contains or not the polyadenylation site. Some of the state-of-the-art machine learning algorithms that have been used are *support vector machines*, *neural networks*, and *classification trees* (*see* **Note 1**).

From the first PolyA-iEP component, the mined E_+/E_- pairs of sets of emerging patterns (*see* Subheading 3.4.3) are used for scoring an instance as being positive or negative. For this reason, pairs of scores for an instance T in transaction format are calculated as described by the following equations:

$$\text{score}(T,+) = \sum_{e \subseteq T, e \in E_+} \text{strength}_{-\rightarrow+}(e)$$

$$\text{score}(T,-) = \sum_{e \subseteq T, e \in E_-} \text{strength}_{+\rightarrow-}(e)$$

The first score indicates if T is positive and the second if it is negative. The final classification could be made by comparing the values of the two scores and assigning the instance to the class with the highest score. The total number of scores that are produced in this step are eight. Two scores (one for positive and one for negative class) are assigned to each of the following discriminations: positives/all negatives, positives/5′ UTR negatives, positives/coding negatives, and positives/intronic negatives.

From the second PolyA-iEP component, the distance of a sequence from a class and subclass (5′ UTR, intronic, or coding) is calculated. For this reason, the sequence is converted into a nucleotide frequency ranking vector using the nucleotide frequency matrix of the class or subclass (*see* Subheading 3.5). Then, the distance from the unary vector is calculated and divided by the length of sequence. For example, given the ranking matrix in Table 2, the ranking vector that corresponds to the sequence "ATGGC" is <4, 1, 2.5, 1, 2>. The distance (Manhattan distance is used in our setup) of this vector from the unary vector <1, 1, 1, 1, 1> is 5.5.

Dividing this distance by the length of the sequence, namely, 5, the mean nucleotide distance is finally calculated to be 1.1. This is the mean nucleotide distance of the above sequence from the category to which the nucleotide frequency matrix in Table 2 belongs. Five distances scores are finally calculated, one for each of the following categories: positives, all negatives, 5′ UTR negatives, coding negatives, and intronic negatives.

3.6.3 Output

The final classification of the input sequences as containing or non-containing a PolyA site.

4 Notes

1. We have mentioned in Subheading 3.6 that the last step of the PolyA-iEP method is the building of a classifier in order to classify unknown sequences. For this purpose, the Weka machine learning library [7] has been used. Weka provides many algorithms implemented in a very efficient way that can handle real-valued attributes. The state-of-the-art machine learning algorithms that have been used are referred to in Subheading 3.6. The implementation of the support vector machines can be found under the tab classify to the path *classifiers/functions/SMO*. The implementation of neural networks can be found also under the tab classify to the path *classifiers/functions/multilayerPerceptrons*. Lastly, in Weka there is an implementation of a classification tree algorithm called C4.5. The implementation of this algorithm can be found under the tab classify to the path *classifiers/trees/J48*.

References

1. Tzanis G, Kavakiotis I, Vlahavas I (2011) PolyA-iEP: a data mining method for the effective prediction of polyadenylation sites. Expert Syst Appl 38:12398–12408
2. Tzanis G, Kavakiotis I, Vlahavas I (2008) Polyadenylation site prediction using interesting emerging patterns. In: Proceedings of the 8th IEEE international conference on bioinformatics and bioengineering, IEEE, Athens, Greece, 2008
3. Agrawal R, Imielinski T, Swami A (1993) Mining association rules between sets of items in large databases. In: Proceedings of the ACM SIGMOD conference on management of data, 1993, Washington, DC, pp 207–216
4. Dong G, Li J (1999) Efficient mining of emerging patterns: discovering trends and differences. In: Proceedings of ACM-SIGKDD'99, 1999, San Diego, CA, pp. 43–52
5. Fan H (2004) Efficient mining of interesting emerging patterns and their effective use in classification. Ph.D. thesis, University of Melbourne, Australia
6. Han J, Pei J, Yin Y (2000) Mining frequent patterns without candidate generation. In: Proceedings of the 2000 ACM SIGMOD international conference on management of data, 2000, Dallas, TX, pp 1–12
7. Hall M, Frank E, Holmes G et al (2009) The weka data mining software: an update. SIGKDD Explor 11(1):10–18

Chapter 12

Detecting Alternative Polyadenylation from Microarray Data

Antonio Lembo and Paolo Provero

Abstract

Gene regulation at the posttranscriptional level is often mediated by trans-acting factors binding the 3′ untranslated region (3′ UTR) of messenger RNAs (mRNAs). Alternative mRNA isoforms differing only in their 3′ UTR can thus be differentially regulated, and it has been recently shown that this mechanism is indeed used by the cell to alter gene regulation effected by microRNAs and RNA-binding proteins, especially in highly proliferating contexts. Here we describe a computational method to analyze alternative 3′ UTR isoforms in gene expression profiling datasets obtained with Affymetrix 3′ IVT microarrays. The approach we describe allows the analysis of 3′ UTR isoform usage in thousands of publicly available gene expression datasets, including many retrospective studies of cancer patients equipped with clinical data.

Key words 3′ UTR, Posttranscriptional regulation, microRNAs, Alternative polyadenylation

1 Introduction

Posttranscriptional regulation mediated by trans-acting factors binding the 3′ untranslated regions (3′ UTR) of transcripts has emerged in the last decade as an important regulatory mechanism in both physiology and pathology. Such trans-acting factors, which include both microRNAs and RNA-binding proteins, bind specific sequences on their mRNA targets and in most cases lead to the depletion of the final gene product, either by causing the degradation of the mRNA or by interfering with its translation.

This type of regulation crucially depends on the presence of binding sites on the target 3′ UTR. Therefore, alternative isoforms of the same mRNA, differing only in their 3′ UTR, can lead to significantly different expression levels of their otherwise identical gene products. The fact that transcripts use alternative polyadenylation (polyA) sites has been known for many years. However, only recently the regulatory consequences of this fact have been systematically investigated.

Joanna Rorbach and Agnieszka J. Bobrowicz (eds.), *Polyadenylation: Methods and Protocols*, Methods in Molecular Biology, vol. 1125, DOI 10.1007/978-1-62703-971-0_12, © Springer Science+Business Media New York 2014

The picture emerging from these studies is that, broadly speaking, highly proliferating cells tend to express shorter 3′ UTRs, thus freeing many transcripts from the negative regulatory action of microRNAs and RNA-binding proteins. This phenomenon has been shown, for example, in activated vs. quiescent T cells [1], in cancer cell lines [2, 3], in early vs. later stages of embryonic development [4, 5], and in more vs. less aggressive human tumors [6].

RNA sequencing is obviously the ideal technique to study this phenomenon. However, thousands of gene expression profiling experiments have been performed in the last decade and deposited in public repositories such as the Gene Expression Omnibus [7], including many retrospective studies of cancer patients, often equipped with rich clinical and histological data. In this work we present a computational approach for the analysis of these datasets specifically aimed at detecting differential polyA site usage.

This is made possible by two features of the Affymetrix 3′ IVT microarrays: (a) the fact that their probes tend to be concentrated toward the 3′ UTR of target transcripts and (b) the possibility of regrouping the probes into custom-defined probesets. Only a subset of all transcripts can be studied in this way, but since they include many genes relevant to both cancer and development, the information obtained, while partial, is of great relevance.

2 Data

- The method can be used on data obtained with any Affymetrix microarrays of the 3′ IVT series.
- It is necessary to have access to the raw data in the form of CEL files.
- A list of known alternative polyA sites must be available, such as the one compiled in the polyA_DB [8, 9] database.
- In the following, we will assume that the aim of the study is to compare the relative usage of alternative polyA sites between two experimental conditions A and B. Generalization to the comparison of more conditions can be designed by straightforward statistical methods.

3 Methods

3.1 Building the Database of Target Sequences

Obtain from a genomic database (such as Ensembl or the UCSC genome browser) the following data for all annotated transcripts (or for the transcripts of interest):

1. Strand (+ or −) where the transcripts is encoded.
   (a) Start and end coordinate of each exon.

2. Obtain from a database of alternative polyA sites (such as polya_DB) the genomic coordinates of all known or predicted polyA sites. Ensure, of course, that the genome version used here and in the previous point match. The easier way to do this is to obtain all information from the same genomic database, for example, the polya_DB database can be obtained as a UCSC track.
3. Using the information obtained above and the sequence data obtained from the genomic database, build a FASTA file containing separately for each transcript T and each polyA site A (*see* **Note 1**).
   (a) The portion of the transcript located 5′ of A (the 5′ portion).
   (b) The portion of the transcript located 3′ of A (the 3′ portion).

3.2 Building an Alternative CDF

1. Use the "altcdfenvs" [10] package of Bioconductor to build an alternative chip description file (CDF) based on the microarray platform at hand and the FASTA file produced above (*see* also **Note 2**). This will produce, for each transcript and each polyA site, two custom probesets, one for the 5′ and one for the 3′ portion. The number of transcripts/polyA sites that can be analyzed depends on two factors:
   (a) The microarray platform.
   (b) The database of alternative polyA sites.

3.3 Computing the Expression Ratio Index

1. The CEL files containing the raw microarray data can be now processed with the custom CDF using any of the normalization methods commonly used for these data, such as RMA [11].
2. This will produce, for each sample and each polyA site, two expression values, one for the 5′ and one for the 3′ portion.
3. The expression ratio index (ERI) is defined as the ratio of the expression values of the 5′ and 3′ portion. Note that:
   (a) What we are interested in biologically is the relative expression of the long and short isoforms (i.e., respectively, the isoform using the alternative polyA site and the isoform using the whole annotated 3′ UTR).
   (b) In ref. 6 we showed that the ERI is a linear function of such relative expression. However, it is not proportional to it (the intercept of the straight line is nonzero). This is simply due to the fact that while the short isoform can only hybridize the 5′ probeset, the long isoform can hybridize both.
   (c) Therefore, the ERI cannot provide information on the value of the relative expression, but can be used, as described below, to compare the relative expression of the two isoform between two experimental conditions.

3.4 Comparing Experimental Conditions

1. To detect differential usage of a polyA site between two experimental conditions, one can use any of the various statistical methods that are used for class comparison in microarray studies, simply replacing the gene expression values with the ERI values.
2. If the ERI value of a polyA site is significantly higher in condition A compared to condition B, it follows that the relative prevalence of the short isoform is higher in condition A or that the transcript is shortened in condition A.

4 Notes

1. Alternative splicing of coding exons can be a source of noise since different isoforms will give a different hybridization signal on the 5′ probeset. There are two possible solutions:
   (a) Consider only one transcript per gene. This is what we did in ref. 6, where we selected for each gene only the longest transcript.
   (b) To build the 5′ probeset, consider (and thus include in the FASTA file) only the exons that are common to all known alternative splicing isoforms.
2. CDF files for human Affymetrix platforms HGU-133A, HGU-133B, and HGU-133_Plus_2, and software to compute ERIs are available as supporting information of ref. 6.

References

1. Sandberg R, Neilson JR, Sarma A et al (2008) Proliferating cells express mRNAs with shortened 3′ untranslated regions and fewer microRNA target sites. Science 320: 1643–1647
2. Mayr C, Bartel DP (2009) Widespread shortening of 3′UTRs by alternative cleavage and polyadenylation activates oncogenes in cancer cells. Cell 138:673–684
3. Fu Y, Sun Y, Li Y et al (2011) Differential genome-wide profiling of tandem 3′ UTRs among human breast cancer and normal cells by high-throughput sequencing. Genome Res 21:741–747
4. Ji Z, Lee JY, Pan Z et al (2009) Progressive lengthening of 3′ untranslated regions of mRNAs by alternative polyadenylation during mouse embryonic development. Proc Natl Acad Sci U S A 106:7028–7033
5. Thomsen S, Azzam G, Kaschula R et al (2010) Developmental RNA processing of 3′UTRs in Hox mRNAs as a context-dependent mechanism modulating visibility to microRNAs. Development 137:2951–2960
6. Lembo A, Di Cunto F, Provero P (2012) Shortening of 3UTRs correlates with poor prognosis in breast and lung cancer. PLoS One 7:e31129
7. Barrett T, Wilhite SE, Ledoux P et al (2013) NCBI GEO: archive for functional genomics data sets – update. Nucleic Acids Res 41: D991–D995
8. Zhang H, Lee JY, Tian B (2005) Biased alternative polyadenylation in human tissues. Genome Biol 6:R100
9. Lee JY, Yeh I, Park JY et al (2007) PolyA_DB 2: mRNA polyadenylation sites in vertebrate genes. Nucleic Acids Res 35:D165–D168
10. Gautier L, Møller M, Friis-Hansen L et al (2004) Alternative mapping of probes to genes for Affymetrix chips. BMC Bioinformatics 5:111
11. Irizarry RA, Bolstad BM, Collin F et al (2003) Summaries of Affymetrix GeneChip probe level data. Nucleic Acids Res 31:e15

Chapter 13

Quantitative Polyadenylation Site Mapping with Single-Molecule Direct RNA Sequencing

Fatih Ozsolak

Abstract

The known regulatory role of 3′ untranslated regions (3′UTRs) and poly(A) tails in RNA localization, stability, and translation, and polyadenylation regulation defects leading to human diseases such as oculopharyngeal muscular dystrophy, thalassemias, thrombophilia, and IPEX syndrome underline the need to fully characterize genome-wide polyadenylation states and mechanisms across normal physiological and disease states. This chapter outlines the quantitative polyadenylation site mapping and analysis strategies developed with the single-molecule direct RNA sequencing technology.

Key words Polyadenylation site mapping, Alternative polyadenylation, Next-generation sequencing, Direct RNA sequencing, Single-molecule sequencing

1 Introduction

The advances in the past decade underlined the complexity of transcriptomes and the existence of many types of RNA species in eukaryotes and prokaryotes [1]. 3′ poly(A)-tail-containing transcripts are arguably the most critical RNA species for a cell's function because the coding subset of 3′ poly(A)-tail-containing RNAs encode for proteins and the noncoding subset have important roles in expression regulation and other functions [2, 3]. Locating the transcript ends of 3′ poly(A)-tail-containing RNAs (e.g., polyadenylation sites) in a genome-wide and quantitative manner has been a long-standing challenge. This information is important not only for the purposes of genome annotation but also to understand transcript functionality, because 3′ end choice and alternative polyadenylation patterns and frequencies determine transcript coding potential, transport, storage, and regulation by RNA binding proteins and factors such as the miRNA machinery [4–7]. It is becoming clear that the regulation of polyadenylation patterns is widespread and the global polyadenylation changes accompany normal biological processes and carcinogenesis [8–10].

Joanna Rorbach and Agnieszka J. Bobrowicz (eds.), *Polyadenylation: Methods and Protocols*, Methods in Molecular Biology, vol. 1125, DOI 10.1007/978-1-62703-971-0_13, © Springer Science+Business Media New York 2014

To extend these recent findings and decipher polyadenylation pattern connections to normal physiology and disease, the scientific community is feeling an urgent need for technologies that can reliably and affordably locate and quantify polyadenylation patterns.

To define polyadenylation sites and to study how polyadenylation machinery functions and alternative polyadenylation patterns lead to biological outcomes, several genome-wide high-throughput approaches have been developed in the past several years [11–13]. Other chapters in this book contain detailed procedures for most of the technological advances in the field. In this chapter, the direct RNA sequencing (DRS)-based quantitative polyadenylation site mapping (QPSM-DRS) technology and associated protocols are outlined [14, 15]. The DRS technology basics are also presented for readers who may not be familiar with this next-generation sequencing approach.

1.1 Direct RNA Sequencing (DRS)

DRS is a single-molecule sequencing-by-synthesis technology developed in 2009 to enable highly quantitative massively parallel RNA measurements with minimal/no bias and artifacts [15]. It is the first and currently the only technology that can sequence RNA directly, without the requirement for prior cDNA conversion and other nucleic acid manipulation steps such as ligation and amplification. The DRS chemistry and operating principles have been described in detail previously in multiple publications [16–19]. Below is a brief summary from the previous presentations of this technology.

The sequencing system consists of two main components [20]: (1) flow cells, which create an environment for template RNA attachment, sequencing chemistry steps, and single-molecule fluorescence imaging, and (2) the HeliScope™ sequencer, which is an integrated fluidics, optics (for automated, fast and high-throughput imaging), and image/data analysis system for introduction of sequencing reagent formulations to the flow cells, taking single-molecule images and analyzing the images in real time (e.g., during the sequencing run) to deduce read sequences. Sequencing flow cell surfaces are coated with poly(dT) oligonucleotides covalently attached at their 5′ amine ends to an epoxide-coated ultraclean glass surface (Fig. 1). These oligonucleotides serve two purposes: (1) the capture of 3′ poly(A)-tail-containing nucleic acids onto surfaces by hybridization and (2) priming and initiation of sequencing steps. The current requirement for RNA sample preparation is the presence of a 3′ poly(A) tail >25–30 nucleotide (nt) in length and "blocked" at its 3′ end against extension by the polymerase used in the sequencing-by-synthesis steps. 3′ polyadenylation and blocking of RNA templates are performed using *E. coli* or yeast poly(A) polymerases with ATP and 3′deoxyATP, respectively. However, for the characterization of RNA species that naturally contain a poly(A) tail for the purposes of gene expression measurements, polyadenylation site mapping, and other applications, such poly(A)-tail addition is not required and direct

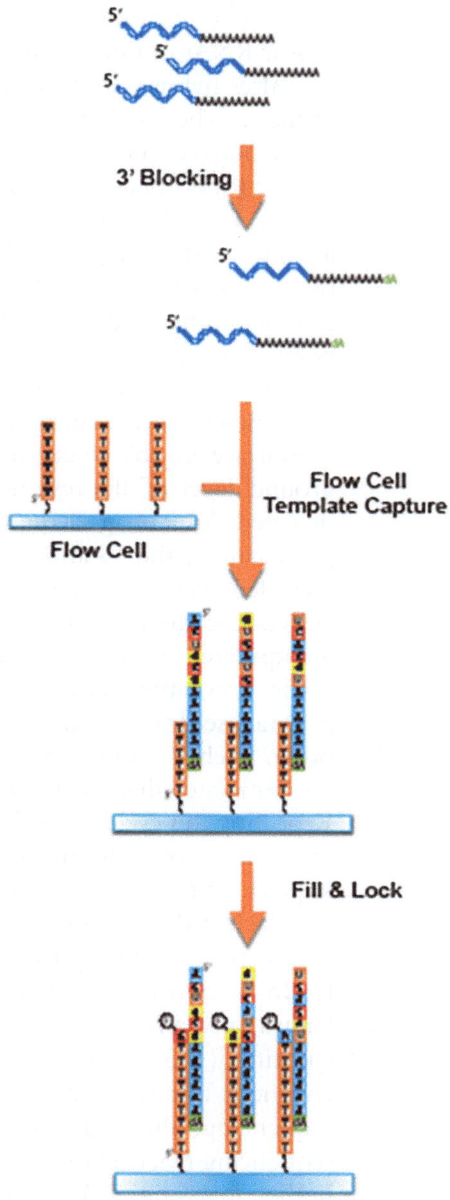

Fig. 1 QPSM-DRS sample preparation and flow cell loading procedure. 3′ blocked RNA (for simplicity, rRNA and other potential RNA species that may be present the input population are not shown) is captured on sequencing surfaces coated with poly(dT) oligonucleotides. The "fill" step is performed with thymidine triphosphate and polymerase. The "lock" step is performed with fluorescent dye-labeled Virtual Terminator™ (VT) A, C, and G analogs and polymerase. The fill and lock steps correct potential misalignments in poly(A/T) stretches and ensure that the sequencing starts at the 3′ end of the RNA template rather than the poly(A) tail. The figure was modified and reprinted with permission from the Springer [18]

hybridization of poly(A)+ RNAs to surfaces can be performed. It is also possible to change the poly(dT) oligonucleotides on the surface to other nucleotides for the capture and sequencing of other targets (for instance, for targeted selection and sequencing of particular RNAs of interest in a single step) [21].

After hybridization of RNA templates to the poly(dT) surface primers, to begin sequencing at the unique template region adjacent to the poly(A) tail, each primer-template pair is "filled" in with thymidine triphosphate by a polymerase and then "locked" in position with A, C, and G Virtual Terminator™ (VT) nucleotides. VTs [22] are nucleotide analogs used for sequencing, containing a fluorescent dye as part of a chemically cleavable group that prevents the addition of another nucleotide. After washing away the excess, unincorporated nucleotides, the surface is irradiated with a laser at an angle that allows total internal reflection at the surface. In such a situation, an evanescent field is generated so that only molecules very close to the surface are able to be excited by the laser. This reduces the background level of fluorescence such that single molecules can be detected. After image acquisition across desired number of positions per channel, the locations of RNA templates on the surface are recorded. The liquid in each channel is replaced with a mixture that cleaves the fluorescent dye and virtual terminator group off the incorporated nucleotide, rendering the strands suitable for further VT incorporation. The sequencing-by-synthesis reaction consists of polymerase-driven cyclic addition of the VTs in the C, T, A, and G order. Each VT addition is followed by rinsing, imaging (to locate the templates that incorporated the particular VT), and cleavage. Repeating this cycle many times provides a set of images from which the base incorporations are detected and then used to generate sequence information for each template molecule.

Each DRS run is currently performed with 120 VT-incorporation cycles and contains up to 50 independent channels, producing up to 25 million aligned reads ≥25 nts in length (up to 55 nts in length, median ~34 nts) per channel depending on the user-defined run time (1–5 days) and throughput (e.g., imaging quantity per channel). Longer reads can be achieved by doing more cycles of sequencing, although the quality and the yield/efficiency of the sequencing reaction may decrease with longer runs. Error rates are in the range of 4–5 %, dominated by missing base errors (~2–3 %), while insertion (~1 %) and substitution (~0.4 %) errors are lower.

2 Materials

2.1 Reagents

Poly(A) Tailing Kit (Ambion AM1350).

3′deoxyATP (Jena Biosciences JBS-NU-1123).

2.2 Equipment

Thermal cycler (Recommended: Bio-Rad DNA Engine Tetrad® 2).

Refrigerated microcentrifuge (Recommended: Eppendorf 5810 R).

Aluminum blocks (VWR 13259-260).

3 Methods

Using poly(dT)-coated sequencing flow cell surfaces, DRS-based quantitative polyadenylation site mapping (QPSM-DRS) technology captures 3′ poly(A)-tail-containing RNA species directly from total or other RNA populations and initiates sequencing at the polyadenylation sites (5′ ends of the reads mark the polyadenylation sites). The only sample preparation step needed is the blocking of the 3′ end of the RNAs against extension with the polymerase used for the DRS chemistry steps. This is currently achieved with a 3′ blocking reaction with 3′ deoxy ATP and *E. coli* poly(A) polymerase prior to the hybridization of RNA to flow cells. Alternative blocking nucleotide analogs and enzymes may be compatible with the system. It is possible to do the template 3′ blocking after RNA capture to flow cell surfaces, before or after the fill and lock step, for the analysis of RNA species present in conditions that disfavor enzymatic manipulation, such as RNAs in cellular lysates. The procedures below list the QPSM-DRS sample preparation protocol that was commercially available when this chapter was written.

3.1 Quantitative Polyadenylation Site Mapping with DRS (QPSM-DRS)

1. Prepare RNA to be 3′ blocked in nuclease-free water in a 29 μL volume in a nuclease-free PCR tube. RNA population to be profiled can be total RNA, poly(A)+-enriched RNA, or other RNA populations (*see* **Notes 1–5**).
2. Heat the RNA at 85 °C for 1 min in a thermocycler with heated lid, followed by rapid cooling in a prechilled aluminum block kept in an ice and water slurry (~0 °C, *see* **Note 6**). Leave the denatured RNA in the prechilled aluminum block for at least 1 min.
3. Add the following reagents in the indicated order while keeping the denatured RNA sample in the chilled aluminum block: 10 μL of 5× *E. coli* Poly(A) polymerase buffer; 5 μL of 25 mM $MnCl_2$; 2 μL of 10 mM 3′deoxyATP; and 4 μL of 2 U/μL *E. coli* Poly(A) polymerase. Mix well by pipetting gently up and down at least three times without vortexing. The final volume for the reaction is 50 μL. Incubate for 1 h at 37 °C in a thermocycler with heated lid.
4. Transfer the sample to a 1.5 mL tube (*see* **Notes 7** and **8**). If the sample cleanup is performed, the sample should be quantified after the cleanup to measure RNA concentration prior to flow cell hybridization (*see* **Note 3**).

3.2 Flow Cell Hybridization and Single-Molecule Direct RNA Sequencing

The detailed procedures for sample loader usage, flow cell rehydration, sample hybridization buffer, fill and lock steps, HeliScope DRS software, and sequencer usage are regularly updated, and the latest versions should be obtained from the manufacturer. This section is intended to describe several relevant details for the purposes of QPSM-DRS.

Hybridization of samples to flow cell channels is performed in 7–100 μL volume. The samples are mixed 50:50 with 2× hybridization buffer provided in the DRS Kit. If an RNA cleanup step is performed after the 3′ blocking reaction, the volume of nuclease-free water to be used to resuspend the RNA sample should be determined considering the input RNA quantity and the volume of hybridization cocktail preferred to be used. In general, 0.5–2 fmol of RNA material is required to optimally load each sequencing channel, although lower quantities can be used for less than optimal aligned read yields per channel. Following hybridization, the RNA molecules are filled and locked, and the flow cells are moved to the HeliScope. The manufacturer-supplied DRS scripts offer the user the opportunity to run one or two flow cells per run, to define the desired number and location of channels to be sequenced, and to request the desired imaging quantity (e.g., throughput) per channel (the system currently allows up to 1,400 fields of view per channel, with up to 25 million aligned reads ≥25 nts in length per channel).

3.3 Data Analysis

The various programs and pipelines for the filtering, alignment, and downstream analyses of the DRS data and the corresponding user manuals can be downloaded freely at http://open.helicosbio.com/. Briefly, an initial filtering step is performed on the raw DRS reads before initiating their alignment to reference sequences. This filtering step involves the following read selection steps:

1. DRS generates reads between 6 and 55 nts in length. Depending on the experimental goals, the reference sequence complexity, and size, a user-defined minimum read length cutoff is employed to remove short reads that cannot be aligned reliably to save computing time and power. ≥25 nts cutoff is routinely used for alignment to human and mouse genomes. However, for smaller and less-complex genomes, DRS reads as short as 15 nts can be used.
2. Any 5′ polyT stretches in DRS reads are trimmed. The likely cause of such T homopolymeric stretches is incomplete fill with dTTP, leading to sequencing initiation within the poly(A) tail. Therefore 5′ polyT trimming is preferred to minimize potential misalignment events.
3. Because of flow cell surface imaging errors, artifactual reads that have a repetition of the VT-addition order sequence (CTAG) may appear. Such reads are eliminated during the filtering step.

Given that the majority of sequencing errors are due to indels, an aligner that is tolerant to these types of errors should be employed. We highly recommend the use of the indexDPgenomic aligner [23] and downstream data analysis, genotyping, and quantification tools freely available at http://open.helicosbio.com.

While multiple aligners are available and can align DRS reads, including Mosaik (http://code.google.com/p/mosaik-aligner/) and SHRiMP [24], the use of these aligners may result in a reduction in actual aligned reads due to their reduced ability to deal effectively with indels.

3.4 QPSM-PCR Applications

The QPSM-DRS has been successfully used to explore polyadenylation patterns in human, yeast, plants, and other species (Fig. 2) [14, 25, 26]. The approach cannot only reveal the locations of alternative polyadenylation sites but also their usage frequency in absolute and relative levels. Two primary considerations need to be kept in mind when analyzing the QPSM-DRS data. First, the resolution of the approach for marking the polyadenylation sites is ±2 nts [14]. This is due to potential incomplete fill and lock steps, causing sequencing initiation location (e.g., 5′ ends of the reads) to vary primarily between the first and second non-A nucleotides after the poly(A) tail. Secondly, given that short reads are produced with DRS, the location of polyadenylation sites identified does not usually directly indicate connectivity to a nearby transcript. Although this assumption is probably true for the vast majority of the cases, there may be rare but important cases where this assumption may fail. This is a common drawback for all next-generation sequencing technologies available today. Even if long-read (>1,000 nts) or paired-end technologies are used, the sequence fragments obtained are partial transcript sequences and thus do not reveal the full transcript structure necessary for proper transcript assignment with full certainty. To eliminate these concerns regarding "connectivity," a sequencing technology needs to generate very long/limitless (e.g., >50,000 nts) read lengths in a massively parallel manner and sequence transcripts fully in a single read. Perhaps the emerging nanopore technologies may allow this capability in the near future [27].

4 Notes

1. RNAs isolated with various techniques can be used. Total RNAs isolated with the Trizol® (Invitrogen) and RNeasy Plus Mini (Qiagen) kits and poly(A)+ RNAs isolated with the Poly(A)Purist™ (Ambion) kit have been successfully used. Since QPSM-DRS allows single-step poly(A)+ RNA selection and sequencing initiated at the polyadenylation sites, users are recommended to start from total or other complex RNA populations to minimize cost and time commitment.
2. Long, undigested DNA hybridizes at minimal/zero levels to poly(dT)-coated sequencing flow cell surfaces due to the need for at least 12–15 nts poly(A) stretches for hybridization and

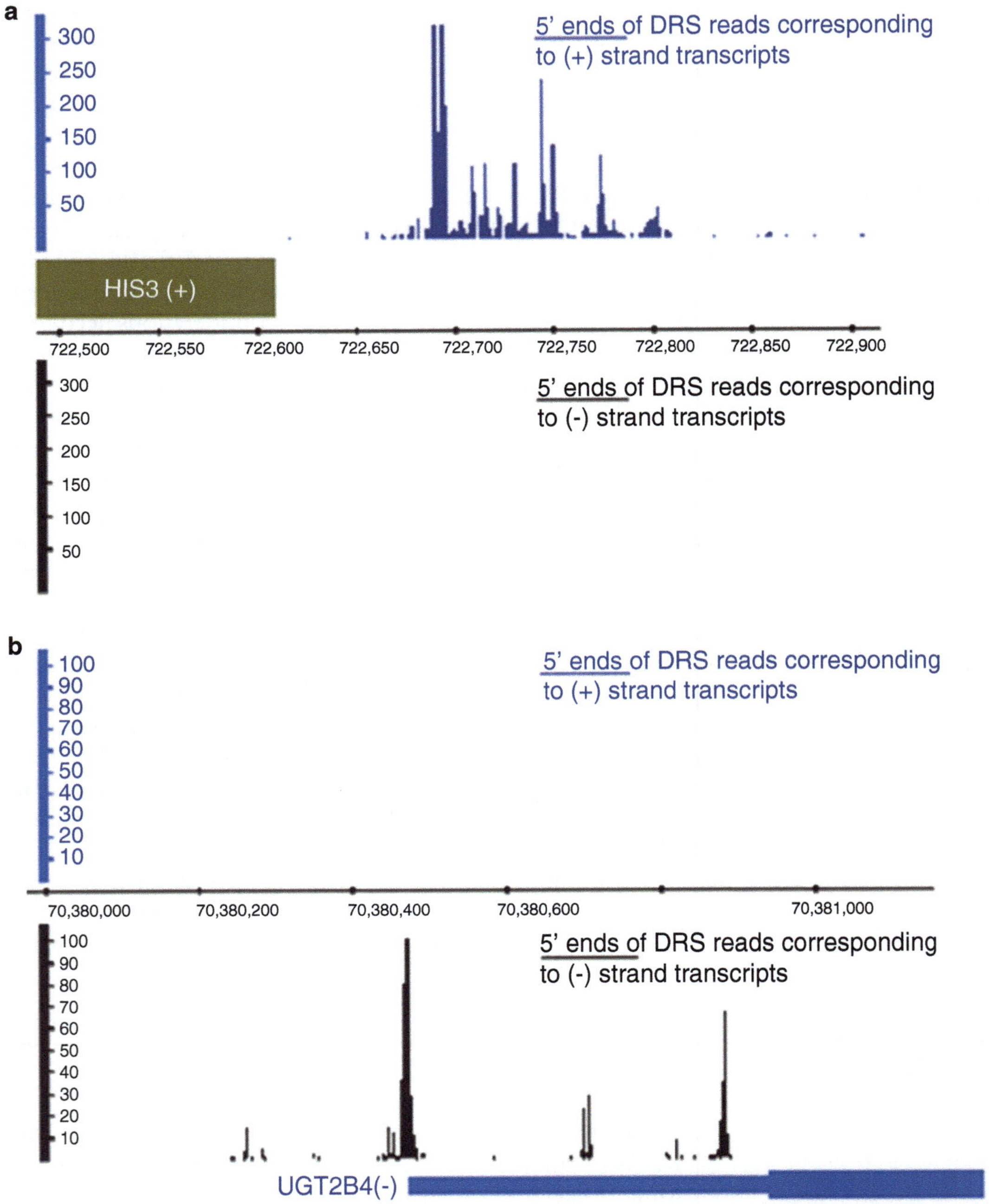

Fig. 2 QPSM-DRS profiles in yeast and human. (**a**) Polyadenylation sites identified downstream of the HIS3 open reading frame in yeast. (**b**) Alternative polyadenylation clusters identified for the UGT2B4 gene in human. *Y*-axis indicates the number of 5′ ends of QPSM-DRS reads marking the polyadenylation site at each X genomic position. The technology reveals polyadenylation site locations and how frequently each location is used. The panels were reprinted with permission from the Elsevier Limited [14]

perhaps due to their shear size and conformation. Even if trace quantities of genomic DNA hybridize to the surface, they are very unlikely to remain stable and sequenced during the course of a 1–5 day sequencing run. Furthermore, DRS chemistry results in high error rates for the sequencing of DNA and complementary DNA (cDNA) templates. Therefore, DNA/cDNA reads generated with the DRS chemistry cannot be aligned reliably to reference genomes. For these reasons, genomic DNA contamination that may be present in RNA samples does not present a significant concern for QPSM-DRS data generation and analysis. The users are encouraged to perform control experiments to revisit these observations with the RNA isolation and processing methods they are employing for their experiments. If the removal of any potential contaminating DNA from the RNA samples is desired, we recommend performing an extensive DNAseI digestion, followed by cleanup with a column-based system that does not preserve short nucleic acid fragments (e.g., <50 nts) in order to minimize short DNA fragments that may remain in the sample due to incomplete DNAseI digestion.

3. RNA quantification can be performed using UV absorbance. For the quantification of low-quantity RNA samples, the Quant-iT™ RiboGreen® reagent (Invitrogen, R11490) tends to provide the most accurate results.

4. The HeliScope generally requires 2–50 ng total RNA to load each of its 50 channels. The total RNA quantity required for optimal loading of sequencing channels is determined by (1) the number of poly(A)+ RNA molecules in each sample, which varies depending on the cell type, cycle stage, and organism being studied, and (2) the flow cell poly(dT) deposition level and performance, which varies between lots depending on poly(dT) concentration and the quality of the epoxide surface used for deposition. HeliScope users are encouraged to perform flow cell hybridization tests to find optimal loading level for the sequencing of each sample. The flow cell lots used for the test hybridization and the actual hybridization for sequencing should be the same. The flow cells, fill and lock kits, and other supplies for test hybridizations can be purchased from the manufacturer.

5. The blocking reaction can be used for RNA sample quantities as high as 10 pmol.

6. It is essential to chill the block to 0 °C in an ice and water slurry and to cool the sample as quickly as possible to 0 °C to minimize reannealing of the denatured RNA strands.

7. A cleanup step is not required if the sample will be used for sequencing immediately after 3′ blocking. After the completion of the blocking step, the sample can be supplemented with 5 μL 100 mM EDTA prior to storage at −80 °C until HeliScope DRS flow cell preparation. For longer-term storage of the 3′ blocked sample, we recommend that the sample be cleaned prior to storage.
8. The cleanup step can be performed using various approaches. We have successfully used the Qiagen RNeasy MinElute Cleanup Kit (Qiagen, 74204) following manufacturer's instructions and the standard phenol/chloroform extraction followed by ethanol/isopropanol precipitation.

Acknowledgments

The author thanks all past and previous Helicos employees for their help and contribution to the development of the technology and the work described here. This work was partly supported by NHGRI grants R01 HG005230 and R44 HG005279 to Patrice M. Milos and Fatih Ozsolak.

References

1. Kapranov P, Willingham AT, Gingeras TR (2007) Genome-wide transcription and the implications for genomic organization. Nat Rev 8:413–423
2. Guttman M, Rinn JL (2012) Modular regulatory principles of large non-coding RNAs. Nature 482:339–346
3. Rinn JL, Chang HY (2012) Genome regulation by long noncoding RNAs. Annu Rev Biochem 81:145–166
4. Andreassi C, Riccio A (2009) To localize or not to localize: mRNA fate is in 3′UTR ends. Trends Cell Biol 19:465–474
5. di Giammartino DC, Nishida K, Manley JL (2011) Mechanisms and consequences of alternative polyadenylation. Mol cell 43:853–866
6. Fu Y, Sun Y, Li Y et al (2011) Differential genome-wide profiling of tandem 3′ UTRs among human breast cancer and normal cells by high-throughput sequencing. Genome Res 21:741–747
7. Proudfoot NJ (2011) Ending the message: poly(A) signals then and now. Genes Dev 25: 1770–1782
8. Ji Z, Lee JY, Pan Z et al (2009) Progressive lengthening of 3′ untranslated regions of mRNAs by alternative polyadenylation during mouse embryonic development. Proc Natl Acad Sci U S A 106:7028–7033
9. Mayr C, Bartel DP (2009) Widespread shortening of 3′UTRs by alternative cleavage and polyadenylation activates oncogenes in cancer cells. Cell 138:673–684
10. Sandberg R, Neilson JR, Sarma A et al (2008) Proliferating cells express mRNAs with shortened 3′ untranslated regions and fewer microRNA target sites. Science 320: 1643–1647
11. Fox-Walsh K, Davis-Turak J, Zhou Y et al (2011) A multiplex RNA-seq strategy to profile poly(A+) RNA: application to analysis of transcription response and 3′ end formation. Genomics 98:266–271
12. Jan CH, Friedman RC, Ruby JG et al (2011) Formation, regulation and evolution of Caenorhabditis elegans 3′UTRs. Nature 469: 97–101
13. Shepard PJ, Choi EA, Lu J et al (2011) Complex and dynamic landscape of RNA polyadenylation revealed by PAS-Seq. RNA 17: 761–772
14. Ozsolak F, Kapranov P, Foissac S et al (2010) Comprehensive polyadenylation site maps in yeast and human reveal pervasive alternative polyadenylation. Cell 143:1018–1029

15. Ozsolak F, Platt AR, Jones DR et al (2009) Direct RNA sequencing. Nature 461:814–818
16. Ozsolak F (2012) Third-generation sequencing techniques and applications to drug discovery. Expert Opin Drug Discov 7:231–243
17. Ozsolak F, Milos PM (2011) Single-molecule direct RNA sequencing without cDNA synthesis. Wiley Interdiscip Rev 2:565–570
18. Ozsolak F, Milos PM (2009) Transcriptome profiling using single-molecule direct RNA sequencing. Meth Mol Biol (Clifton, NJ) 733:51–61
19. Ozsolak F, Milos PM (2012) RNA sequencing: advances, challenges and opportunities. Nat Rev 12:87–98
20. Harris TD, Buzby PR, Babcock H et al (2008) Single-molecule DNA sequencing of a viral genome. Science 320:106–109
21. Thompson JF, Reifenberger JG, Giladi E et al (2012) Single-step capture and sequencing of natural DNA for detection of BRCA1 mutations. Genome Res 22:340–345
22. Bowers J, Mitchell J, Beer E et al (2009) Virtual terminator nucleotides for next-generation DNA sequencing. Nat Meth 6: 593–595
23. Giladi E, Healy J, Myers G et al (2010) Error tolerant indexing and alignment of short reads with covering template families. J Comput Biol 17:1397–1411
24. Rumble SM, Lacroute P, Dalca AV et al (2009) SHRiMP: accurate mapping of short color-space reads. PLoS Computat Biol 5(5): e1000386
25. Li Z, Lin Y, Ozsolak F et al (2012) An in-depth map of polyadenylation sites in cancer. Nucleic Acids Res 40:8460–8471
26. Sherstnev A, Duc C, Cole C et al (2012) Regulated 3′ end formation revealed by direct sequencing of Arabidopsis thaliana RNA. Nat Struct Mol Biol 19:845–852
27. Schneider GF, Dekker C (2011) DNA sequencing with nanopores. Nat Biotech 30: 326–328

Chapter 14

RHAPA: A New Method to Quantify Alternative Polyadenylation

Ashley L. Cornett and Carol S. Lutz

Abstract

3′ end formation of eukaryotic messenger RNAs (mRNAs) is an essential process that influences mRNA stability, turnover, and translation. Polyadenylation is the process by which mRNAs are cleaved at specific sites in response to specific RNA sequence elements and binding of trans-acting protein factors; these cleaved mRNAs subsequently acquire non-templated poly(A) tails at their 3′ ends. Alternative polyadenylation occurs when multiple poly(A) signals are present in the primary mRNA transcript, in either the 3′ untranslated region (3′UTR) or other sites within the mRNA, resulting in multiple transcript variants of different lengths. We demonstrate here a new method, termed RHAPA (*R*Nase *H* *a*lternative *p*olyadenylation *a*ssay), that employs conventional RT–PCR with gene-specific oligonucleotide hybridization and RNase H cleavage to directly measure and quantify alternatively polyadenylated transcripts. This method gives an absolute quantified expression level of each transcript variant and provides a way to examine poly(A) signal selection in different cell types and under different conditions. Ultimately, it can be used to further examine posttranscriptional regulation of gene expression.

Key words Polyadenylation, Alternative polyadenylation, 3′ end formation, RNase H, Oligonucleotide, 3′ untranslated region

1 Introduction

Polyadenylation is an essential processing event that forms the 3′ ends of all eukaryotic mRNAs, with the exception of histone mRNAs. The process of polyadenylation is accomplished by the actions of multi-subunit protein complexes which bind to specific RNA sequences in the 3′ untranslated region (3′UTR) of mRNAs. Alternative polyadenylation, that is, the presence and use of more than one polyadenylation signal, is now understood to occur in more than half of all human mRNAs [1]. Recent studies have highlighted that alternative polyadenylation is an important regulatory mechanism of gene expression [2–7]. Alternative polyadenylation is important not only in defining the 3′UTR length but also in inclusion or exclusion of additional RNA elements, such as AU-rich stability elements or microRNA (miRNA) binding sites.

Joanna Rorbach and Agnieszka J. Bobrowicz (eds.), *Polyadenylation: Methods and Protocols*, Methods in Molecular Biology, vol. 1125, DOI 10.1007/978-1-62703-971-0_14, © Springer Science+Business Media New York 2014

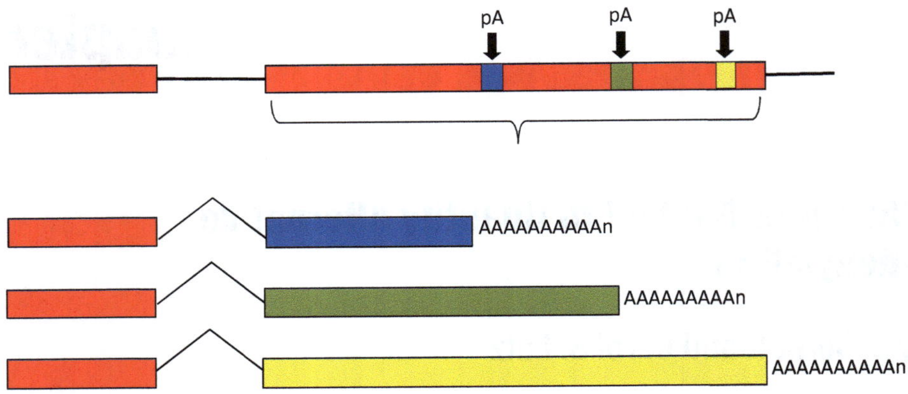

Fig. 1 Schematic representation of alternative polyadenylation. In this example, multiple poly(A) signals are all located in the 3′-most exon. As a result, RNA transcripts generate 3′ untranslated regions (UTRs) of varying lengths. Exons are represented as *red boxes*; introns are represented by *black lines*. Varying lengths of 3′ UTRs are represented by *blue*, *green*, and *yellow boxes*. pA, poly(A) signal

As shown in Fig. 1, cleavage and subsequent poly(A) tail addition can result in RNAs of varying 3′UTR lengths. Polyadenylation at the 3′ most poly(A) signal ("distal" with respect to the stop codon) would result in an mRNA containing a 3′UTR of greatest length (shown in Fig. 1 as yellow), while polyadenylation at a more proximal site would lead to a shortened 3′UTR (blue or green).

Analysis of alternative polyadenylation has been limited to two techniques until now: Northern blotting and RT–PCR. Northern blotting analysis requires use of a labeled probe to hybridize to a given complementary sequence within the RNA molecule of interest; however, it provides only a relative analysis of alternatively polyadenylated transcript expression. RT–PCR provides a more quantitative method of analyzing RNA expression; however, this quantitation is still a relative measure. Figure 2 depicts the mechanism of action of a standard RT–PCR reaction. In addition, RT–PCR does not account directly for use of the proximal polyadenylation signal as the primer set F1:R1 will not only amplify the short RNA transcript, but it will also amplify the long RNA transcript.

Although RT–PCR does not provide an absolute method to quantify alternatively polyadenylated transcripts, it was a starting point for our design of RHAPA, or *R*Nase *H a*lternative *p*olyadenylation *a*ssay, as depicted in Fig. 3. RHAPA employs the principles of RT–PCR but requires the RNA to undergo hybridization with a gene-specific DNA oligonucleotide that will anneal to the RNA between the polyadenylation signals. RNase H treatment ensues, which will specifically cleave the DNA:RNA hybrid. Next, the cleaved RNA is converted into cDNA by reverse transcription (RT) with an oligo d(T) primer, which will only amplify transcripts that have a poly(A) tail. The cDNA is then used for quantitative PCR analysis.

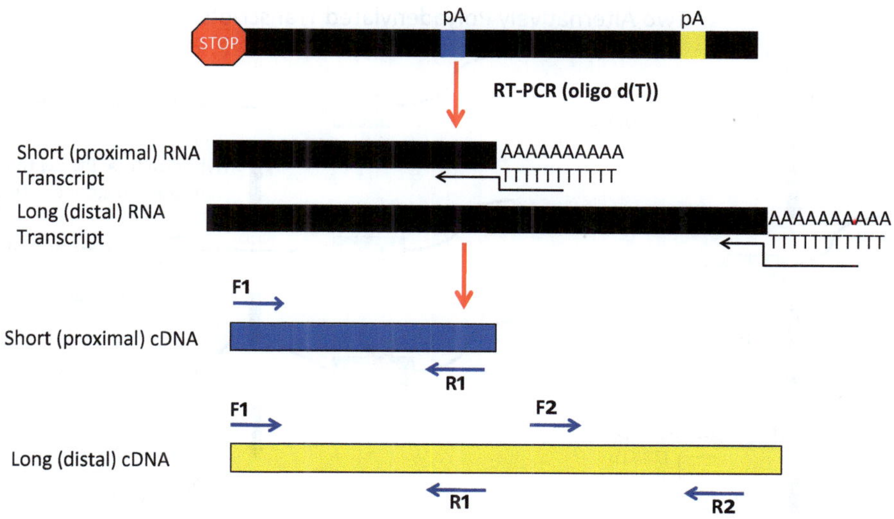

Fig. 2 Schematic of RT–PCR and problems with quantitative analysis. This diagram is a schematic representation of why standard RT–PCR is not sufficient to quantify transcripts that result from alternative polyadenylation. A specific primer set (F1:R1) cannot be used to amplify the short RNA transcript alone, because those primers will also cause amplification of the long mRNA. The short RNA transcript refers to mRNA polyadenylated at the proximal poly(A) signal, with respect to the stop codon; the long RNA transcript refers to mRNA polyadenylated at the distal poly(A) signal

Our new RHAPA method described here has several advantages. The method not only provides accurate and absolute quantification of alternatively polyadenylated transcripts but is fast (~1–1.5 days) and reliable. It can be applied to any gene of interest and can be adapted to be used for multiple treatments or time-course experiments. It can also be easily coupled with global methods such as RNASeq or microarray analysis to get a broad picture of gene expression regulation.

2 Materials

2.1 Tissue Culture

1. DMEM complete: Dulbecco's Modified Eagle's Medium (DMEM) with 4,500 mg/mL glucose/L, L-glutamine, $NaHCO_3$, pyridoxine HCl (Sigma Aldrich, St. Louis, MO, USA), 5 % fetal bovine serum (FBS,), 4 mM additional L-glutamine.
2. 10× phosphate-buffered saline (PBS), pH 7.2.
3. 1× PBS: Dilution of 100 mL of 10× PBS described above into 900 mL of sterile autoclaved dH_2O.
4. Trypsin-ethylenediaminetetraacetic acid (EDTA) (1×).

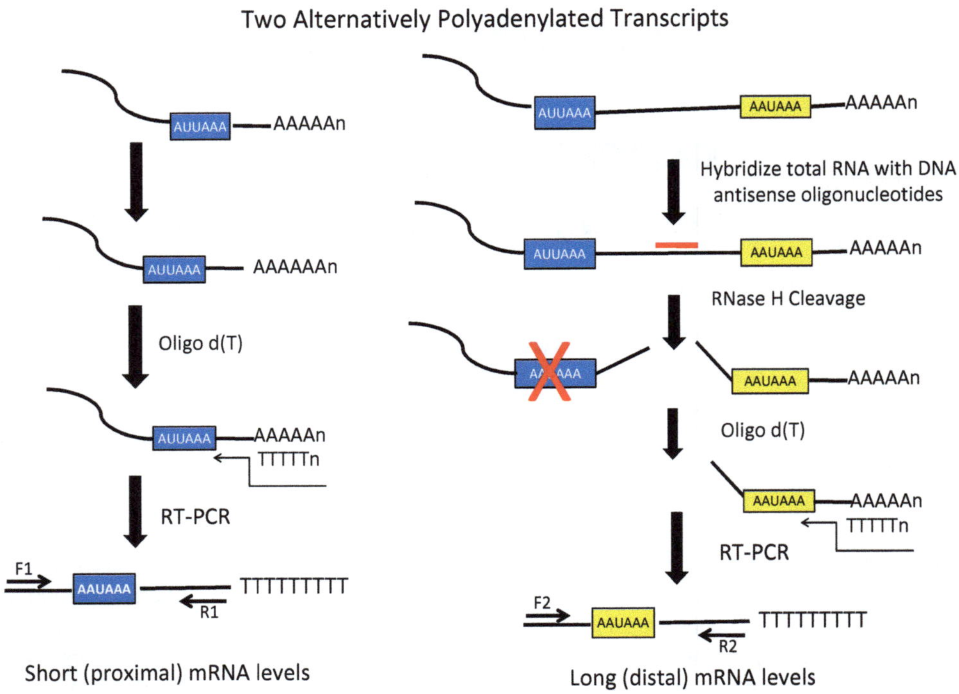

Fig. 3 RHAPA mechanism of action. This assay is designed as an answer to accurate and faithful quantification of alternatively polyadenylated transcripts. This method employs RNase H digestion of RNA:DNA hybrids to cleave RNA sequences between two given polyadenylation signals, which allow the separate quantification of alternatively polyadenylated transcripts. The first step involves the hybridization of the DNA oligonucleotide to its complimentary RNA sequence, followed by subsequent RNase H treatment. The 3′ cleaved RNA is then converted into cDNA by reverse transcription (RT) with an oligo d(T) primer, thereby only amplifying transcripts with a poly(A) tail. The cDNA is then used for PCR analysis. Poly(A) signals are shown in *blue* (proximal) and *yellow* (distal)

2.2 Total RNA Isolation

1. 1× PBS.
2. TRIzol reagent (Invitrogen Corp.).
3. Chloroform.
4. Isopropanol.
5. 80 % Ethanol: To make 500 mL, dilute 400 mL of 200-proof absolute, anhydrous ethyl alcohol in 100 mL of sterile RNase-free double-distilled H_2O.
6. Sterile RNase-free double-distilled H_2O: Place double-distilled water in RNAse-free containers that have been used only for RNase-free double-distilled H_2O. Autoclave the bottles filled with double-distilled H_2O to sterilize them.

2.3 Gene-Specific Oligonucleotide Hybridization and RNase H Cleavage

1. 1× DNA oligonucleotide annealing buffer: 10 mM Tris–HCl, 50 mM NaCl, 1 mM EDTA, pH 7.5–8.0.
2. 500 ng/μL gene-specific antisense DNA oligonucleotide (*see* **Note 1**).

3. RNase H (Invitrogen Corp).
4. Heat block: set at 70 °C.

2.4 Reverse Transcription (RT)

1. ThermoScript RT (15 U/μL) (Invitrogen Corp.).
2. 5× cDNA synthesis buffer (Invitrogen Corp.).
3. 50 μM oligo $(dT)_{20}$.
4. 0.1 M DTT (Invitrogen Corp.).
5. RNaseIN RNase inhibitor (Promega Corp., Madison, WI, USA).
6. Sterile RNase-free double-distilled H_2O.
7. Template RNA (500 ng–1 μg/reaction).
8. 10 mM 2′deoxynucleoside-5′-triphosphate (dNTP): Dilute 10 μL 100 mM deoxycytidine triphosphate (dCTP), 10 μL of 100 mM deoxyadenosine triphosphate (dATP), 10 μL of 100 mM deoxyguanosine triphosphate (dGTP), and 10 μL of 100 mM thymidine triphosphate (dTTP) in 60 μL distilled dH_2O.

2.5 Polymerase Chain Reaction (PCR)

1. 10× PCR buffer: 200 mM Tris–HCl (pH 8.4), 500 mM KCl.
2. 50 mM $MgCl_2$.
3. *Taq* polymerase (5 U/μL).
4. 10 mM dNTP.
5. cDNA from first-strand (RT) reaction.
6. 10 μM gene-specific isoform 1 forward primer (F1).
7. 10 μM gene-specific isoform 1 reverse primer (R1).
8. 10 μM gene-specific isoform 2 forward primer (F2).
9. 10 μM gene-specific isoform 2 reverse primer (R2).
10. 10 μM gene-specific control for cleavage forward primer (F3).
11. 10 μM gene-specific control for cleavage reverse primer (R3).
12. 10 μM housekeeping gene forward primer (such as primer specific for GAPDH).
13. 10 μM housekeeping gene reverse primer (such as primer specific for GAPDH).
14. PCR Cycler: PTC-100, Programmable Thermal Controller (MJ Research™ Inc., Boston, MA, USA).

2.6 Agarose Gel Electrophoresis Components

1. Sterile RNase-free double-distilled dH_2O.
2. Agarose (Invitrogen Corp.).
3. 10× Tris–Borate–EDTA buffer (TBE): 1 M Tris, 0.9 M boric acid, and 0.01 M EDTA.
4. 1× TBE buffer: Dilute 100 mL of 10× TBE buffer into 900 mL sterile RNase-free double-distilled H_2O.

5. Ethidium bromide.
6. DNA gel loading dye: 15 % glycerol, 0.05 % w/v bromophenol blue, 0.05 % w/v xylene cyanol.
7. DNA 100 bp ladder (Invitrogen).
8. Gel box.
9. Combs.

2.7 Detection and Analysis Equipment and Software (See Note 2)

1. GeneGenius bio-imaging system (Syngene, Frederick, MD, USA).
2. GeneSnap image acquisition software (Syngene).
3. GeneTools analysis software (Syngene).

3 Methods

Use of the TRIzol method as per the manufacturer's protocol for RNA extraction is provided below. However, other RNA isolation techniques can be employed.

3.1 RNA Isolation

1. Cells are grown to 80 % confluency in DMEM + 5 % FBS.
2. Media is removed and cells are then washed with 1× PBS.
3. TRIzol reagent (1 mL) is added to each 60 mm dish and left on the cells for 5 min at room temperature, allowing cell lysis (*see* **Note 3**).
4. Cell mixture is added to 1.5 mL microfuge tubes for further steps.
5. 200 μL chloroform is added (1/5 of the original volume of TRIzol used) to tubes. Tightly secure microfuge tube caps, and vigorously shake by hand for 15 s. The mixed solution is then incubated at room temperature for 10 min followed by centrifugation (~10,000 × *g*) in a microcentrifuge for 15 min.
6. After centrifugation, the resulting mixture separates into a lower, red phenol-chloroform phase, an interphase, and a colorless upper aqueous phase. The RNA-containing aqueous phase is removed and transferred to a fresh microfuge tube.
7. Precipitate the RNA from the aqueous phase by mixing with isopropanol. Use 0.5 mL isopropanol for every 1 mL of TRIzol reagent used at the initial step. Incubate for 10 min at room temperature and centrifuge for 10 min at 10,000 × *g*. RNA will precipitate into a pellet at bottom of the tube. Gently decant supernatant to avoid disrupting the RNA pellet.
8. To wash the RNA pellet, add an equal volume (1 mL) of 80 % ethanol. Mix the sample gently and centrifuge at 7,500 × *g* for 5 min at 2–8 °C.

9. Briefly dry RNA pellet, and resuspend in 60 μL RNase-free H_2O.
10. Store RNA at −80 °C.

3.2 Gene-Specific DNA Oligonucleotide Hybridization and RNA:DNA Hybrid Cleavage

1. Incubate 1.5 μg template RNA with 2 μg of gene-specific DNA oligo in 50 μL 1× oligonucleotide annealing buffer in a microfuge tube.
2. Mix briefly, then centrifuge for 15 s at 7,500 × *g*.
3. Place tube in a standard heat block set at 70 °C for 10 min, then remove heat block from the apparatus, allowing tube and heat block to slowly cool to room temperature on the work bench (*see* **Note 4**).
4. After the tube and mixture therein cools to ~25 °C, add 1 μL (2 U/μL) of RNase H (Invitrogen) and 0.5 μL (40 U/μL) of RNaseIN (Promega). Mix briefly, and centrifuge briefly for 10 s to collect contents to the bottom of the tube.
5. Incubate at 30 °C for 60 min (*see* **Note 5**).

3.3 RNA Purification

RNA purification is carried out by QIAGEN RNeasy RNA cleanup protocol, and those steps are briefly provided below. Other methods to purify RNA can be applied (*see* **Note 6**).

1. Adjust the sample volume to 100 μL with RNase-free water. Add 350 μL buffer RLT and mix.
2. Add 250 μL ethanol (100 %) to the dilute RNA, and mix by pipetting.
3. Transfer the sample to an RNeasy mini spin column placed in a 2 mL collection tube (supplied by the QIAGEN kit). Close lid, and centrifuge for 15 s at ≥8,000 × *g*. Discard flow through (*see* **Note 7**).
4. Add 500 μL buffer RPE to the RNeasy spin column. Close lid, and centrifuge for 15 s at ≥8,000 × *g* to wash the spin column membrane. Discard flow through.
5. Add 500 μL buffer RPE to the RNeasy spin column. Close lid, and centrifuge for 2 min at ≥8,000 × *g* to wash the spin column membrane. Discard flow through.
6. Place RNeasy spin column in a new 1.5 mL collection tube (supplied by QIAGEN kit). Add 12 μL RNase-free water directly to the spin column membrane. Close the lid, and centrifuge for 1 min at ≥8,000 × *g* to elute the RNA.
7. Keep eluted RNA sample on ice and immediately proceed to Subheading 3.4 (*see* **Note 8**).

3.4 Synthesis of cDNA and Polymerase Chain Reaction (RT–PCR)

Complimentary DNA (cDNA) is produced following the Invitrogen ThermoScript Reverse Transcriptase (RT) protocol. PCR is carried out by the Invitrogen Taq polymerase as per the manufacturers' protocol.

1. Incubate ~1 μg RNA (from Subheading 3.3), 1 μL 50 μM oligo$(dT)_{20}$, 2 μL 10 mM dNTP mix in 12 μL reaction, and add RNase-free H_2O to reach volume. Incubate mixture at 65 °C for 5 min.
2. To reaction mixture, add 4 μL 5× cDNA synthesis buffer, 1 μL 0.1 M DTT, 1 μL RNaseIN, 1 μL sterile distilled water, and 1 μL Thermoscript RT (15 U/μL). Mix gently and incubate at 50 °C for 60 min. Terminate reaction by heating at 85 °C for 5 min.
3. Proceed to **step 4** or cDNA can be stored at 4 °C–20 °C.
4. Incubate 5 μL of 10× PCR buffer, 1.5 μL 50 mM $MgCl_2$, 1 μL 10 mM dNTP mix, 1 μL of a sense and antisense GAPDH primer, or other housekeeping gene primer set, 1 μL of a sense and antisense gene of interest primer, and 0.5 μL Taq polymerase (1 U/1 μL). Dilute reaction to 50 μL in RNase-free H_2O.

3.4.1 PCR Cycling Conditions

Step 1: Incubate PCR reaction at 95 °C for 5 min.

Step 2: Denature 94 °C for 30 s.

Step 3: Anneal 55 °C for 1 min.

Step 4: Extend 72 °C for 30 s.

Repeat **steps 2–4** for a total of 30 cycles.

Step 5: Final extension 72 °C for 5 min.

Step 6: Keep reaction at 4 °C after cycling.

Figure 4 diagrams the primer sets for the gene of interest and also how these primers are used to ensure RNAse H digestion (*see* **Note 9**).

3.5 Agarose Gel Electrophoresis

1. Prepare 1–1.5 % agarose gel. Add 1–1.5 g agarose to 90 mL of H_2O in a 250 mL Erlenmeyer flask. Microwave agarose mixture until agarose dissolves in water (*see* **Note 10**).
2. Allow to cool briefly. Add 10 mL 10× TBE to flask.
3. Then add 10 μL ethidium bromide and immediately pour gel into gel apparatus (*see* **Note 11**).
4. While gel is solidifying, add 5 μL of 6× DNA gel loading dye to each PCR reaction.
5. Each PCR reaction is subsequently separated by electrophoresis on a 1.5 % agarose gel for approximately 2 h at 100 V.

3.6 Analysis and Quantification of Data

1. The agarose gel is placed on a gel imaging software docking station.
2. Bands are quantified using GeneSnap tool.
3. Bands are normalized to the GAPDH mRNA control and are expressed as a percent of total (*see* **Note 12**). Figure 4 depicts an example of a gel revealing the results of RHAPA analysis.

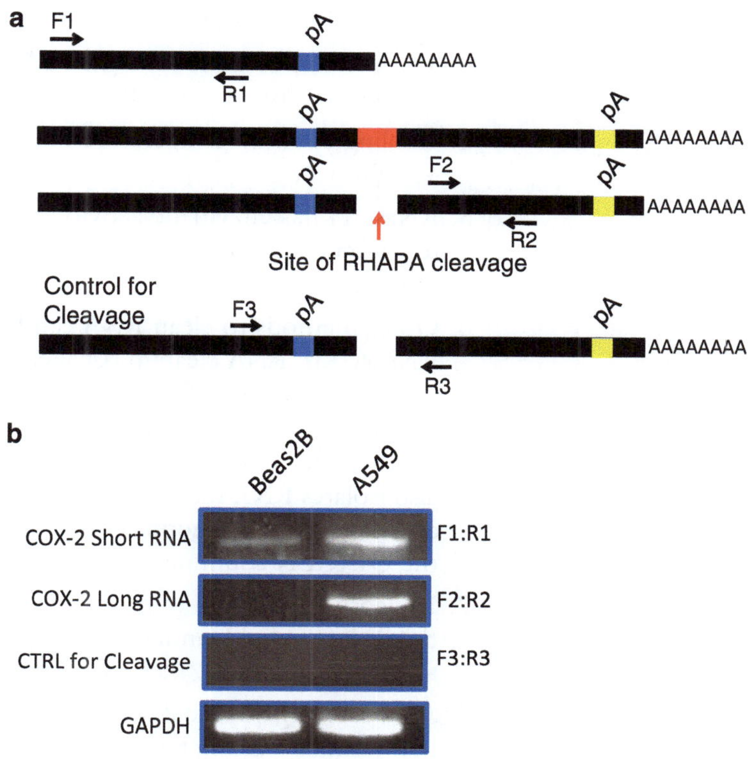

Fig. 4 RHAPA analysis. (**a**) Diagram depicting primer location and how the assay works. *Blue* pA, proximal polyadenylation signal; *Yellow* pA, distal polyadenylation signal; and *red box*, site of DNA oligonucleotide annealing. The primer set F1:R1 will amplify a short, or proximal, RNA transcript. The primer set F2:R2 will amplify a long, or distal, RNA transcript. The primer set F3:R3 will test for completeness of the RNase H cleavage reaction. Amplification of a PCR fragment resulting from F3:R3 would indicate that the RNase H cleavage reaction did not occur to completion, and therefore, the quantification would not be absolute. (**b**) Total RNA was isolated from Beas2B (normal lung cell line) and A549 (lung cancer) cells. RNA was subjected to COX-2-specific RHAPA analysis. Shown here are lanes from a representative ethidium bromide stained agarose gel

4 Notes

1. Design the gene-specific antisense oligonucleotide under similar conditions as considered when designing a PCR primer:
   (a) The oligonucleotide should be roughly 18–22 nucleotides in length.
   (b) The GC content should be roughly 40–60 %.
   (c) The melting temperature (T_m) should be approximately 60 °C.

2. Gel imaging software.
3. The amount of TRIzol reagent added is based on the area of the culture dish (1 mL/10 cm^2).
4. Cooling to room temperature on a work bench takes roughly 45 min–1 h.
5. RNase H has enzymatic activity up to 37 °C. Heat blocks set to a temperature in the range of 30–37 °C are appropriate. Do not exceed 37 °C.
6. Other methods to clean RNA can be employed; however, the RNeasy Kit: RNA cleanup (Qiagen) was selected based on the principle of purpose for this protocol. This method is designed to clean up RNA previously isolated by different methods and/or after enzymatic reactions, e.g., RNase H digestion. It also isolates RNA that is of a size larger than 200 nt; therefore, any excess DNA oligonucleotide in the reaction will remain in the QIAGEN column.
7. In depth description of protocol specification can be found in the RNeasy Mini Handbook. The protocol can be downloaded at www.qiagen.com/literature/protocols/RNeasyMini.aspx.
8. After carrying out the RNA cleanup protocol, it can be assumed that residual starting template RNA is not eluted through each treatment; therefore, as a fail-safe, the RHAPA method requires 1.5 μg starting template RNA to compensate for RNA lost from step to step. With this assumption, at the point of the reverse transcription step, 1 μg of RNA is assumed to remain to serve as the template for cDNA.
9. PCR cycling conditions are based on melting temperature of the primer sets. The annealing temperature should be roughly 5 °C lower than melting temperature for the given primer. Also the extension time is based on the size of the PCR fragment, e.g., for every 1 kb the extension time should be 1 min.
10. Plug the 250 mL Erlenmeyer flask with a paper towel or tissue to avoid bubbling over of the agarose mixture.
11. Ethidium bromide (EtBr) is an intercalating hazardous agent. This material should be handled according to the material safety data sheet (MSDS).
12. In order to quantify PCR fragment bands, GeneTools has a "manual band quantification" option. By manually drawing rectangular boxes around each band and setting each box to be the same size, this provides a way to quantify intensity of each band, using the intensity of the GAPDH PCR fragment band to normalize expression of the gene of interest.

Acknowledgements

The authors wish to thank Anita Antes for careful reading of the manuscript. The authors also wish to thank the Lung Cancer Research Foundation and the American Heart Association, #12GRNT9120029 for grants awarded to C.S.L. Cheers to all.

References

1. Tian B, Hu J, Zhang H et al (2005) A large-scale analysis of mRNA polyadenylation of human and mouse genes. Nucleic Acids Res 33:201–212
2. Lutz CS, Moreira AM (2011) Alternative polyadenylation of eukaryotic mRNAs: an effective regulator of gene expression. WIREs RNA 2: 22–31
3. di Giammartino DC, Nishida K, Manley JL (2011) Mechanisms and consequences of alternative polyadenylation. Mol Cell 43:853–866
4. Sandberg R, Neilson JR, Sarma A et al (2008) Proliferating cells express mRNAs with shortened 3′ untranslated regions and fewer microRNA target sites. Science 320: 1643–1647
5. Mayr C, Bartel DP (2009) Widespread shortening of 3′ UTRs by alternative cleavage and polyadenylation activates oncogenes in cancer cells. Cell 138:673–684
6. Ji Z, Lee JY, Pan Z et al (2009) Progressive lengthening of 3′ untranslated regions of mRNAs by alternative polyadenylation during mouse embryonic development. Proc Natl Acad Sci U S A 106:7028–7033
7. Shi Y (2012) Alternative polyadenylation: new insights from global analyses. RNA 18: 2105–2117

Chapter 15

Multiplex Analysis of PolyA-Linked Sequences (MAPS): An RNA-Seq Strategy to Profile Poly(A+) RNA

Yu Zhou, Hai-Ri Li, Jie Huang, Ge Jin, and Xiang-Dong Fu

Abstract

We summarize 12 experimental methods that have been developed for profiling gene expression by focusing on the 3′-end of poly(A+) mRNA, distilling both common and unique features. Of this family of methods, we provide detailed protocol for MAPS, a method we believe is the simplest and most cost-effective for profiling gene expression and quantifying alternative polyadenylation events by oligo-dT priming followed by random priming and deep sequencing. This method also enables library multiplexing by using a set of bar coding primers during PCR amplification. We also provide a general guideline for analysis of the data generated by MAPS by using the software package *maps3end*.

Key words Gene expression profiling, RNA-seq, Alternative polyadenylation, Multiplexing strategies

1 Introduction

Gene expression profiling is a key strategy to study regulated gene expression, to classify disease samples based on gene expression signatures, and to discover potential drug targets. Full-length RNA-seq is now a widely used technology to measure gene expression and detect structural variations in processed transcripts genome-wide. However, the cost is still prohibitive in cases where there are lots of samples and multiple replicates to be analyzed, emphasizing the need for efficient and simplified methods.

Polyadenylation is a conserved process in eukaryotic cells, which proceeds in two essential steps: cleavage at near the 3′-end of the primary transcripts and addition of a polyA tail of 200–250 nt in length [1]. Importantly, more than 40 % genes have been shown to undergo alternative polyadenylation (APA), as detected in zebrafish [2], mice, and humans [3], indicating APA as a critical regulatory step in gene expression. APA can cause 3′ UTR lengthening or shortening, resulting in different platforms for regulation by miRNAs, nonsense-mediated RNA decay (NMD), mRNA localization, and translational control [4, 5].

Joanna Rorbach and Agnieszka J. Bobrowicz (eds.), *Polyadenylation: Methods and Protocols*, Methods in Molecular Biology, vol. 1125, DOI 10.1007/978-1-62703-971-0_15, © Springer Science+Business Media New York 2014

APA is dynamically regulated during cell proliferation and development via cis-elements, including PAS (Poly(A) Signal), USE (Upstream Sequence Element) and DSE (Downstream Sequence Element), and trans-factors, such as core 3′-end processing factors like CPSF and CstF. APA is coupled with other layers of gene expression and is subject to modulation by RNA-binding proteins like HuR and Nova2 [5]. Mis-regulated APA has been linked to various human diseases, particularly cancers [6]. Researchers are still early in deciphering the polyadenylation code, similar to the splicing code, an effort that began 10 years ago [7].

Motivated by the requirement for gene expression profiling and pA site usage quantification, a family of RNA-seq methods named 3′-end RNA-seq has been developed with the basic idea to only count reads in the proximity of pA sites. Sequencing reads mapped at the end of genes reflect the levels of gene expression and pA usage, but do not provide information on other differences in the transcripts, such as alternative promoter usage and alternative splicing. Up to now, there are at least 12 deep sequencing-based methods that have been described in literature, as summarized in Fig. 1. All these methods utilize the unique signature of the long polyA tail for pulling down mRNAs with complementary oligo-dT. One major confounding factor is the internal A-rich sequences, which may be annealed to oligo-dT to generate unwanted signals. Some methods have been specifically designed to address this problem.

DRS [8] permits direct sequencing of the captured mRNAs from 50 oligo-dT coated on the slide, which is exemplified with several successful applications [6, 9]. However, this method is only amenable on the Helicos platform, which has not been widely available to the research community. All other methods generally involve several common steps that include extraction of small RNA fragments close to pA sites, conversion of RNA to cDNA by RT, ligation to sequencing adaptors and amplification by PCR, and finally, sequencing on Illumina or 454 sequencers. The main differences lie at the treatment at the 5′-end and 3′-end of short RNA fragments. To remove most upstream sequences from the pA site, three strategies are used: random fragmentation by heating, enzymatic cleavage with DpnII/RNase T1/RNase I, or random priming with octamer. Several creative strategies have been developed to treat the 3′-end. NV-anchor oligo-dT are used in 3-SEQ [10], PolyA capture [11], PAS-Seq [3], MAPS [12], SAPAS [13], PolyA-Seq [14], 3′-end-seq [15], 3′Seq [16], and A-seq [17]. All of these methods require bioinformatics treatment to remove reads from potential internal priming [4]. In contrast, 3P-seq [18] is designed to remove internal polyA priming experimentally by attaching splint oligos to the end of polyA tail followed by selection with biotin. 3′READS [19] uses a composed oligo (U_5T_{45}) coupled with stringent wash conditions to enrich for polyadenylated RNA fragments.

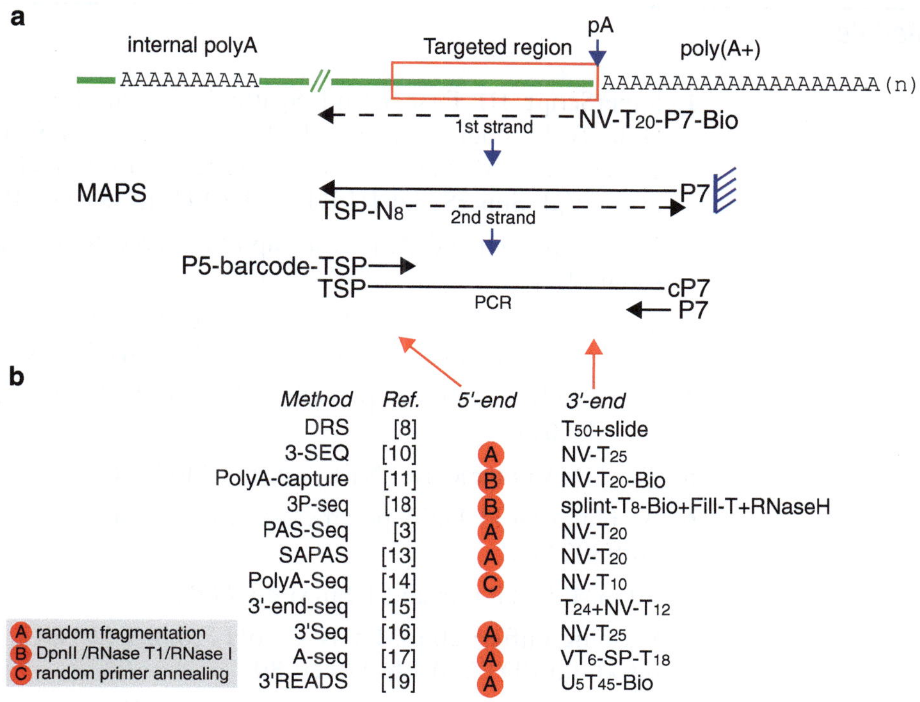

Fig. 1 MAPS and other strategies in 3′-end RNA-seq. (**a**) Schematic of MAPS procedure. *Green line* represents an mRNA with polyadenylated tail and one internal polyA stretch. *Red box* highlights the targeted region where the sequencing reads are supposed to be mapped. (**b**) Summary of other strategies. The methods are sorted by the publication date, with abbreviated description of experimental designs. *TSP* target sequencing primer, *cP7* reverse complement of P7 primer, *Bio* biotin, *Ref.* reference number, *SP* sequencing primer in a stem-loop, *N* A/C/T/G, *V* A/C/G

We believe that the MAPS method we have developed is the simplest and most cost-effective for profiling the 3′-end of mRNA by RNA-seq, as illustrated in Fig. 1. In this method, oligo-dT linked to the 3′ adaptor is first used to prime reverse transcription from mRNAs. After removing free priming oligos and biotin selection, the second-strand synthesis is initiated on beads with random octamers directly anchored to the 5′ adaptor. The released second-strand products are PCR amplified, which takes advantage of PCR bias in combination with size selection to enrich small fragments near the 3′-end. This method provides strand-specific information. A set of bar-coded primers is individually used in the PCR reactions for multiplexing libraries to be sequenced, permitting genome-wide gene expression profiling of up to 12 libraries per lane on the Illumina HiSeq 2000 platform [12]. Here we describe a detailed protocol for this Multiplex Analysis of PolyA-linked Sequences (MAPS) method and provide a general guide to data analysis by using the software package *maps3end*.

2 Materials

1. SuperScript III First-Strand Synthesis System for RT-PCR (Life Technologies, cat. no. 18080-051), including 10× RT buffer, 25 mM $MgCl_2$, 0.1 M DTT, 10 mM dNTP mix, 200 U/μL SuperScript III RT, and 40 U/μL RNaseOUT.
2. NucleoSpin gel and PCR clean-up (Macherey-Nagel, cat. no. 470609).
3. Terminal transferase (NEB, cat. no. M0315S), including 10× buffer, 2.5 mM $CoCl_2$, and 20 U/μL Terminal Transferase.
4. Dynabeads Myone Streptavidin C1 (Life Technologies, cat. no. 650.01).
5. Taq DNA polymerase (NEB, cat. no. M0237L).
6. Amplitaq Gold DNA polymerase (Life Technology, cat. no. N8080241).
7. 10 mM dNTP mix and 10 mM ddNTP mix.
8. Washing buffer: 20 mM Tris–Cl, pH 7.6, 0.1 M NaCl, 1 mM disodium EDTA, 0.1 % Tween-80.
9. Library construction primers – first-strand RT primer: biotin-CAAGCAGAAGACGGCATACGAGT$_{(20)}$VN; second-strand primer: GCTGATGCTACGACCACAGGN$_{(8)}$.
10. PCR amplification primers – bar coding primers: AATGATAC GGCGACCACCGAGATN$_{(4)}$GCTGATGCTACGACCAC AGG; P7 primer: CAAGCAGAAGACGGCATACGAG.
11. Sequencing primers – P5 primer (for sequencing barcodes): AATGATACGGCGACCACCGAGAT; target sequencing primer: GCTGATGCTACGACCACAGG.

3 Methods

3.1 Reverse Transcription and Blocking

1. Add the following components to denature RNA:

   1 μg of Trizol-isolated total RNA.

   1 μL of 50 μM first-strand RT primer.

   1 μL of 10 mM dNTP mix.

   RNase-free H_2O to 10 μL.
2. Incubate at 65 °C for 5 min and then place on ice for at least 1 min.
3. Add the following components to the above mixture to initiate the RT reaction:

   2 μL of 10× RT buffer.

   4 μL of 25 mM $MgCl_2$.

2 μL of 0.1 M DTT.

1 μL of 40 U/μL RNaseOUT.

1 μL of 200 U/μL SuperScript III.

4. Incubate at 50 °C for 50 min and terminate the reaction at 85 °C for 5 min.
5. Purify the first-strand cDNA with NucleoSpin gel and PCR clean-up kit following manufacturer's instruction to remove free RT primer.
6. Add the following components to block the 3′-end of the cDNA:

   19 μL of purified cDNA (supplement with water if necessary).

   2.5 μL of 10× terminal transferase buffer.

   2.5 μL of 2.5 mM $CoCl_2$.

   0.25 μL of 10 mM ddNTP.

   0.25 μL of 20 U/μL terminal transferase.
7. Incubate at 37 °C for 30 min and terminate reaction at 70 °C for 10 min.
8. Wash magnetic beads (5 μL per reaction) twice with washing buffer and resuspend in washing buffer (equivalent to the original volume).
9. Add the beads in washing buffer to the above mixture and incubate at room temperature for 30 min, agitating a few times during the incubation.
10. Collect beads on a magnetic stand and discard the supernatant.
11. Briefly denature the cDNA by suspending beads in 100 μL of 0.1 N NaOH, followed by incubation at room temperature for 5 min, and washing the beads twice with H_2O.

3.2 Synthesis of the Second-Strand cDNA and PCR

1. Add the following components to washed beads to synthesize the second strand:

   1 μL of 100 μM AD primer.

   5 μL of 10× Taq DNA Polymerase Buffer.

   1 μL of 10 mM dNTP mix.

   1 μL of 5 U/μL Taq DNA polymerase (NEB).

   Add H_2O to 50 μL.
2. Incubate at 25 °C for 60 min, heat at 68 °C for 30 s, 75 °C for 5 min, and immediately put the tubes on magnetic stand and wash the beads twice with washing buffer.
3. Resuspend beads with 20 μL of H_2O.

4. Heat at 95 °C for 5 min and immediately put the tubes on magnetic stand to collect supernatant containing the released second-strand cDNA.
5. Add the following components for PCR amplification (*see* **Note 1**):

   5–20 μL of the second-strand cDNA.

   5 μL of 10× PCR buffer.

   3 μL of 25 mM MgCl2.

   1 μL of 10 mM dNTP mix.

   1.5 μL of 20 μM P7 primer.

   1.5 μL of 20 μM bar coding primer.

   0.3 μL of 5 U/μL Amplitaq Gold DNA polymerase (Life Technologies).

   Add H_2O to 50 μL.
6. Run PCR 20–25 cycles in the thermal cycler using the following program: 94 °C for 10 min (initial denaturation and polymerase activation), 94 °C for 30 s (denaturation), 58 °C for 30 s (annealing), and 72 °C for 30 s (extension).
7. Examine PCR products in 2 % agarose gel.
8. Cut band corresponding to 200–400 nt and purify PCR product with NucleoSpin gel and PCR clean-up kit following manufacturer's instruction.
9. Quantify the amount of purified PCR products by using a Qubit fluorometer, and pool purified PCR products for multiplex sequencing (*see* **Note 2**).

3.3 Sequencing

1. Use a Qubit fluorometer to quantify the pooled PCR products to make an estimate for loading onto individual lanes of Illumina sequencer (*see* **Note 3**).
2. Load 10 pM of multiplexed libraries to individual lanes of flow cell.
3. Use the target sequencing primer to sequence a desired length (35–75 nt).
4. Strip off the sequenced products (using the procedure described in the Illumina sequencing manual) and then prime the flow cell with the P5 primer to sequence the barcode region.

3.4 Data Analysis

Here we describe a basic pipeline for processing MAPS data and also provide some example scripts at https://code.google.com/p/maps3end/

1. Decoding of multiplexed samples.

   The output from a MAPS sequencing run is a FASTQ file containing the reads for all samples. The first step is to decode the samples according to specific barcodes incorporated in different libraries. The program *decode.py* in the *maps3end* package is designed for this task, which can be run in the following way: *decode.py* [*--startpos= 37 –mismatch= 0 –outprefix lane1*] *f_fastq f_barcode*. The *f_barcode* is a tab-delimited text file and contains bar coding information for all libraries sequenced in the given lane. In this text file, each line represents a library where the first column is the barcode sequence and the second column is the unique library name. The option *startpos* is for setting the start position of the barcode in the read sequence. The option *mismatch* is the maximum number of mismatches allowed in assigning read to the corresponding barcode (we normally allow 0 or 1 mismatch, as long as the read can be assigned to a single barcode). The decoded reads are written into individual files for each library and any reads that failed to be decoded are saved in the file starting with *outprefix* for diagnosis of potential sequencing problems.

2. Mapping of sequencing reads.
   MAPS generates strand-specific reads from mRNAs, which should be only mapped to the sense strand of genes. Generally we map reads directly to the reference genome (e.g., mm9 for mouse, hg18 for human, *see* **Note 4**), which could be downloaded from the UCSC genome browser FTP site [20]. There are multiple programs, such as Bowtie [21] and BWA [22], for mapping the reads to genomes. We normally use Bowtie with the following parameters: *bowtie -l25 -n2 -e 200 -m1 –best –strata –trim5 4 EBWT sample.fastq* (*see* **Note 5**). *EBWT* is the reference index built from genome sequence in FASTA format by the *bowtie-build* program. As the beginning nucleotides from random primers are not required for perfect matching with primed sequences, we skip the first 4 nt of reads in mapping. Two mismatches are allowed in the remaining first 25 nt seed sequences. Only the reads uniquely mapped to the genome are kept for further analysis.

3. Filtering reads resulting from internal priming.
   Potential internal priming is checked for each read against the downstream sequence up to 300 nt to determine the presence of one or more polyA stretches (consecutive 8 As or 9 As in a 10 nt window) [23]. The polyA stretches can be scanned by the script *scan4astretch.py* across the sequences of all annotated genes or the genome and saved for later use. The output BED file can be compared with the mapped reads by the *intersectBed* program in BEDTools [24] to filter out those potential internal priming events.

4. Quantification of gene expression and APA usage.
   Even though the current MAPS protocol cannot accurately assign the pA position in each gene (*see* **Note 6** for an

alternative sequencing strategy to assign pA sites), the data can be used to quantify gene expression without confounding effect of gene length. For the purpose of gene expression profiling, we sum up all reads mapped to the targeted regions (300 nt upstream of the 3′-end, *see* **Note 7**) in annotated genes with more than one 3′-ends. To calculate differential APA usage, we use reads mapped to individual targeted regions that belong to the same gene. We take two steps in computing the quantitative information at each pA site. We first count uniquely assignable reads at individual pA sites and then use the frequency at each site to assign a specific fraction of reads that are mapped to multiple 3′-ends. We provide the scripts *gene2land.py* and *land2exp.py* to assist such computation, which report the quantitative information in terms of reads per kb per million mapped reads (RPKM) for each gene or pA site.

4 Notes

1. A single bar coding primer from a synthetic set of primers each containing a specific sequence in the barcode region is used in combination with the common primer (P7) to amplify each library. We normally take a small aliquot (2 μL) of the second-strand cDNA to perform preliminary amplification to estimate the yield, based on which to use the right amount of the second-strand cDNA for final PCR amplification in order to obtain detectable products with a minimal PCR cycle number to minimize PCR biases.
2. For transcriptome profiling in mammalian genomes, we take an equal amount of PCR-amplified materials from each library to make a pool for multiplex sequencing. We normally pool 12 libraries for sequencing on a single HiSeq 2000 lane, which will yield at least ten million reads per library.
3. Before sequencing, it is important to make an estimation of the materials for loading, which can be determined by various quantitative methods, such as Qubit or real-time PCR. Although 10 pM is recommended for loading into each lane, careful adjustment is required based on previous sequencing results to load the right amount to obtain the optimal density of ~200 millions per lane on HiSeq 2000.
4. The genome version choice, like mm9 or mm10 both for mouse, depends on user's preference, because of other existing data for comparative purposes.
5. Alternatively, to increase the number of assignable reads, we can first map the reads to the mRNA reference sequences and assign remaining reads to the genome reference. We also use TopHat [25] to map reads that span exon-exon junctions.

6. The library can be sequenced from the 3′-end by using a different sequencing primer to determine the accurate pA sites as described [26].
7. The 300 nt interval is a rough choice based on the size selection during the preparation of sequencing library. The sequencing data can be used to estimate the average distance between the positions of mapped reads and the actual 3′-ends of transcripts that contain a single pA site.

Acknowledgments

We acknowledge early contribution of Kristi Fox-Walsh to this method. This work was supported by NIH grants (HG004659) to XDF.

References

1. Proudfoot NJ (2011) Ending the message: poly(A) signals then and now. Genes Dev 25(17):1770–1782
2. Li Y, Sun Y, Fu Y et al (2012) Dynamic landscape of tandem 3′ UTRs during zebrafish development. Genome Res 22(10):1899–1906
3. Shepard PJ, Choi EA, Lu J et al (2011) Complex and dynamic landscape of RNA polyadenylation revealed by PAS-Seq. RNA 17(4): 761–772
4. Wang L, Dowell RD, Yi R (2013) Genome-wide maps of polyadenylation reveal dynamic mRNA 3′-end formation in mammalian cell lineages. RNA 19(3):413–425
5. Shi Y (2012) Alternative polyadenylation: new insights from global analyses. RNA 18(12):2105–2117
6. Lin Y, Li Z, Ozsolak F et al (2012) An in-depth map of polyadenylation sites in cancer. Nucleic Acids Res 40(17):8460–8471
7. Fu XD (2004) Towards a splicing code. Cell 119(6):736–738
8. Ozsolak F, Platt AR, Jones DR et al (2009) Direct RNA sequencing. Nature 461(7265): 814–818
9. Ozsolak F, Kapranov P, Foissac S et al (2010) Comprehensive polyadenylation site maps in yeast and human reveal pervasive alternative polyadenylation. Cell 143(6):1018–1029
10. Beck AH, Weng Z, Witten DM et al (2010) 3′-end sequencing for expression quantification (3SEQ) from archival tumor samples. PloS One 5(1):e8768
11. Mangone M, Manoharan AP, Thierry-Mieg D et al (2010) The landscape of C. elegans 3′UTRs. Science 329(5990):432–435
12. Fox-Walsh K, Davis-Turak J, Zhou Y et al (2011) A multiplex RNA-seq strategy to profile poly(A+) RNA: application to analysis of transcription response and 3′ end formation. Genomics 98(4):266–271
13. Fu Y, Sun Y, Li Y et al (2011) Differential genome-wide profiling of tandem 3′ UTRs among human breast cancer and normal cells by high-throughput sequencing. Genome Res 21(5):741–747
14. Derti A, Garrett-Engele P, Macisaac KD et al (2012) A quantitative atlas of polyadenylation in five mammals. Genome Res 22(6): 1173–1183
15. Haenni S, Ji Z, Hoque M et al (2012) Analysis of C. elegans intestinal gene expression and polyadenylation by fluorescence-activated nuclei sorting and 3′-end-seq. Nucleic Acids Res 40(13):6304–6318
16. Jenal M, Elkon R, Loayza-Puch F et al (2012) The poly(A)-binding protein nuclear 1 suppresses alternative cleavage and polyadenylation sites. Cell 149(3):538–553
17. Martin G, Gruber AR, Keller W et al (2012) Genome-wide analysis of pre-mRNA 3′ end processing reveals a decisive role of human cleavage factor I in the regulation of 3′ UTR length. Cell Rep 1(6):753–763
18. Jan CH, Friedman RC, Ruby JG et al (2011) Formation, regulation and evolution of Caenorhabditis elegans 3′UTRs. Nature 469(7328):97–101
19. Hoque M, Ji Z, Zheng D, Luo W et al (2013) Analysis of alternative cleavage and polyadenylation by 3′ region extraction and deep sequencing. Nat Methods 10(2):133–139

20. Meyer LR, Zweig AS, Hinrichs AS et al (2013) The UCSC Genome Browser database: extensions and updates 2013. Nucleic Acids Res 41((Database issue)):D64–D69
21. Langmead B, Trapnell C, Pop M et al (2009) Ultrafast and memory-efficient alignment of short DNA sequences to the human genome. Genome Biol 10(3):R25
22. Li H, Durbin R (2009) Fast and accurate short read alignment with Burrows-Wheeler transform. Bioinformatics 25(14):1754–1760
23. Beaudoing E, Gautheret D (2001) Identification of alternate polyadenylation sites and analysis of their tissue distribution using EST data. Genome Res 11(9):1520–1526
24. Quinlan AR, Hall IM (2010) BEDTools: a flexible suite of utilities for comparing genomic features. Bioinformatics 26(6):841–842
25. Trapnell C, Pachter L, Salzberg SL (2009) TopHat: discovering splice junctions with RNA-Seq. Bioinformatics 25(9):1105–1111
26. Wilkening S, Pelechano V, Jarvelin AI et al (2013) An efficient method for genome-wide polyadenylation site mapping and RNA quantification. Nucleic Acids Res 41(5):e65

Chapter 16

Global and Quantitative Profiling of Polyadenylated RNAs Using PAS-seq

Chengguo Yao and Yongsheng Shi

Abstract

mRNA alternative polyadenylation (APA) has been increasingly recognized as a widespread and evolutionarily conserved mechanism for eukaryotic gene regulation. Here we describe a method called poly(A) site sequencing that can not only map RNA polyadenylation sites on a transcriptome level but also provide quantitative information on the relative abundance of polyadenylated RNAs. This method has been successfully used for both global APA analysis and digital gene expression profiling.

Key words Alternative polyadenylation, mRNA 3′ processing, Genomics, High-throughput sequencing, PAS-seq

1 Introduction

In eukaryotes, a large portion (30–70 %) of genes have been estimated to produce mRNAs bearing multiple alternative 3′ ends, a phenomenon known as alternative polyadenylation (APA) [1, 2]. To globally characterize APA, early experimental studies used various microarray platforms [3–5]. In these studies, APA changes are monitored by calculating the ratio between the average signal intensities of the probes targeting the extended regions found only in the longer APA isoforms and those of the probes for the common regions. Although these microarray-based methods can be used to detect APA changes, there are several serious limitations. For example, microarrays cannot be used to map polyadenylation sites (PASs), and the quantification is challenging, especially for genes with more than two APA isoforms.

In the past several years, many high-throughput sequencing (HTS)-based methods have been introduced for global characterization of mRNA polyadenylation [1], which can be generally classified into three distinct types. The first type, called direct RNA sequencing (DRS) [6], is based on the Helicos single-molecule sequencing platform. As polyadenylated RNAs are directly captured

Joanna Rorbach and Agnieszka J. Bobrowicz (eds.), *Polyadenylation: Methods and Protocols*, Methods in Molecular Biology, vol. 1125, DOI 10.1007/978-1-62703-971-0_16, © Springer Science+Business Media New York 2014

and sequenced by synthesis without library construction, DRS is believed to be more quantitative. However, when compared to other more commonly used HTS systems such as the Illumina platform, the disadvantages of DRS include lower read counts, shorter read length, higher error rate, and lack of multiplexing capacity [7]. The second method, called 3P-seq, utilizes a series of enzymatic steps designed to map the true 3′ ends of polyadenylated RNAs [8]. However, 3P-seq is labor intensive, and experimental bias may be introduced at various steps. The third type and most commonly used HTS method, including poly(A) site sequencing (PAS-seq), is based on oligo(dT)-primed reverse transcription [9–11]. The advantages of this method include its simplicity and quantitative performance. One limitation is the possibility of oligo(dT) primers hybridizing to internal A-rich RNA sequences, thereby identifying false-positive PASs. Computational methods can be applied to identify and remove the majority of these sites. In addition, various modifications have been introduced to this basic method to reduce internal priming and facilitate library construction and sequencing. For example, PAS-seq takes advantage of the SMART reverse transcription system in library construction [12]. Using this method, reverse transcription and linker addition on both ends are accomplished in one step, thereby significantly simplifying library construction. Additionally, HTS is carried out on the Illumina platform using a custom sequencing primer, which allows sequencing to start at the poly(A) junction and avoid the problematic A stretch at the beginning of the reads. Below we describe the detailed protocol for PAS-seq and offer technical advice on optimization and troubleshooting.

2 Materials

2.1 Solutions

1. Ammonium acetate (10 M).
2. 100 % Ethanol.
3. 1×TBE buffer (89 mM Tris base, 89 mM boric acid, 2 mM EDTA).

2.2 Enzymes, Reagents, Equipment

1. Trizol (Life Technologies).
2. Dynal Beads $(dT)_{25}$ (Life Technologies).
3. 10× RNA Fragmentation Buffer (10× Fragmentation Reagent, Life Technologies).
4. 10× Stop Buffer (10× Stop Solution, Life Technologies).
5. 6× DNA Loading Dye (Thermo Scientific).
6. Glycogen (Life Technologies).
7. RNAseOUT (Life Technologies).
8. Superscript III reverse transcriptase (Life Technologies).

9. QIAquick PCR purification kit (Qiagen).
10. Phusion DNA polymerase (New England Biolabs).
11. 25 bp DNA ladder (Promega).
12. QIAquick Gel Extraction kit (Qiagen).
13. NanoDrop 1000 (Thermo Scientific).
14. PCR Thermal Cycler (Eppendorf).

2.3 Primer Sequences

HITS-5′:

CGGTCTCGGCATTCCTGCTGAACCGCTCTTCCGATCTr (GGG).

HITS-3′:

ACACTCTTTCCCTACACGACGCTCTTCCGATCTTTTTT TTTTTTTTTTTTTTVN (V:A/C/G; N:A/T/C/G).

PE 1.0:

AATGATACGGCGACCACCGAGATCTACACTCTTTCCC TACACGACGCTCTTCCGATCT.

PE 2.0:

CAAGCAGAAGACGGCATACGAGATCGGTCTCGGCAT TCCTGCTGAACCGCTCTTCCGATCT.

PAS-seq:

ACACTCTTTCCCTACACGACGCTCTTCCGATCTTTTT TTTTTTTTTTTTTTT.

3 Methods

3.1 Poly(A+) RNA Purification

1. Purify total RNAs from cells/tissues using Trizol or other reagent as per manufacturer's instructions.
2. Purify poly(A+) RNAs from total RNAs using Dynal Beads $(dT)_{25}$ as per manufacturer's instructions.

3.2 Poly(A+) RNA Fragmentation

1. Prepare the following mixture:

   9 μl poly(A+) RNA (0.5–1 μg).

   1 μl 10× RNA fragmentation buffer.

   Incubate at 70 °C for 10 min (minutes).
2. Add 1 μl STOP buffer (10×), and leave on ice for 2 min.
3. Add:

   190 μl H_2O.

   50 μl ammonium acetate (10 M).

   750 μl 100 % ethanol.

0.5 μl glycogen (20 μg/μl).

Incubate on dry ice for 10 min, and spin at top speed in a microfuge for 15 min at 4 °C.

4. Rinse the pellet with 70 % ethanol, air-dry the pellet, and resuspend it in 22 μl of H_2O.

3.3 Reverse Transcription (RT)

1. Prepare the following mixture:

   22 μl RNA.

   2 μl HITS-3′ (12 μM).

   2 μl dNTP mix (10 mM).

   Incubate at 65 °C for 5 min and then on ice for 5 min.

2. Add:

   8 μl 5× First-strand reverse transcriptase buffer.

   2 μl DTT (0.1 M).

   2 μl HITS-5′ (10 μM).

   2 μl RNaseOUT (40 U/μl).

   2 μl Superscript III reverse transcriptase (200 U/μl).

   Incubate at 50 °C for 30 min, then 42 °C for 30 min, and 90 °C for 5 min, and cool down to room temperature.

3.4 First-Round PCR Amplification of cDNAs

1. Add 60 μl of H_2O to the 40 μl RT reaction, and isolate cDNAs using the QIAquick PCR purification kit. Elute cDNA from the column with 40 μl of H_2O.
2. Prepare the following PCR mixture (50 μl total volume):

   36.4 μl cDNA.

   10 μl 5× High-fidelity phusion polymerase buffer.

   1 μl PE-1.0 primer (12 μM).

   1 μl PE-2.0 primer (12 μM).

   1 μl dNTP mix (10 mM).

   0.5 μl High-fidelity phusion polymerase (2 U/μl).

   Run the following PCR program:

   1. 98 °C 30 s.
   2. 98 °C 10 s.
   3. 60 °C 30 s.
   4. 72 °C 30 s.
   5. Go to **step 2** twice.
   6. 72 °C 5 min.
   7. 4 °C HOLD.

3.5 Size Selection of cDNAs

1. Add 10 μl of 6× DNA loading dye to the PCR reaction, and run the above PCR reaction on a 2 % agarose gel.
2. Cut out 200–300 bp band.
3. Extract DNA using QIAquick Gel Extraction kit, and elute DNA from the columns with 40 μl H_2O.

3.6 Second-Round PCR Amplification of cDNA Fragments

1. Prepare the following PCR mixture (50 μl total volume):

   36.5 μl cDNA (eluate from the previous step).

   10 μl 5× High-fidelity Phusion polymerase buffer.

   1 μl PE-1 primer (12 μM).

   1 μl PE-2 primer (12 μM).

   1 μl dNTP mix (10 mM).

   0.5 μl High-fidelity phusion polymerase (2 U/μl).

   Run PCR program:

   1. 98 °C 30 s.
   2. 98 °C 10 s.
   3. 65 °C 30 s.
   4. 72 °C 30 s.
   5. Go to **step 2** 14 times.
   6. 72 °C 5 min.
   7. 4 °C HOLD.
2. Purify DNA using QIAquick PCR purification kit.
3. Run 10 % of the PCR product on a 2 % agarose gel to examine the size. A typical gel picture of the library is shown in Fig. 1.
4. Measure the DNA concentration using a Nanodrop spectrophotometer, and adjust the concentration according to the requirement of your sequencing facility.
5. Submit the library for sequencing on an Illumina Genome Analyzer or Hi-Seq using a custom sequencing primer (PAS-seq).

3.7 Bioinformatic Analysis

1. The PAS-seq raw reads are first trimmed by removing Ts from the beginning of the reads.
2. The remaining sequences are mapped to the proper reference genome using Bowtie with the setting (-n 2 -m 1 -s 1) (up to two nucleotide mismatches and one unique match to the reference genome allowed) [13].
3. To address the internal poly(A) priming issue, mapped reads that have six or more consecutive As or seven As in total within the ten nucleotides downstream of mapped poly(A) junction are removed.

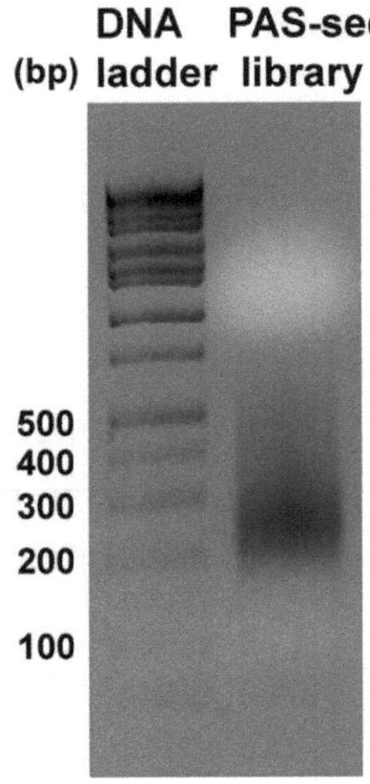

Fig. 1 PAS-seq library. 10 % of the second-round PCR reaction (Subheading 3.6) is resolved on a 2 % agarose gel to check the size of DNA fragments

4. PASs that are found within 40 nt of one another are likely due to heterogeneity of the cleavage reaction and are thus pooled together, and the weighed average position of all the PASs are designated as the final PAS. This process is repeated iteratively until there is no more PAS found within 40 nt.
5. For further bioinformatics analysis, please refer to references 9–11, 14.

4 Trouble-Shooting Tips

1. It is important to purify the cDNAs after the RT step and remove unused RT primers. Otherwise the RT primers may be used in the PCR amplification, and produce DNAs with incomplete linker sequences.
2. The purpose for the first short PCR step is to convert cDNAs into double-stranded DNAs. Gel purification helps to make sure that the DNAs are relatively homogenous in size, therefore minimizing the possible amplification bias.
3. Be extra careful with each DNA purification step. Maximize the recovery by eluting twice.
4. The low-cycle-number PCR is important for quantification. If the starting RNA amount is low, PCR cycle number may be increased to obtain the necessary DNA amount. However this may negatively influence the quantification performance.
5. If sequencing using the custom primer is not possible at your sequencing facility, one can change the orientation of the

library by redesigning the primer sequences such as sequencing reads start at the 5′ end of the corresponding RNA sequence and read into the poly(A) tail. For this approach, we usually gel purify 175–225 bp DNA fragments and perform single-end 100 nt sequencing.

6. PAS-seq can be used for digital gene expression profiling (DEG) analysis. RNA-seq, one of the most commonly used methods for DEG analysis, relies on the average read counts over the entire genes for quantification. As each mRNA has only one 3′ end, PAS-seq and related methods should be at least as accurate as RNA-seq for DEG analysis [14].

Acknowledgements

This work was supported by National Institutes of Health Grant R01 GM090056 and American Cancer Society Grant RSG-12-186.

References

1. Shi Y (2012) Alternative polyadenylation: new insights from global analyses. RNA 18: 2105–2117
2. Di Giammartino DC, Nishida K, Manley JL (2011) Mechanisms and consequences of alternative polyadenylation. Mol Cell 43:853–866
3. Sandberg R, Neilson JR, Sarma A et al (2008) Proliferating cells express mRNAs with shortened 3′ untranslated regions and fewer microRNA target sites. Science 320: 1643–1647
4. Flavell SW, Kim TK, Gray JM et al (2008) Genome-wide analysis of MEF2 transcriptional program reveals synaptic target genes and neuronal activity-dependent polyadenylation site selection. Neuron 60:1022–1038
5. Ji Z, Lee JY, Pan Z et al (2009) Progressive lengthening of 3′ untranslated regions of mRNAs by alternative polyadenylation during mouse embryonic development. Proc Natl Acad Sci U S A 106:7028–7033
6. Ozsolak F, Kapranov P, Foissac S et al (2010) Comprehensive polyadenylation site maps in yeast and human reveal pervasive alternative polyadenylation. Cell 143:1018–1029
7. Ozsolak F, Milos PM (2011) RNA sequencing: advances, challenges and opportunities. Nat Rev Genet 12:87–98
8. Jan CH, Friedman RC, Ruby JG et al (2010) Formation, regulation and evolution of *Caenorhabditis elegans* 3′UTRs. Nature 469: 97–101
9. Shepard PJ, Choi E, Lu J et al (2011) Complex and dynamic landscape of RNA polyadenylation revealed by PAS-Seq. RNA 17:761
10. Fox-Walsh K, Davis-Turak J, Zhou Y et al (2011) A multiplex RNA-seq strategy to profile poly(A+) RNA: application to analysis of transcription response and 3′ end formation. Genomics 98:266–271
11. Fu Y, Sun Y, Li Y et al (2011) Differential genome-wide profiling of tandem 3′ UTRs among human breast cancer and normal cells by high-throughput sequencing. Genome Res 21:741–747
12. Zhu YY, Machleder EM, Chenchik A et al (2001) Reverse transcriptase template switching: a SMART approach for full-length cDNA library construction. Biotechniques 30:892–897
13. Langmead B, Trapnell C, Pop M et al (2009) Ultrafast and memory-efficient alignment of short DNA sequences to the human genome. Genome Biol 10:R25
14. Derti A, Garrett-Engele P, Macisaac KD et al (2012) A quantitative atlas of polyadenylation in five mammals. Genome Res 22:1173–1183

Chapter 17

High-Throughput Sequencing of RNA Isolated by Cross-Linking and Immunoprecipitation (HITS-CLIP) to Determine Sites of Binding of CstF-64 on Nascent RNAs

Petar N. Grozdanov and Clinton C. MacDonald

Abstract

Genome-wide analysis of gene expression has changed the RNA world. Recent techniques leading to this revolution have been the use of cross-linking and immunoprecipitation (CLIP) combined with high-throughput sequencing (HITS-CLIP) to determine sites on nascent mRNAs to which RNA-binding proteins bind. Several researchers (including us) have been examining the role of RNA-binding proteins in polyadenylation, including the role of the 64,000 M_r component of the cleavage stimulation factor, CstF-64. In this chapter, we present our optimizations of the CLIP procedure for examination of CstF-64 binding to nascent pre-mRNAs expressed in testis. For CstF-64 CLIP, we use a well-characterized monoclonal antibody (3A7) that recognizes CstF-64. Rather than optimizing tricky but essential RNA fragment cloning schemes, we illustrate the use of the proprietary Illumina TruSeq Small RNA Sample Preparation kit for this step. Other techniques such as SDS-PAGE and the transfer to the nitrocellulose membrane techniques follow the original Illumina protocol (though we point out potential pitfalls). Finally, we discuss the options for high-throughput sequencing and some general suggestions for bioinformatic analysis of the data.

Key words Polyadenylation, CstF-64, RNA-binding protein, UV cross-linking, Immunoprecipitation, CLIP, RNA-seq

1 Introduction

Polyadenylation site choice is governed by the coordinated action of a swarm of RNA-binding proteins [1–3] that bind to signals in the nascent mRNA [4]. Molecular techniques have been used to map some of these interactions to individual nascent mRNAs [5–7], but newer techniques are required to examine protein–RNA interactions that encompass genes expressed from the entire genome. The approach that combines ultraviolet cross-linking and immunoprecipitation (CLIP, [8]) with high-throughput sequencing of cDNAs generated from the bound RNA fragments (HITS-CLIP, [9]) is ideal for global analysis of these protein–RNA interactions to reveal functional insights [10].

Joanna Rorbach and Agnieszka J. Bobrowicz (eds.), *Polyadenylation: Methods and Protocols*, Methods in Molecular Biology, vol. 1125, DOI 10.1007/978-1-62703-971-0_17, © Springer Science+Business Media New York 2014

Immunoprecipitation can have greater or lesser specificity, depending on the antibody and conditions. Therefore, optimization steps should precede HITS-CLIP. In this chapter, we present our optimizations for the 3A7 monoclonal antibody that recognizes the 64,000 M_r subunit of the cleavage stimulation factor (CstF-64, [11]) in HITS-CLIP from mouse testis. Because 3A7 is highly specific for the somatic form of CstF-64 in mice [12] and shows low nonspecific background immunoprecipitation [5], it has proven to be ideal for HITS-CLIP [13]. Other antibodies may work, as well [10, 14]. We further give tips on high-throughput sequencing of the resulting cDNA libraries.

2 Materials

2.1 Reagents

All reagents, buffers, and other solutions were prepared using double-deionized water and stored as indicated. General laboratory reagents (e.g., Tris base, NaCl, SDS, MgCl2), if not specifically indicated, were purchased from one of the major bioscientific vendors located in the United States: Fisher Scientific or Sigma-Aldrich. Other items purchased included the following:

(a) NuPAGE® Novex 10 % Bis–Tris Gel 1.0 mm, 10 Well SDS-PAGE gels (NP0301Box, Life Technologies).
(b) NuPAGE® Sample Reducing Agent (10×, NP0009, Life Technologies).
(c) NuPage Antioxidant (NP0005, Life Technologies).
(d) Protran BA-85, 0.45 μm (10 402 594, Whatman) nitrocellulose membrane (Sigma-Aldrich).
(e) Water-saturated phenol, pH 6.6 (AM9710, Life Technologies).
(f) Agencourt AMPure XP beads (A63880, Beckman Coulter).
(g) Corning® Costar® Spin-X® centrifuge tube filters (CLS8161, Sigma-Aldrich).
(h) SequaGel–UreaGel Concentrate (EC-830, National Diagnostic).
(i) 30 % acrylamide and bis-acrylamide solution, 37.5:1 ratio (161-0158, Bio-Rad).
(j) 20 mg/mL UltraPure™ Glycogen (10814-010, Life Technologies).
(k) Dynabeads® Protein G for Immunoprecipitation (100-04D, Life Technologies).
(l) EasyTide [γ-^{32}P]-ATP, 3,000 Ci/mmol, 10 mCi/mL (BLU502A100UC, PerkinElmer).
(m) Precision Plus Protein™ Dual Color Standards (161-0374, Bio-Rad).

(n) Kaleidoscope Prestained Standards (161-0324, Bio-Rad).

(o) Low Molecular Weight DNA Ladder (N3233S, New England Biolabs).

(p) Adenosine 5′-triphosphate (ATP, R0441, Fermentas) was prepared as a 10 mM solution in aliquots, which were stored at −80 °C and discarded after each use.

2.2 Buffers

The buffers were made from stock solutions with double-deionized water prior to the beginning of the experiment. The following buffers and solutions were obtained from commercial vendors:

(a) 10× Dulbecco's phosphate-buffered saline (DPBS), no calcium, no magnesium (14200-075, Life Technologies).

(b) Buffers in a Box, Premixed TBE Buffer, 10× (1666703001, Roche Applied Science).

(c) NuPAGE® MOPS SDS Running Buffer (for Bis–Tris Gels only, 20×, NP0001, Life Technologies).

(d) NuPAGE® Transfer Buffer (20×, NP0006-1, Life Technologies).

(e) NuPAGE® LDS Sample Buffer (4×, NP0007, Life Technologies).

(f) 3 M Sodium Acetate, pH 5.5 (AM9740, Life Technologies).

(g) Blue/Orange Loading Dye, 6× (G1881, Promega).

(h) SequaGel–UreaGel Diluent (EC-830 National Diagnostic).

(i) Gel Loading Buffer II (Denaturing PAGE, AM8547G, Life Technologies).

Whenever possible we used the buffers supplied with the corresponding enzymes or systems. Some of the general laboratory stock supply buffers are not listed. However, the following buffers were always prepared "in-house":

(a) Low-salt immunoprecipitation buffer—1× DPBS, 0.1 % SDS (w/v), 0.5 % NP-40 (v/v), 0.5 % deoxycholic acid sodium salt (w/v).

(b) High-salt immunoprecipitation buffer—5× DPBS, 0.1 % SDS, 0.5 % NP-40, 0.5 % deoxycholic acid, sodium salt.

(c) Dephosphorylation buffer—50 mM Tris–HCl pH 7.9, 100 mM NaCl, 10 mM $MgCl_2$, 1 mM dithiothreitol (added fresh before use).

(d) Phosphatase wash buffer—50 mM Tris–HCl pH 7.5, 20 mM EGTA, 0.5 % NP-40 (v/v).

(e) Polynucleotide kinase buffer—50 mM Tris–HCl pH 7.9, 50 mM NaCl, 10 mM $MgCl_2$.

(f) Proteinase K reconstitution buffer—100 mM Tris–HCl pH 7.9, 50 mM NaCl.

(g) Proteinase K dilution buffer—100 mM Tris–HCl pH 7.9, 50 mM NaCl, 10 mM EDTA.

(h) Proteinase K urea buffer—6 M Urea, 100 mM Tris–HCl pH 7.9, 50 mM NaCl, 5 mM EDTA.

(i) RNA PAGE elution buffer—1.66 mg/mL proteinase K, 50 mM Tris–HCl pH 7.5, 25 mM NaCl, 0.3 M sodium acetate pH 5.5, 0.5 % SDS.

(j) DNA PAGE elution buffer—50 mM Tris–HCl pH 7.5, 25 mM NaCl/0.3 M sodium acetate pH 5.5, 0.5 % SDS.

2.3 DNA and RNA Adapters

Below are listed the RNA and DNA adapters and primers used in the CLIP protocol. For the RNA PCR primer (RPI1) containing index 1, the index sequence is underlined. Illumina's oligonucleotides are modified and purified in a proprietary manner (*see* **Note 1**):

(a) RNA 5′ adapter (RA5): 5′-GUUCAGAGUUCUACAGUC CGACGAUC-3′.

(b) RNA 3′ adapter (RA3): 5′-TGGAATTCTCGGGTGCCAAGG-3′.

(c) Stop oligo (STP): 5′-GAAUUCCACCACGUUCCCGUGG-3′.

(d) RNA RT primer (RTP): 5′-GCCTTGGCACCCGAGAATT CCA-3′.

(e) RNA PCR primer (RP1): 5′-AATGATACGGCGACCACCG AGATCTACACGTTCAGAGTTCTACAGTCCGA-3′.

(f) RNA PCR primer, index 1 (RPI1): 5′-CAAGCAGAAGACGGC ATACGAGATCGTGATGTGACTGGAGTTCCTTGGCAC CCGAGAATTCCA-3′.

2.4 Enzymes

The following enzymes and kits were used as recommended by the suppliers:

(a) SUPERase • In (AM2696, Life Technologies).

(b) RQ1 RNase-Free DNase (M610A, Promega).

(c) RNase I (Cloned), 100 U/µL (AM2294, Life Technologies).

(d) Calf intestinal alkaline phosphatase (CIP, M0290S, New England Biolabs).

(e) T4 polynucleotide kinase (M0201L, New England Biolabs).

(f) T4 RNA Ligase 2, truncated (M0242L, New England Biolabs).

(g) SuperScript® III Reverse Transcriptase (18080-044, Life Technologies).

Proteinase K was prepared as follows: 25 mg of Proteinase K, recombinant, PCR Grade Lyophilizate from *Pichia pastoris* (03115836001, Roche Applied Science) was dissolved in 6.25 mL of Proteinase K reconstitution buffer to produce a 4 mg/mL solution, and 0.25 mL aliquots were frozen at −80 °C. Prior to use, a working solution was prepared by mixing 0.25 mL of the frozen aliquot with 0.25 mL Proteinase K dilution buffer and incubating for 10 min at 37 °C.

2.5 Equipment and Kits

(a) UV Crosslinker CL-1000 (UVP LLC.) or equivalent.

(b) Thermomixer® R (Eppendorf, North America).

(c) Sonic Dismembrator Model 60 (Fisher Scientific).

(d) DynaMag™-2 Magnetic Particle Concentrator (Invitrogen) or equivalent.

(e) XCell SureLock™ Electrophoresis Cell/Novex Mini-Cell (Invitrogen).

(f) TruSeq small RNA prep kit (Indexes 1–12, RS-200-0012, Illumina).

(g) PicoFuge® microcentrifuge (Stratagene) or equivalent.

2.6 Animals

25-day-old male C57BL/6 mice were obtained from litters bred in-house. All animal protocols were performed as approved by the Texas Tech University Health Sciences Center Institutional Animal Care and Use Committee.

3 Methods

3.1 Tissue Disruption and UV Cross-Linking

The amount of material required for CLIP varies depending on the tissue source, protein abundance, and efficacy of the antibody. For this protocol (Fig. 1), we describe CLIP using the 3A7 monoclonal antibody to isolate CstF-64–RNA complexes from mouse testes. Guidelines to optimize the procedure for other tissues and conditions are discussed in **Note 2**.

1. Sacrifice two male mice in the manner approved by your Institutional Animal Care and Use Committee. Collect the seminiferous tubules by making incision into the tunica albuginea of the testes and expressing the content in 3–5 mL ice-cold 1× DPBS in a 15 mL conical tube. Disrupt the seminiferous tubules using vigorous shaking. Let the material settle down for ~3 min on ice. Remove the supernatant. Add 3–5 mL fresh ice-cold 1× DPBS. Repeat the procedure for a total of three times.
2. Resuspend the seminiferous tubules in 1 mL of 1× DPBS and transfer to a single well of 12-well plate, pre-chilled on ice.
3. Place the 12-well plate in the UVP UV Crosslinker CL-1000, and irradiate the tissues with 200 mJ/cm^2 at 254 nm three times (for a total of 600 mJ/cm^2). Mix between irradiations by swirling the plate. Transfer the samples to 2 mL microcentrifuge tubes, and rinse the well with an additional 1 mL of 1× DBPS.

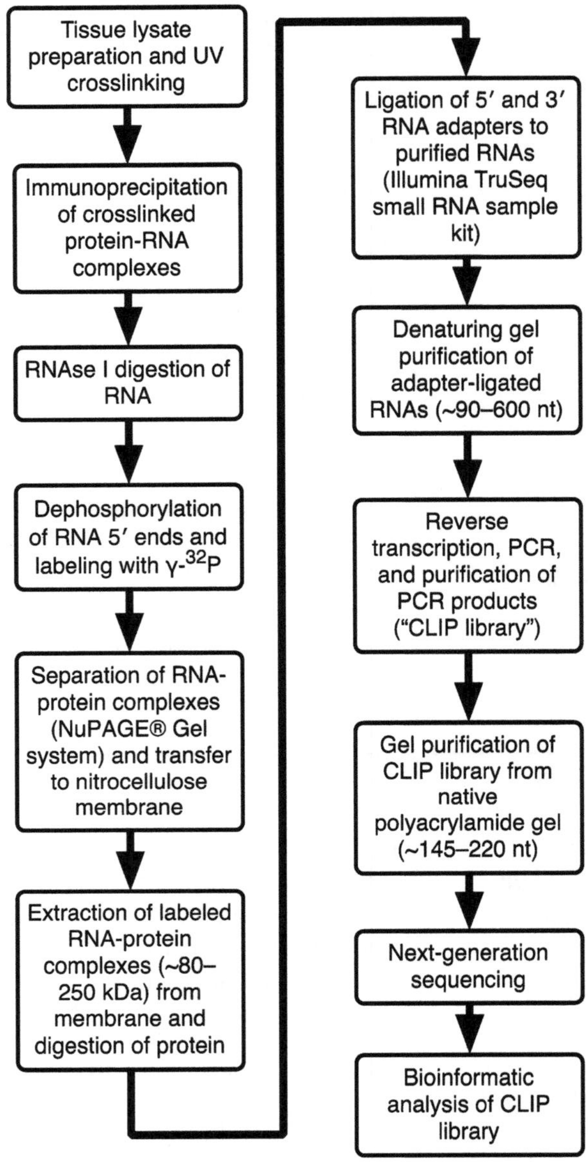

Fig. 1 Summary of steps in CLIP protocol. Shown is a flow chart of the steps described in this chapter

4. Collect the samples by low-speed centrifugation (1,000 × *g*) for 5 min at 4 °C.
5. At this stage, you can store the tissue at −80 °C or continue directly with tissue lysis. For freezing, snap-freeze the samples in liquid nitrogen or dry ice; slower freezing (say, by transferring to a −80 °C freezer) is less desirable, since it might promote proteolysis.

3.2 Antibody Coupling to the Paramagnetic Dynabeads Protein G

1. Transfer 100 μL of Dynabeads Protein G in each of the four 1.5 mL microcentrifuge tubes and resuspend by vortexing. The 3A7 mouse monoclonal antibody we use here is of the IgG_1 isotype, which binds best to Protein G. If your antibody is of a different isotype, choose the appropriate Dynabeads.
2. Wash Dynabeads twice with 1 mL low-salt immunoprecipitation buffer:
   (a) Add 1 mL of low-salt immunoprecipitation buffer.
   (b) Vortex briefly.
   (c) Clear liquid quickly by microcentrifugation.
   (d) Place tube in the magnetic particle concentrator for 30–60 s.
   (e) Remove supernatant by vacuum aspiration.
   (f) Repeat.
3. Resuspend Dynabeads in 500 μL low-salt immunoprecipitation buffer.
4. Add 20 μg of IgG_1 (~500 μL of tissue culture supernatant) of mouse monoclonal anti-CstF-64 clone 3A7 antibody.
5. Incubate the tubes on a Thermomixer R (1,100 rpm, shaking for 20 s and resting for 30 s) for 2 h at 23 °C.
6. Wash three times with 1 mL low-salt immunoprecipitation buffer. *See* **Note 3** for antibody-coupled Dynabeads storage conditions.

3.3 Total Cell Lysate Preparation, RNAse I and DNase I Digestions, and Immunoprecipitation of the CstF-64 Protein

1. Prepare and chill (4 °C) the low-salt immunoprecipitation buffer, at least 1 mL/testis. Usually, we have the testes from one animal collected in one 2 mL tube (*see* Subheading 3.1). Immediately prior to use, add 2 μL SUPERase·In per testis and ~5 mg of phenylmethylsulfonyl fluoride (PMSF) powder to 10 mL of buffer (*see* **Note 4**).
2. Prepare 1:500 (low RNase) and 1:50 (high RNase) dilutions of RNase I in low-salt immunoprecipitation buffer and store on ice (*see* **Note 5**).
3. If tissue/cell pellets have been frozen at −80 °C, thaw on ice.
4. As soon as the tissue is thawed, add 0.7 mL ice-cold low-salt immunoprecipitation buffer. Pipette up and down several times until most of the tissue cells are lysed. Add an additional 0.8 mL buffer, and mix by pipetting. The lysate should become extremely viscous.
5. Incubate on ice for 5 min.
6. Sonicate using the Sonic Dismembrator small probe three times on ice with five pulses each time. Avoid foaming or warming the lysate.

7. Clear the lysates by centrifugation at maximum speed (15,800 × *g*) at tabletop microcentrifuge for 3 min at 4 °C.
8. Transfer the supernatant (~1.5 mL) to a fresh 2 mL tube. Avoid disturbing the pellet.
9. *Optional*: Incubate the lysate for 5 min at 75 °C and then cool on ice (*see* **Note 6**).
10. Add 40 μL RQ1 RNase-free DNase to each of the samples and vortex.
11. Split the volume of each tissue lysate into two tubes, ~700 μL each. Save a small amount of lysate for examination by Western blot (*see* Fig. 2a). For mice, we split the lysate from both testes of a single animal into two tubes, one of which will be digested

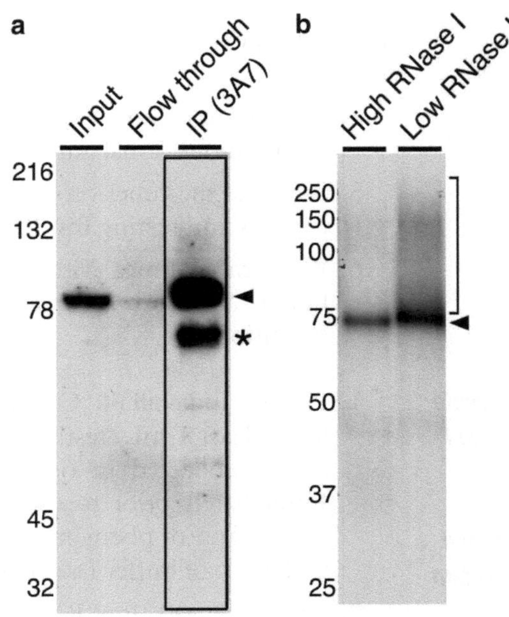

Fig. 2 The 3A7 monoclonal antibody recognizes CstF-64 cross-linked to RNA. (**a**) A Western blot of CstF-64 from testis lysate showing total extract (input), unbound (flow-through), and immunoprecipitate-bound material [IP (3A7)], corresponding to Subheading 3.3. The *arrowhead* indicates the mobility of the CstF-64 protein; the *asterisk* indicates a breakdown product we have observed previously (cf., [23]); the *boxed lane* highlights the immunoprecipitated material. Approximate molecular weights in kilodaltons are indicated to the *left*. (**b**) Representative autoradiograph after the CLIP procedure using 3A7 antibody against CstF-64 protein (Subheading 3.7). Shown are radioactively labeled RNA oligomers cross-linked to the immunoprecipitated CstF-64 protein. The *arrowhead* indicates the mobility of CstF-64 after digestion with either high or low RNase I (see text). The region of the membrane that was excised from the low-RNase I lane is indicated by the *bracket* on the *right*. Approximate molecular weights in kilodaltons are indicated to the *left*. Note the slight difference in the mobility of the CstF-64 protein resolved on 10 % SDS-PAGE (**a**) and NuPAGE® Novex 10 % Bis–Tris gels (**b**)

with the high-RNase dilution ("tube 1") and the other digested with the low-RNase dilution ("tube 2").

12. Add 7 μL of high-RNase dilution to tube 1 and of the low-RNase dilution to tube 2.
13. Mix by vortexing, microcentrifuge briefly to clear droplets, and incubate for 5 min at 37 °C.
14. Centrifuge the lysates at 15,800 × *g* for 3 min at 4 °C.
15. Transfer the supernatant (~700 μL) to the antibody-coupled Dynabeads Protein G. Vortex and spin down briefly to collect the liquid from the sides and the lid of the tubes.
16. Incubate on a Thermomixer R (1,100 rpm shaking for 20 s and resting for 30 s) for 1 h at 4 °C (*see* **Note 7**).
17. Collect the antibody-coupled Dynabeads on the magnetic particle concentrator.
18. Save the supernatant as flow-through for a Western blot (*see* Fig. 2a).
19. Wash the beads once with 1 mL low-salt immunoprecipitation buffer as for **step 2** in Subheading 3.2. Take care not to remove Dynabeads from the tube.
20. Wash the antibody-coupled Dynabeads Protein G four times with 1 mL high-salt immunoprecipitation buffer.
21. Wash the beads once with 1 mL low-salt immunoprecipitation buffer, and leave them in the buffer until all necessary reagents are ready to be used in the next step.

3.4 Dephosphorylation of the 5′ RNA Ends with Calf Intestinal Alkaline Phosphatase

1. Wash Dynabeads twice with 1 mL of dephosphorylation buffer as in **step 2** in Subheading 3.2. After the second wash, remove the supernatant with a pipette and discard it.
2. Prepare the dephosphorylation reaction mix by adding 2 μL/tube CIP (10 U/μL) to 50 μL dephosphorylation buffer (supplied by the manufacturer, 45 μL of water and 5 μL of 10× buffer 3).
3. Resuspend beads in dephosphorylation reaction mix.
4. Incubate the suspension for 10 min at 37 °C in the Thermomixer R (1,100 rpm shaking for 20 s and resting for 30 s).
5. Wash the antibody-coupled Dynabeads Protein G twice with 1 mL phosphatase wash buffer as in **step 2** in Subheading 3.2.

3.5 γ-^{32}P ATP Labeling of the RNA 5′ Ends Using T4 Polynucleotide Kinase

1. Wash the antibody-coupled Dynabeads Protein G twice in 1 mL polynucleotide kinase buffer, and discard the supernatant.
2. Working behind a Plexiglas shield and taking precautions for radioactive use, prepare the T4 polynucleotide kinase mix by adding 3 μL of T4 polynucleotide kinase (10 U/μL) to 47 μL

of 5 μL 10× T4 polynucleotide kinase buffer, 0.5 μL SUPERase · In, 0.5 μL γ-^{32}P-ATP, and 41 μL of water.

3. Add 50 μL of the mix to the tube and mix by vortexing.
4. Incubate the tubes for 30 min at 37 °C in the Thermomixer R (1,100 rpm shaking for 20 s and resting for 30 s).
5. Add 0.75 μL of 10 mM ATP to each tube and incubate for an additional 5 min at 37 °C.
6. Collect the antibody-coupled Dynabeads Protein G on a magnetic collector.
7. Remove the T4 polynucleotide kinase reaction according to the safety regulations for disposal of radioactive materials.
8. Wash the antibody-coupled Dynabeads Protein G five times with 0.8 mL polynucleotide kinase buffer, and discard the supernatant as radioactive waste.
9. The beads can be stored at −20 °C overnight in a Plexiglas-shielded container.

3.6 Separation of Protein–RNA Complexes by SDS-PAGE Using the NuPAGE® Novex Bis–Tris Gel System

In order to prevent alkaline hydrolysis of RNA oligonucleotides, it is necessary to use an electrophoresis system that buffers to pH 7.0. Therefore, Ule et al. [15] recommend the Novex NuPAGE 10 % Bis–Tris system (for homemade alternatives, *see* **Note 8**).

1. Prepare 1× NuPAGE® MOPS SDS Running Buffer by diluting the 20× buffer. Add 0.5 mL NuPAGE® antioxidant to 200 mL pre-chilled 1× MOPS SDS running buffer before use. Remove the sealing sticker from the bottom of the NuPAGE® Novex 10 % Bis–Tris Gel 1.0 mm, 10-well polyacrylamide gel. Remove the comb, and wash the wells two to three times with water and three to four times with the 1× NuPAGE® MOPS SDS running buffer supplemented with the NuPAGE Antioxidant. Assemble the electrophoretic unit as recommended. If you run only a single gel, use the "dam" to block one of the sides. Pour the 1× NuPAGE® MOPS SDS running buffer supplemented with the NuPAGE® Antioxidant in the "top" inner chamber of the assembled unit.
2. Resuspend the Dynabeads Protein G from the T4 polynucleotide kinase labeling step in 25 μL of 2× NuPAGE® LDS Sample Buffer supplemented with 1× NuPAGE® Sample Reducing Agent.
3. Incubate for 10 min at 75 °C to elute the ^{32}P-labeled CstF-64–oligoribonucleotide complexes from the antibody. Mix twice during the heating by gently flicking the tubes (do not invert the tube, as that can spread the radioactive material to the tube lid). Allow the samples to cool briefly on ice.
4. Spin the tubes briefly, and collect the Dynabeads on the magnetic particle collector.

5. Load 10 μL of Precision Plus Protein™ Dual Color Standard and 15 μL of LDS sample buffer to the first gel well. Load subsequent wells with supernatants from each sample (~25 μL). Take care to prevent sample spillover between the wells. Fill all "empty" lanes with 25 μL of 2× NuPAGE® LDS Sample Buffer.
6. Run the gel at 180 V for 1–2 h with cooling at 4 °C (possibly in a cold room).

3.7 Transfer the Separated ^{32}P-Labeled Protein–RNA Complexes to a Nitrocellulose Membrane

1. Pre-chill 1 L of 1× transfer buffer (50 mL 20× NuPAGE® Transfer buffer, 100 mL methanol, 850 mL water). Add 1 mL NuPAGE® Antioxidant for efficient transfer to the 1× transfer buffer just before use.
2. Cut a nitrocellulose membrane (BA-85) to the size of the gel (7.5×9 cm) and pre-wet in 1× transfer buffer in a separate container. Wet the transfer sponges and assemble on a wet transfer electrophoresis cassette. While submerged, squeeze the sponges to remove any trapped air.
3. Make a "transfer sandwich" by stacking three sheets of pre-wetted thick chromatographic paper (7.9×9.4 cm) and the nitrocellulose membrane on top of the transfer sponge. Be sure to eliminate bubbles at every stage. If necessary, assemble the "transfer sandwich" while submerged to eliminate the possibility of trapped air bubbles between the layers.
4. Disassemble the electrophoresis unit, and release the gel from the glass plates. Be aware that the gel is radioactive, so take appropriate precautions.
5. Wet your gloves in the 1× transfer buffer to prevent sticking to the gel. Using the gel releaser wedge as carrying and support tool, place the gel on top of the nitrocellulose membrane in the transfer cassette.
6. Cover the gel with three sheets of pre-wetted thick chromatographic paper, and place a wet transfer sponge on top of the paper. To eliminate residual bubbles, roll a small pipette or 15 mL tube over the "sandwich" and close the transfer cassette.
7. Place the "sandwich" in the electrophoretic transfer unit. *Make note of the cassette orientation to ensure correct direction of transfer.*
8. Fill the electrophoresis tank with ~1 L of cold 1× transfer buffer.
9. Transfer at 60 V for 2.5–3 h at 4 °C in a cold room on ice.
10. At the end of the transfer, disassemble the electrophoretic cassette and discard the buffer and gel as radioactive waste.
11. Rinse the membrane once with 1× DPBS.

12. Wrap the wet membrane with clear plastic wrap. Expose it to X-ray film at −80 °C. We include luminescent stickers (such as Glogos, Agilent) on the plastic backing to correctly align the developed film to the membrane.
13. Expose to film for 6–24 h (*see* **Note 9**).

3.8 Extract the Labeled RNAs, Which Are Bound to the CstF-64 Protein from the Membrane Using Proteinase K

1. Operating behind a Plexiglas shield, align the developed film over the membrane using the Glogos as a guide (Fig. 2b). Identify the band that corresponds to the size of RNA-binding protein in the sample digested with the high concentration of RNase I (high RNase, arrow, Fig. 2b), and mark the plastic wrap. The position of the peak of radioactive signal in the high-RNase lane will serve to indicate the approximate size of the CstF-64–oligoribonucleotide complex.
2. Compare the low-RNase lane to the high-RNase lane, and determine where the CstF-64–oligoribonucleotide complex will migrate (right bracket, Fig. 2b). Mark the plastic wrap. Using a clean razor or scalpel, cut out the bracketed region of the membrane corresponding to the low RNase (right bracket, Fig. 2b). Transfer excised membrane to a sterile microcentrifuge tube.
3. Dilute a 250 μL aliquot of Proteinase K (PK, 4 mg/mL) with a 250 μL of Proteinase K dilution buffer to yield a 2 mg/mL PK solution in 100 mM Tris–HCl pH 7.5, 50 mM NaCl, and 5 mM EDTA.
4. Incubate the PK solution at 37 °C for 10 min.
5. Add 200 μL of PK solution to each tube containing the excised membranes. Make sure that the membrane is soaked in the PK solution.
6. Incubate at 37 °C in the Thermomixer R (1,100 rpm shaking for 20 s and resting for 30 s) for 20 min.
7. Add 200 μL Proteinase K urea buffer; incubate at 55 °C in the Thermomixer R (1,100 rpm shaking for 20 s and resting for 30 s) for 20 min.
8. Reset the Thermomixer R to 37 °C and allow to cool (~6–7 min). Add 400 μL of water-saturated phenol (pH 6.6) and 130 μL chloroform and incubate at 37 °C in the Thermomixer R (1,100 rpm shaking for 20 s and resting for 30 s) for 5 min.
9. Separate the aqueous and organic phases by brief centrifugation at maximum speed in a tabletop centrifuge.
10. Transfer 400–450 μL of the aqueous phase to a fresh microcentrifuge tube, add 50 μL of 3 M sodium acetate pH 5.2, 0.5 μL of glycogen (20 mg/mL), and 1 mL of ice-cold 100 % ethanol. Allow the precipitate to form at −80 °C for at least 20 min. Alternatively, you could split the ethanol precipitation in two tubes.

11. Centrifuge the samples at maximum speed at 4 °C for 15 min.
12. Wash the pellet twice with 0.5 mL of cold 75 % ethanol.
13. Let the pellet dry at room temperature for 5–10 min or until no ethanol can be detected. Do not over-dry the pellet.
14. Add 5 μL of RNase-free water to the pellet.
15. Incubate at 37 °C for 5 min with some occasional vortexing and pipetting to resuspend the pellet.
16. Spin the tube briefly to settle the droplets.

3.9 Adapter Ligation to the Isolated RNAs Using Illumina's TruSeq Small RNA Sample Prep Kit

In this section, we present an abbreviated version of a procedure, which is outlined in detail in the Illumina TruSeq small RNA prep kit manual (available through the Illumina, Inc. website [16]).

1. Dilute the original adapters 5′ (RA5) and 3′ (RA3) 1:5 (4 μL RNAse-free water and 1 μL of each adapter).
2. Ligation of the 3′ adapter:
   (a) Transfer 5 μL of the dissolved RNA from the extracting step into a 0.2 mL PCR tube.
   (b) Add 1 μL diluted RA3 to the isolated RNA.
   (c) Incubate the tube for 2 min at 70 °C and then cool on ice.
   (d) Prepare the following reaction mix in a fresh tube (scale up as needed for multiple reactions):
      - 2 μL ligation buffer (HML, Illumina).
      - 1 μL RNase inhibitor.
      - 1 μL T4 RNA ligase 2, truncated.
   (e) Transfer the above mixture to the tube containing the RNA and RA3 adapter.
   (f) Mix by pipetting up and down eight to ten times.
   (g) Briefly spin down the tube to collect all the droplets.
   (h) Incubate for 1 h at 28 °C.
   (i) Add 1 μL of the Stop Oligo (STP, Illumina); continue to incubate for an additional 15 min at 28 °C.
   (j) Place the tube on ice.
3. Ligation of the 5′ adapter (RA5).
   (a) Add 1 μL of diluted 5′ adapter in a new tube.
   (b) Incubate the tube at 70 °C for 2 min.
   (c) Immediately place the tube on ice.
   (d) Add to the RA5:
      - 1 μL 10 mM ATP.
      - 1 μL T4 RNA ligase.

(e) Mix by pipetting up and down.

(f) Transfer 3 μL of the RA5 adapter mix to the RNA/RA3 reaction mix from **step 2j**; mix by pipetting up and down eight to ten times.

(g) Incubate at 28 °C for 1 h.

(h) Place on ice.

(i) Add 15 μL of 2× gel loading buffer II.

(j) Incubate at 75 °C for 10 min.

(k) Place on ice.

3.10 Denaturing 9.3 % Polyacrylamide Gel Electrophoresis

1. Prepare a 9.3 % denaturing polyacrylamide gel (20 mL) using 16 × 17 cm glass plates with 0.75 mm thick spacers and the corresponding 15-well comb:
   (a) 2 mL 10× TBE.
   (b) 7.5 mL SequaGel–UreaGel Concentrate.
   (c) 10.5 mL SequaGel–UreaGel Diluent.
   (d) 0.12 mL 10 % ammonium persulfate.
   (e) 0.02 mL *N*,*N*,*N′*,*N′*-tetramethylethane-1,2-diamine (TEMED).
2. Assemble the gel in a vertical gel apparatus; fill cathode and anode reservoirs with 1× TBE running buffer. Remove the comb, and pre-run the gel at 10 W for ~1 h until glass plates are warm to touch.
3. Immediately before loading your sample, clean accumulated urea from the wells of the gel using a syringe and a fine-gauge needle.
4. Incubate the samples at 75 °C for 10 min, spin down briefly, and place on ice. Load samples into each lane. Include the radioactively pre-labeled low-molecular-weight DNA ladder.
5. Run the gel with the loaded samples until the bromophenol blue dye nears the bottom of the gel.
6. Wrap the gel in plastic wrap and expose to X-ray film overnight at −80 °C (*see* Subheading 3.7, **step 12**).

3.11 Extract the RNA with Ligated Adapters from Denaturing Polyacrylamide Gel

1. Develop the film, and mark the sizes of the radioactively pre-labeled low-molecular-weight DNA ladder on the film using a permanent marker pen. Align the film to the gel using the Glogo dots. Mark on the plastic wrap the position of the DNA ladder and migration of the loading dyes. Cut out the radioactive material above 90 nt (usually above the xylene cyanol FF dye) all the way to the beginning of the well (~600 nt, *see* Fig. 3a). You may also verify the size of your RNA fragments using as a guide the sizes of the radioactively pre-labeled low-molecular-weight DNA ladder (*see* **Note 10**).

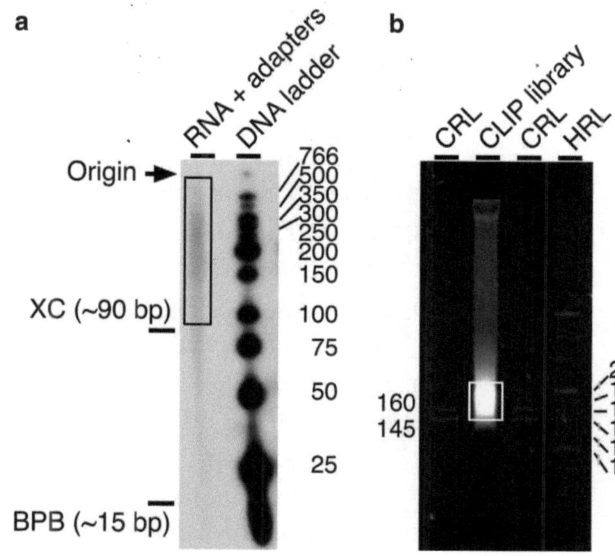

Fig. 3 Purification of RNA oligomers and DNA amplimers after the CLIP procedure. (**a**) Autoradiograph of the ^{32}P-labeled RNA oligonucleotides with various sizes isolated as described in the text and subsequently ligated to the RA3 and RA5 adapters using the Illumina's TruSeq small RNA prep kit (Subheading 3.11). The *box* indicates radioactive RNA with the ligated adapters (RNA + adapters) that is extracted to produce the CLIP DNA library. Shown are sizes in nucleotides of the radioactively labeled low-molecular-weight DNA ladder (DNA ladder, *right*) and approximate mobilities of the xylene cyanol (XC) and bromophenol blue (BPB, *left*). (**b**) Representative fluorescent image of ethidium bromide-stained native PAGE of the CLIP library (Subheading 3.14). The *box* indicates the region of the final CLIP library that was extracted from the gel. Lanes are the custom RNA ladder (CRL), CLIP library, and high-resolution ladder (HRL). Ladder fragment sizes are indicated in base pairs. Please note that the theoretical size of a PCR product without an insert is 117 bp

2. Prepare a "gel crushing tube," by making two holes on the bottom of 0.5 mL microfuge tube with an 18-gauge needle.
3. Place the gel crushing tube inside a fresh 2 mL microfuge tube.
4. Place the cutout gel fragment inside the gel crushing tube and centrifuge at 6,000 × *g*. At the end of this process, the gel fragment will be crushed and extruded into the 2 mL tube.
5. Discard the 0.5 mL crushing tube as radioactive waste. Add 600 μL of RNA PAGE elution buffer to the 2 mL tube containing the crushed gel fragments. Incubate with end-over-end tumbling for 3–4 h at room temperature to elute the adapter-ligated RNA fragments.
6. Transfer the slurry to a Spin-X tube and centrifuge at 16,000 × *g* for 5 min or until the majority of the solution enters the collection tube.

7. Phenol/chloroform extract the flow-through material: Add 400 μL of water-saturated phenol, pH 6.6 and 130 μL of chloroform. Vigorously shake the samples for 15 s, collect the aqueous phase, add 1 μL glycogen (20 mg/mL), and split the volume equally into two tubes (~300 μL/tube). Discard phenol/chloroform as radioactive organic waste.
8. Add 900 μL of ice-cold 100 % ethanol. Precipitate the samples overnight at −20 °C.
9. Collect the precipitated ligated RNA mixture by centrifugation at maximum speed at 4 °C for 15 min.
10. Discard the supernatant as radioactive waste. Wash the pellet twice with 0.5 mL of ice-cold 75 % ethanol.
11. Let the pellet dry for 5–10 min at room temperature or until no more ethanol can be detected. Do not over-dry the pellet. Proceed immediately to Subheading 3.12.

3.12 Reverse Transcriptase (RT) Reaction and PCR Using Illumina's TruSeq Small RNA Sample Prep Kit

1. Resuspend the RNA pellets in 6 μL water. Transfer the solution in a 0.2 mL tube.
2. Perform the RT reaction as follows:
   (a) Add 1 μL of the RTP to 6 μL of the RNA, mix, and incubate for 2 min at 70 °C. Place immediately on ice.
   (b) Prepare 12.5 mM dNTPs by mixing 0.5 μL of water with 0.5 μL 25 mM dNTPs provided with the Illumina kit.
   (c) In a separate tube mix the following ("enzyme mix"):
      - 2 μL 5× Buffer (Illumina).
      - 1 μL 0.1 M DTT.
      - 0.5 μL of 12.5 mM dNTPs.
      - 1 μL SUPERase · In.
      - 1 μL SuperScript III Reverse Transcriptase.
   (d) Add 5.5 μL of the enzyme mix to the RNA/RTP sample. The total volume will be now 12.5 μL.
   (e) Incubate the samples as follows using a thermocycler: 25 °C for 5 min; 42 °C for 20 min; 50 °C for 1 h; and 4 °C unlimited.
3. Prepare PCR reactions using the indexed RPI1 and RP1 primers. Be sure to record the RPI1 indices for each corresponding sample. A "master mix" can be prepared that includes the PCR master mix from the Illumina kit and RP1 primer (prepare at least 10 % extra "master mix" to account for pipetting errors):

(a) TruSeq small RNA kit PCR mix	25 µL
(b) 10 µM RP1 primer	2 µL
(c) 10 µM RPI1 primer	2 µL
(d) cDNA total reaction	12.5 µL
(e) Water	8.5 µL

PCR cycle conditions:

1×	98 °C, 1 min
25–28×	98 °C, 15 s 60 °C, 30 s 72 °C, 30 s
1×	72 °C, 10 min
1×	Hold at 10 °C

3.13 Purify the PCR Product Using Agencourt AMPure XP Beads

The purification step is important for concentrating the sample and reducing the salt concentration, which otherwise would interfere with the electrophoretic mobility of the DNA. Alternatively, this step may be omitted and the samples can be treated as recommended in the Illumina manual for the TruSeq small RNA prep kit. However, we recommend performing this concentration and purification step.

1. Bring the AMPure XP magnetic beads to room temperature (about 30 min).
2. Resuspend the beads by vortexing briefly.
3. Transfer the PCR reaction mix (Subheading 3.12) to a 1.7 mL tube, add 90 µL of AMPure beads, and mix by pipetting gently ten times.
4. Incubate for 10 min at room temperature.
5. Place the tube on the magnetic concentrator stand for 2 min.
6. Gently remove the clear solution using a pipette; dispose of as radioactive waste.
7. Add 200 µL of 80 % ethanol, being careful not to disturb the beads. Remove the wash solution using a pipette.
8. Repeat the 80 % ethanol wash for a total of two times.
9. Allow the pellet to dry with an open lid on the magnetic stand (about 5–10 min or until the pellet looks dry).
10. Resuspend the dried beads in 10 µL of 10 mM Tris–HCl, pH 8.0. Ensure that the beads are in suspension by vortexing or pipetting up and down.
11. Incubate at room temperature for 2 min. Spin briefly to settle the liquid.

12. Place the tube on a magnetic stand for 2 min.
13. Remove 8.5–10 μL of the clear solution and place in a new tube.
14. Add 3 μL of 6× Blue/Orange Loading Dye Buffer (Promega).

3.14 Resolve the PCR Product Using Native PAGE

1. Prepare a 6 % native polyacrylamide gel (*see* the recipe for 10 mL below). We prepare these gels using the same glass plates as for the protein mini gels (8.6×6.8 cm) with 1.00 mm thick spacers and the corresponding 10-well comb:

1 mL	10× TBE
2 mL	30 % acrylamide and bis-acrylamide solution, 37.5:1 ratio
120 μL	10 % ammonium persulfate
20 μL	TEMED
6.88 mL	Water

2. Prepare the custom RNA and high-resolution DNA ladders (supplied with the Illumina TruSeq small RNA prep kit) as recommended in the Illumina manual (Fig. 3b).
3. Load the ladders on the side and your sample in the middle (Fig. 3b). Leave empty lanes between samples to minimize cross-contamination. Perform electrophoresis at 110 V for ~1 h or until the bromophenol blue dye reaches the bottom of the gel. The majority of the PCR product will be between 145 and 220 bp, approximately 1.5–2.5 cm slower than the bromophenol blue dye when it reaches the bottom of the gel (Fig. 3b).

3.15 Extract the PCR Product with Molecular Weight ~145–220 nt from the Native Polyacrylamide Gel

1. Stain the native polyacrylamide gel with 0.5 μg/mL ethidium bromide in water.
2. Use a UV transilluminator to visualize the DNA (*see* Fig. 3b).
3. Excise the gel slices using a clean scalpel or razor blade for each sample. For CLIP products in the 35–45 nt range, excise a broad region between 145 and 220 bp. Try to make the gel piece as small as possible without sacrificing yield.
4. Prepare a "gel crushing tube" as for Subheading 3.11.
5. Place the crushing tube inside a fresh 2 mL microfuge tube.
6. Place the cutout gel fragment inside the gel crushing tube and centrifuge at 6,000 × *g* to crush the gel and elute the fragments into the larger tube.
7. Discard the 0.5 mL crushing tube. Add 400 μL of DNA PAGE elution buffer to the 2 mL tube containing the crushed gel and incubate with end-over-end tumbling for 3–4 h at room temperature.
8. Transfer the slurry to a Spin-X tube and centrifuge at maximum speed (~16,000 × *g*) until majority of the solution enters the collection tube.

9. Add 1 μL of glycogen (20 mg/mL) and 900 μL 100 % ethanol. Precipitate overnight at −20 °C.
10. Collect the precipitated PCR products by centrifugation at maximum speed at 4 °C for 15 min.
11. Wash the pellet twice with 0.5 mL cold 75 % ethanol.
12. Let the pellet dry for 5–10 min at room temperature or until no more ethanol can be detected. Do not over-dry the pellet.
13. Dissolve the pellet in 20 μL of 10 mM Tris–HCl, pH 7.9. This is the final CLIP library.
14. Assess the size and quantity of the CLIP library using an Agilent 2100 Bioanalyzer with a DNA 1000 kit. To obtain good-quality data from the sequencing on Illumina platform, a concentration of at least 1.5 nM and total of 35 ng is needed.

3.16 Illumina Sequencing of the CLIP Library

We recommend single-end (SE) sequencing with the Illumina platform for the CLIP library, performing 50 cycles of sequencing of the inserts ("50SE"). In the case of multiple samples, a multiplexing (use of several short sequence tags in the adapters to uniquely identify different samples) can be performed. 1 Gb (~20 million reads) of clean data (de-multiplexed and adapter sequences removed) is usually sufficient for analysis of a single CLIP experiment. Currently, there are many commercial vendors who will perform the sequencing for a reasonable fee. Before choosing a specific vendor, consider the following: the platform that they are using for sequencing, the physical location of their facility (which will affect shipping costs), the turnaround time, and, last but not least, the price.

3.17 Bioinformatic Analysis of the CLIP Library

A detailed workflow of bioinformatic analysis of CLIP data is beyond the scope of this chapter and requires specialized skills [13, 17]. As a general outline, the sequencing CLIP data is first aligned to a reference genome (i.e., of the organism from which you obtained your samples, the *Mus musculus* genome in our example) and any ambiguous localizations of reads removed. Next, filter all identical RNA sequences, thus eliminating duplicates that may have arisen during the PCR steps. Thus filtered, sequence reads should be analyzed with regard to their annotated locations in the genome—e.g., in exons, introns, intergenic sequences, and 5′ or 3′ UTRs. Any further analysis probably needs to be tailored to the characteristic of your specific RNA-binding protein and to your hypotheses. Also (and way beyond the scope of this chapter), you should consider comparing your HITS-CLIP data to RNA-seq [17, 18] or high-throughput technologies to identify polyadenylation site [13, 19] data from the same samples. Those correlations can be used to address functional aspects and suggest new hypotheses to test.

4 Notes

1. For more information please refer to the Illumina custom sequence letter as of September 7, 2012, available through their website for free after registration (https://www.illumina.com/). Oligonucleotide sequences © 2007–2012 Illumina, Inc. All rights reserved [20].
2. Generally, the amount of tissue to consider sufficient for a CLIP procedure needs to be in the range that the radioactivity from the ^{32}P-labeled RNAs can be visualized in an overnight exposure on X-ray film as indicated in Subheading 3.7. In our experience, 1–2 testes (about 80 mg each) obtained from a single 25-day-old male animal were sufficient for the CLIP procedure with 3A7 antibody. However, pilot experiments should be performed to optimize the amount of the starting material.
   (a) In our experience the physical disruption of the tissue by vigorous shaking, triturating, and pipetting techniques is not problematic and lead to efficient cross-linking between the CstF-64 protein and RNA. Since the UV can penetrate several layers of cells obtaining single-cell suspension may not be necessary. Some tissues (e.g., liver, heart, muscle) may be more resistant to physical disruption. Therefore, an enzymatic digestion (e.g., by collagenase) of the tissue may be considered. If you work with a tissue requiring enzymatic digestion, use appropriate techniques to prevent uncontrolled digestion of the RNA.
   (b) We determined that irradiating thrice with the 200 mJ/cm^2 254 nm UV source was optimal by performing pilot experiments to examine patterns of cross-linked ^{32}P-labeled proteins (*see* Fig. 2b). RNA-binding proteins may have different RNA-interacting surfaces and therefore different cross-linking efficiencies. Therefore, we recommend determining the optimal cross-linking conditions for each RNA-binding protein empirically.

   We also recommend optimizing the initial immunoprecipitation conditions by trying different detergents (RIPA, NP-40, etc.), salts, and times. Determine the yield by Western blot. Once the immunoprecipitation conditions are determined, we suggest proceeding and performing a pilot CLIP experiment to verify that the conditions are indeed optimal for CLIP. If you have any doubts about the conditions, you can further optimize them by performing CLIP up to Subheading 3.8, which visualizes the protein–RNA complexes (Fig. 2b). In this protocol, we exclusively describe immunoprecipitation conditions that work best with the 3A7 antibody, which recognizes the mammalian CstF-64 protein.

3. You can prepare the antibody-coupled Dynabeads Protein G a few days ahead by washing once with 1 mL low-salt immunoprecipitation buffer and storing the coupled antibody in low-salt immunoprecipitation buffer supplemented with 0.05 % sodium azide at 4 °C. Prior to the immunoprecipitation wash the Dynabeads Protein G twice with 1 mL low-salt immunoprecipitation buffer.
4. PMSF has a short half-life in aqueous solution. Therefore, rather than preparing a stock PMSF solution, we find that adding a "pinch" of powder (enough to cover the tip of a metal spatula, ~5–10 mg) is sufficient to saturate a 10 mL aqueous solution. In our experience PMSF was sufficient to prevent degradation of CstF-64 in our testis samples. In tissues with greater endogenous proteases, consider using a cocktail of proteinase inhibitors.
5. RNase dilutions were prepared as follows: to 499 and 49 μL of low-salt immunoprecipitation buffer 1 μL of RNase I (*see* Subheading 2.4) was added to create the low- and high-RNase solutions, respectively. Tubes with the corresponding dilutions are vortexed and briefly spun down to clear the droplets.
6. Heating at 75 °C is an antibody-specific step and many antibody–protein interactions will not benefit from it. Because the 3A7 monoclonal antibody recognizes an epitope [21] that is revealed upon denaturation [11], we find that heating the lysate prior to immunoprecipitation increases yield while decreasing background.
7. We recommend the use of the Thermomixer. In lieu of using the Thermomixer, one could agitate the samples on a rocker platform at 37 °C.
8. We recommend using Novex® NuPAGE® SDS-PAGE Gel System from Life Technologies. The use of the NuPAGE system will greatly reduce variability during the SDS-PAGE electrophoresis. However, if you cannot obtain the Novex® NuPAGE® SDS-PAGE Gel System, homemade gels and buffers can be prepared and used according to the protocols published by the laboratory of Dr. Robert T. Sauer at the Department of Biology at the Massachusetts Institute of Technology [22].
9. Exposure times will vary greatly depending on the amount of the starting material, specific activity of the radioactivity, and immunoprecipitation efficiency of the antibody. However, if it takes longer than overnight at −80 °C to visualize the signal on an X-ray film with a fresh radioactivity, you should consider increasing the amount of the starting tissue/cells (*see* **Note 2**).

10. The size of the RA3 and RA5 adapters ligated to each other is 47 nt. If one wants to obtain mainly insert sizes of 13 nt one should cut a band between 58 and 62 nt. However, in practice it is much better to collect as much material as possible above 90–100 nt. In our experience, if we do not perform this step, majority of our products contain adapters ligated to each other. In addition, nucleic acid sequences smaller than 20–30 nt are difficult to align unequivocally to the reference genome.

References

1. Änko ML, Neugebauer KM (2012) RNA-protein interactions in vivo: global gets specific. Trends Biochem Sci 37:255–262
2. Darmon SK, Lutz CS (2012) mRNA 3′ end processing factors: a phylogenetic comparison. Comp Funct Genomics 2012:876893
3. Millevoi S, Vagner S (2010) Molecular mechanisms of eukaryotic pre-mRNA 3′ end processing regulation. Nucleic Acids Res 38: 2757–2774
4. Tian B, Graber JH (2012) Signals for pre-mRNA cleavage and polyadenylation. Wiley Interdiscip Rev RNA 3:385–396
5. MacDonald CC, Wilusz J, Shenk T (1994) The 64-kilodalton subunit of the CstF polyadenylation factor binds to pre-mRNAs downstream of the cleavage site and influences cleavage site location. Mol Cell Biol 14:6647–6654
6. Murthy KGK, Manley JL (1995) The 160-kD subunit of human cleavage-polyadenylation specificity factor coordinates pre-mRNA 3′-end formation. Genes Dev 9:2672–2683
7. Brown KM, Gilmartin GM (2003) A mechanism for the regulation of pre-mRNA 3′ processing by human cleavage factor I_m. Mol Cell 12:1467–1476
8. Ule J, Jensen KB, Ruggiu M et al (2003) CLIP identifies Nova-regulated RNA networks in the brain. Science 302:1212–1215
9. Licatalosi DD, Mele A, Fak JJ et al (2008) HITS-CLIP yields genome-wide insights into brain alternative RNA processing. Nature 456:464–469
10. Martin G, Gruber AR, Keller W et al (2012) Genome-wide analysis of pre-mRNA 3′ end processing reveals a decisive role of human cleavage factor I in the regulation of 3′ UTR length. Cell Rep 1:753–763
11. Takagaki Y, Manley JL, MacDonald CC et al (1990) A multisubunit factor CstF is required for polyadenylation of mammalian pre-mRNAs. Genes Dev 4:2112–2120
12. Wallace AM, Dass B, Ravnik SE et al (1999) Two distinct forms of the 64,000 M_r protein of the cleavage stimulation factor are expressed in mouse male germ cells. Proc Natl Acad Sci U S A 96:6763–6768
13. Hoque M, Ji Z, Zheng D et al (2012) Analysis of alternative cleavage and polyadenylation by 3′ region extraction and deep sequencing. Nat Methods 10(2):133–139
14. Yao C, Biesinger J, Wan J et al (2012) Transcriptome-wide analyses of CstF64-RNA interactions in global regulation of mRNA alternative polyadenylation. Proc Natl Acad Sci U S A 109:18773–18778
15. Ule J, Jensen K, Mele A et al (2005) CLIP: a method for identifying protein-RNA interaction sites in living cells. Methods 37:376–386
16. Illumina Inc (2012) Illumina website. Available from http://www.illumina.com/ Accessed 15 Jan 2013
17. Li W, Yeh HJ, Shankarling GS et al (2012) The τCstF-64 polyadenylation protein controls genome expression in testis. PLoS One 7:e48373
18. Wang Z, Gerstein M, Snyder M (2009) RNA-Seq: a revolutionary tool for transcriptomics. Nat Rev Genet 10:57–63
19. Derti A, Garrett-Engele P, Macisaac KD et al (2012) A quantitative atlas of polyadenylation in five mammals. Genome Res 22:1173–1183
20. Illumina Inc (2012) Customer letter. Available from https://www.illumina.com/ Accessed 15 Jan 2013
21. Dass B, McDaniel L, Schultz RA et al (2002) The gene CSTF2T encoding the human variant CstF-64 polyadenylation protein τCstF-64 is intronless and may be associated with male sterility. Genomics 80:509–514
22. OpenWetWare contributors (2009) Sauer: bis-Tris SDS-PAGE, the very best. Available from http://openwetware.org/index.php?title=Sauer:bis-Tris_SDS-PAGE%2C_the_very_best&oldid=300293 Accessed 15 Jan 2013
23. Dass B, Tardif S, Park JY et al (2007) Loss of polyadenylation protein τCstF-64 causes spermatogenic defects and male infertility. Proc Natl Acad Sci U S A 104:20374–20379

Part II

Bacteria and Organelles

Chapter 18

Polyadenylation in Bacteria and Organelles

Joanna Rorbach, Agnieszka Bobrowicz, Sarah Pearce, and Michal Minczuk

Abstract

Polyadenylation is a posttranscriptional modification present throughout all the kingdoms of life with important roles in regulation of RNA stability, translation, and quality control. Functions of polyadenylation in prokaryotic and organellar RNA metabolism are still not fully characterized, and poly(A) tails appear to play contrasting roles in different systems. Here we present a general overview of the polyadenylation process and the factors involved in its regulation, with an emphasis on the diverse functions of 3′ end modification in the control of gene expression in different biological systems.

Key words Polyadenylation, PAP I, Bacteria, Chloroplasts, Mitochondria, RNA degradation

1 Introduction

Polyadenylation is a posttranscriptional modification found at the 3′ end of RNA molecules that involves the non-templated extension of RNA consisting of primarily adenosine monophosphate (AMP) residues, known as poly(A) tails. Whilst such poly(A) tails are often homopolymeric, heteropolymeric tails have also been characterized that are primarily adenine rich but also contain other ribonucleotides.

The process of polyadenylation is present across the kingdoms of life in almost all species studied. However, the function that poly(A) tails play in RNA metabolism can vary. To date, perhaps the best characterized poly(A) tails are those found on messenger RNAs (mRNAs) in the eukaryotic cytoplasm. Almost all eukaryotic nuclear-encoded mRNAs rely on the presence of a homopolymeric poly(A) tail for nuclear export, stability, and ability to be translated. In contrast, RNA polyadenylation in prokaryotes has been linked primarily to signalling RNA molecules, including mRNAs, to undergo degradation, although this process remains largely uncharacterized.

Joanna Rorbach and Agnieszka J. Bobrowicz (eds.), *Polyadenylation: Methods and Protocols*, Methods in Molecular Biology, vol. 1125, DOI 10.1007/978-1-62703-971-0_18, © Springer Science+Business Media New York 2014

Mitochondria and chloroplasts are thought to have originated from endosymbiotic relationships between early eukaryotes and ancestors of alpha-proteobacteria or cyanobacteria, respectively [1]. Both organelles have retained a much reduced form of their own genomes and rely on their own transcription and translation systems, independent of those of the cytoplasm, for synthesis of certain key proteins. Although polyadenylation of RNA species in these organelles has been shown, exactly how poly(A) tails control RNA levels appears to be contrasting between organelles and far from fully understood. Whilst our knowledge into the diverse roles that polyadenylation plays in coordinating RNA metabolism in prokaryotes and organelles of bacterial origin remains limited compared to that of the eukaryotic cytoplasm, we discuss here the current understanding of the field and the protein factors involved in the production and degradation of poly(A) tails in these different systems.

2 Enzymes Involved in Polyadenylation and Deadenylation

2.1 Poly(A) Polymerases

The best characterized bacterial enzyme responsible for synthesizing RNA poly(A) tails is the poly(A) polymerase I (PAP I) from *Escherichia coli*. PAP I is encoded by the *pcnB* gene and belongs to the tRNA nucleotidyltransferase/poly(A) polymerase family. It is characterized by two specific domains: poly A polymerase head domain (Pfam: PF01743) and polymerase A arginine-rich C-terminus (Pfam: PF12626). Bioinformatic analyses actually suggest that bacterial PAP descends from tRNA CCA-adding enzymes (tRNA nucleotidyltransferases) [2, 3]. It catalyzes the template-independent addition of AMP moieties to 3′-OH groups of RNA using ATP as a substrate. In vitro PAP I is active as a monomer [4] and requires neither RNA *cis* elements nor auxiliary factors for its polymerization activity, although other proteins are likely to regulate PAP I function in vivo (see other sections). Most of the interactions of PAP I with other proteins are transient in nature [5]. PAP I can inefficiently incorporate nucleotides other than AMP in vitro [6]; however, in vivo the enzyme exclusively synthesizes homopolymeric poly(A) tails [7].

Homologues of bacterial PAP I belonging to the tRNA nucleotidyltransferase/poly(A) polymerase family have been shown to localize to chloroplasts of flowering plants and green algae, e.g., AtNtr1 and CrPAP4, respectively [8]. These enzymes display polyadenylation activity in vitro and when expressed in *E. coli* deletion strains lacking PAP I (Δ*pcnB*) produce homopolymeric poly(A) tails. In addition, the presence of homopolymeric poly(A) tails in *Arabidopsis* and other plant chloroplasts suggests the activity of a PAP I-like enzyme [9]. However, the physiological role of these extensions remains to be elucidated.

In mitochondria, the nature of enzymes responsible for adding homopolymeric poly(A) tails varies considerably among the tree of life. Yeast *Saccharomyces cerevisiae* mitochondria lack a PAP homologue, consistent with no poly(A) tails being detected on mitochondrial RNAs [10]. Some homologues of bacterial-type tRNA nucleotidyltransferase/poly(A) polymerase-like PAP have been localized to plant mitochondria, but nothing is known about their roles in the organelle. Despite the postulated common endosymbiotic origin of mitochondria, the organelles found in protists and animals contain a poly(A) polymerase that differs from bacterial-type PAP and belongs to the Cid1-like family. This family is named after Cid1, the first member of this class of noncanonical PAPs discovered in fission yeast [11]. Family members are characterized by the presence of the nucleotidyltransferase domain (Pfam: PF01909) and the PAP-associated domain (Pfam: PF03828). Similarly to the bacterial enzymes, Cid1-like proteins catalyze the addition of AMP to 3′-hydroxyl termini of RNA in a template-independent manner using ATP as a substrate. However, many members of this family, including the predecessor Cid1, also have a poly(U) polymerase activity. Both *Trypanosoma* and human mitochondrial poly(A) polymerases, kinetoplast poly(A) polymerase 1 (KPAP1, also known as TUT5) and mitochondrial poly(A) polymerase (MTPAP, also known as PAPD1 and TUTase1), respectively, have been identified based on homology searches of sequence databases [12–14]. In vitro KPAP1 exhibits great specificity towards ATP and is unable to perform polymerization using UTP as a substrate. MTPAP can utilize all four ribonucleotides for polymerization, although it is more active with ATP or UTP [15]. Structural studies of MTPAP revealed that its N-terminus contains an RNA recognition motif (RRM)-like (RL) domain. The RL domain contributes to dimerization of MTPAP, which is required for its catalytic activity. It is unknown whether the RL domain itself contributes to RNA binding or whether a separate protein is required to mediate this interaction [15, 16].

2.2 Polynucleotide Phosphorylases

Polynucleotide phosphorylase (PNPase) was first discovered as an RNA-synthesizing enzyme [17]. However, PNPases can act either as 3′-terminal RNA polymerases or as phosphorolytic exoribonucleases, degrading RNA in the 3′–5′ direction. The type of reaction performed by PNPases depends on the availability of inorganic phosphate (Pi). At high NDP and low Pi concentrations the synthesis of untemplated heteropolymeric tails is favored, whereas low NDP and high Pi concentrations favor the exoribonucleolytic degradation of transcripts [18]. PNPases generally contain two tandem copies of the RNase PH domain (Pfam: PF01138), the S1 (Pfam: PF00575) and KH (Pfam: PF00013) RNA-binding domains. The crystal structure of bacterial PNPase [19] shows that its homotrimer forms a ring of six RNase PH domains, one side of which has a cap that contains the RNA-binding domains. This ringlike structure of RNase PH domains is also conserved in RNA-degrading machines from eukaryotic and archaeal exosomes [20, 21]. It has been well documented that PNPase participates in the degradation, processing, and polyadenylation of RNA in bacteria and organelles. Most bacterial species have homologues of PNPase [9]. PNPase is also directly involved in RNA metabolism in plant organelles [9]. However, the involvement of PNPase in degradation/polyadenylation of RNA in animal mitochondria remains controversial [22–24].

2.3 Other Ribonucleases

Several endo- and exoribonucleases assist the poly(A) metabolism in bacterial and organellar systems. The main enzymes are briefly characterized in this section.

Endonuclease RNase E is an important player in the polyadenylation-stimulated degradation pathway in bacteria and chloroplasts. Bacterial RNase E, together with PNPase, is an important part of the macromolecular bacterial complex involved in RNA processing and decay known as the RNA degradosome. RNase E is a large multidomain protein with an N-terminal catalytic region and a C-terminal non-catalytic part that contains sites for binding RNA and for protein–protein interactions with other components of the RNA degradosome [25]. Other endoribonucleases, e.g., RNase J and CSP41a/b, also participate in bacterial and chloroplast poly(A)-assisted RNA decay, respectively. However, no RNase E- or RNase J-type enzymes have been identified in plant mitochondria, despite RNA degradation system resembling that of bacteria and chloroplasts. In addition, no such proteins have been identified in animal mitochondria.

The RNR-type exoribonucleases (e.g., *E. coli* RNase II and RNase R) that hydrolytically degrade RNA in the 3′–5′ direction are also involved in polyadenylation-stimulated RNA decay pathways [26]. The RNR exoribonuclease family members have been identified in bacteria, plant organelles, and also mitochondria of protists [9]. However, no homologue of RNase II/R has been identified in animal mitochondria thus far.

Poly(A)-specific ribonucleases (deadenylases) control changes to the length of mRNA poly(A) tails in response to various stimuli. No deadenylases have been described in bacteria and chloroplasts, likely owing to the unstable character of poly(A) tails. Thus far, PDE12 (2′-PDE) is the only identified mitochondrial deadenylase [27, 28]. PDE12 belongs to the endonuclease/exonuclease/phosphatase (EEP) family (Pfam: PF03372). In vitro PDE12 displays specificity for RNA 3′-ends containing a free hydroxyl group and does not possess 5′ → 3′ exoribonucleolytic or endoribonucleolytic activity. The enzyme prefers homopolymeric substrates containing adenosine or uracil, whereas cytosine- or guanine-homopolymers are poorly degraded. Incubation of recombinant PDE12 with polyadenylated mRNA results in the removal of the poly(A) tail, and the reaction stops 6–8 nt upstream from the mRNA–poly(A) boundary [27].

2.4 Other Factors

Several regulatory proteins are involved in the poly(A) metabolism in bacteria and organelles. Hfq, an RNA-binding protein of ring-like oligomeric architecture, is one of the key bacterial factors that influence turnover rates of specific transcripts. Several mechanisms by which Hfq can control regulatory responses in vivo have been proposed, including stimulation of PAP I activity, and they are described in detail elsewhere [29]. Eukaryotic cytosolic poly(A)-binding proteins are often involved in the protection of transcripts and interaction with other proteins. Several prokaryotic proteins, including Hfq and CspE, have been reported to bind poly(A) in vitro; however, there is no evidence for their poly(A)-binding activity in vivo [30, 31]. A member of the poly(A)-binding family of proteins, RB47, has been identified in chloroplasts. However, it has been shown that RB47 binds to the structured 5′ UTR, rather than the poly(A) tail, in order to regulate translation of the psbA mRNA (coding for the D1 polypeptide of the photosystem II reaction center) [32]. Recently, we have demonstrated that an isoform of PABPC5, a member of the poly(A)-binding protein family, is present in human mitochondria. There is no data regarding whether or not PABPC5 binds to polyadenylated mitochondrial transcripts in vivo. However, the protein co-immunoprecipitates with MTPAP, suggesting that it plays a role in RNA metabolism in the organelle [33].

Several proteins of the pentatricopeptide repeat (PPR) family have been shown to play important regulatory roles in polyadenylation-related events in organelles. The PPR family is characterized by the presence of a 35-amino acid structural motif that is tandemly repeated 2–26 times per protein; the PPR domains appear to be involved in RNA–protein interactions. The vast majority of members of the PPR family are plant proteins, and nearly all localize to chloroplasts or mitochondria. Many PPR proteins are known to play various roles in RNA metabolism [34]. It has recently been shown that a heterodimer of PPR proteins, named kinetoplast polyadenylation/uridylation factors (KPAFs) 1 and 2, stimulates polyadenylation by KPAP1 and regulates mitochondrial translation in protists [16]. It has also been recently suggested that a PPR protein, LRPPRC, is important for RNA polyadenylation in mammalian mitochondria [35, 36].

3 Role of Polyadenylation

3.1 Regulation of RNA Stability

3.1.1 Bacteria

Although the bacterial polymerase activity was described over 50 years ago [37], for many years it was believed that polyadenylation was unique to eukaryotic cells, with only a minor role in bacterial RNA metabolism. The facts that prokaryotic poly(A) tails are much shorter, that mRNA turnover is rapid, and that only a small fraction of the total mRNA population bear poly(A) tails made the detection and analysis of polyadenylation in bacteria much more difficult than in eukaryotic cells. The first indication of the function of bacterial polyadenylation came from studies in *E. coli* where mutations in the PAP I-coding gene (*pcnB*) resulted in reduction of the copy number of ColE1 plasmid. It was shown that polyadenylation facilitated regulation of the plasmid copy number by controlling the steady-state levels of RNAI, a noncoding RNA repressor of plasmid replication. Lack of RNAI polyadenylation in the *pcnB* mutant led to stabilization of RNAI [38, 39]. Later, it was discovered that when exoribonucleases responsible for degradation of the poly(A) tail, such as PNPase and RNAse II, are absent from the cell polyadenylated RNA could accumulate to significant levels [40–42]. Polyadenylation has since been established to be involved in the degradation of various transcripts in *E. coli* [43, 44].

Addition of poly(A) tails can occur not only on the mature mRNAs but also on RNA fragments that are products of exo- and endonucleolytic cleavage, creating a single-stranded extension to facilitate binding of ribonucleases to the 3′ end of the RNA molecule. Exonucleases, like PNPase, can degrade RNA molecules containing weak secondary structure. However, when the enzyme encounters a strong secondary structure its action is blocked and it dissociates from the substrate. The tailless RNAs can then be readenylated, and the exonuclease can reinitiate its action. It is now evident that polyadenylation is a dynamic process and cycles of polyadenylation and exonucleolytic degradation are needed to overcome accumulation of the structured decay intermediates [45–48].

Thus far, most studies into the role of bacterial polyadenylation have been performed in *E. coli* models, and only limited data is available for other bacterial species. Still, studies on Gram-positive *Streptomyces coelicolor* and *Bacillus subtilis* showed association of the poly(A) and heteropolymeric tails with RNA degradation intermediates, confirming a common role of polyadenylation in RNA decay in bacteria [49, 50]. Polyadenylation activity has also been detected in cyanobacteria [51] that are considered to be related to the chloroplast's ancestor [52]. In this case, heterogeneous tails were found on mRNAs as well as rRNA and the single intron located at the tRNA(fmet) [51].

3.1.2 Chloroplasts

Organelles such as mitochondria and chloroplasts are believed to originate from prokaryotes [1]. However, the nature and polyadenylation patterns of the 3′ end of their respective RNA species are very diverse, suggesting different evolutionary pathways. In the chloroplast, most of the polyadenylation sites that were found by RT-PCR of oligo(dT)-primed chloroplast RNA were localized within the amino acid-coding region of the mRNA rather than at the mature 3′ end [53, 54]. This implies that, similar to bacteria, polyadenylation is part of the RNA decay system and targets endonucleolytic cleavage products for degradation. In contrast to the relatively short tails of bacterial RNA (40–60 nucleotides), tails in spinach chloroplast RNA are much longer (reaching almost 300 nucleotides). They usually contain 70 % adenosines, 25 % guanosines, and 5 % cytidines and uridines, suggesting that PNPase rather than PAP is the major polyadenylation enzyme in spinach chloroplast [54, 55].

3.1.3 Mitochondria

In plant mitochondria, polyadenylation also plays an important role in RNA decay. The process has been studied in the mitochondria of several different plant species, revealing that most of the mature mRNAs are not polyadenylated, while the polyadenylated transcripts are of very low abundance, possibly due to their rapid degradation [9, 56, 57]. The observed poly(A) tails consist mostly of adenosines, suggesting that PAP is the enzyme responsible for their addition.

Yeast mitochondria are the only organelles where mRNAs are not polyadenylated. Only oligo(A) tails of up to ~8 nt are present on yeast mitochondrial transcripts, but the role of this modification and the enzyme that produces such tails remain uncharacterized [58]. An AU-rich dodecamer sequence at the 3′ end of the transcripts is required for their stability [59]. There is no PNPase present in the organelle, and mRNAs are degraded by the mitochondrial degradosome (mtEXO), a protein complex consisting of SUV3 helicase and DSS1 exoribonuclease [60].

In trypanosomes, polyadenylation is related to the editing status of the transcripts and has different effects at various stages of processing. Mitochondrial mRNAs can either have short (20–50) or long (200–300) tails. Pre-edited mRNAs contain only short tails, while edited and never-edited molecules can both either have short or long A/U heteropolymeric tails. The studies of mRNA stability *in organello* [61, 62] and RNA degradation activities in submitochondrial fractions [63, 64] indicated that the short poly(A) tails protect edited mRNAs against 3′–5′ degradation. On the contrary, the short tails stimulate degradation of pre-edited mRNAs in vitro [63, 64]. In the following studies, it was shown that inactivation of mitochondrial (kinetoplast) poly(A) polymerases (KPAP1) leads to a rapid decay of never-edited and fully edited mRNAs, whereas pre-edited mRNAs are not destabilized by the lack of 3′ poly(A) tails [12]. These findings indicate that short poly(A) tails can "switch" their function from destabilizing or neutral in pre-edited mRNAs to stabilizing upon initiation of editing [65].

The situation is even more complicated in mammalian mitochondria, where the effect of polyadenylation on RNA stability varies between individual transcripts. The mammalian mitochondrial genome encodes only 13 proteins, all of which are essential components of the oxidative phosphorylation system. Stable poly(A) tails with the average length of 45 nt are observed on 12 of the 13 transcripts [66]. Inactivation of MTPAP has led to varied effects on the stability of the transcripts. When MTPAP gene silencing was performed via RNAi, the poly(A) tail of each mRNA was shortened. However, the steady-state levels of several mRNAs (e.g., ND1, ND2, ND3, and CYTB) were unchanged or increased, whereas steady-state levels of others, including COX1 and COX2, were reduced [13, 14, 23]. In the case where deadenylation of mitochondrial mRNAs was induced by targeting cytosolic deadenylase (PARN) to mitochondria or by overexpressing mitochondrial deadenylase PDE12, the steady-state level of some transcripts (COX1 and COX2) mRNAs decreased, while mRNA levels of others (e.g., ND1, ND2) increased [27, 67]. Interestingly, in addition to the stable 3′-end poly(A) tails, the existence of internally polyadenylated RNAs that may represent transient degradation intermediates has been detected at very low levels in human mitochondria [68]. Recently, a study using deep sequencing-based method for quantitative and global analysis of RNA polyadenylation (PAS-Seq) confirmed that polyadenylated truncated mRNAs and rRNA are widespread in human mitochondria [69]. These observations suggest that both stabilizing and destabilizing types of polyadenylation can coexist in the mammalian organelles.

3.2 RNA Quality Control

As demonstrated by many studies, addition of poly(A) tails in *E. coli* promotes degradation of transcripts [44, 47, 70]. This function is not unique to mRNA, and polyadenylated forms of many stable RNA precursors could be found in *E. coli* when RNA maturation was slowed due to the absence of processing exoribonucleases [71]. Moreover, later it was shown that polyadenylation is directly responsible for elimination of defective stable RNAs [72]. A temperature-sensitive mutant tRNATrp does not accumulate to normal levels in *E. coli* as its precursor is rapidly degraded. The cells lacking poly(A) polymerase accumulate large amounts of the defective precursor, suggesting a quality control mechanism for defective tRNAs [72]. Interestingly, a recent study has shown that only a small fraction of the polyadenylated tRNA species appears to be degraded via a general quality control process. In the absence of both RNase T and RNase PH, the two major tRNA 3′ processing ribonucleases, the majority of the transcripts become polyadenylated by PAP I, generating significant pool of nonfunctional tRNAs. Most of these species are not degraded but instead very slowly processed by the remaining tRNA-processing enzymes such as RNase D, RNase BN, or RNase II, suggesting a role of polyadenylation in regulating the intracellular levels of functional tRNAs in *E. coli* I [73].

3.3 Regulation of Translation

In contrast to the eukaryotic cytosol, where polyadenylation has been proven to play an essential role in translation initiation, there is still little known about the effects of poly(A) addition on protein synthesis in the bacterial system. It is possible that polyadenylation may directly control the level of expression of a gene by modulating the stability of a functional transcript. It was shown, for example, that inactivation of PAP I causes accumulation and stabilization of the glmS transcript, which leads to the overexpression of glucosamine-6-phosphate synthase (GlmS) [74]. A recent study by Mohanty et al. suggests that polyadenylation helps to regulate levels of functional tRNAs in *E. coli* [75]. Increased expression of PAP I in cells was shown to lead to rapid polyadenylation of mature tRNAs, reducing the fraction of aminoacylated tRNAs necessary for protein synthesis. This caused a dramatic inhibition of protein production and cell death. This discovery suggests that a tight regulation of PAP I levels seems to be critical not for preventing mRNA degradation, but rather for maintaining normal levels of functional tRNAs and protein synthesis in *E. coli*.

In trypanosomes, edited transcripts bearing 200–300 nucleotide-long A/U tails, but not short A tails, were found in translating ribosomal complexes. Addition of long A/U tails by KPAP1 and RET1 terminal uridyltransferase is coordinated by the heterodimer of PPR proteins KPAF1-2 and facilitates mRNA binding to the small ribosomal subunit, leading to translation initiation [16].

For 7 out of 13 protein-coding ORFs in mammalian mitochondria, stop codons encoded by mtDNA are incomplete and polyadenylation is necessary to create functional UAA stop codons. However, whether poly(A) tail addition is directly involved in regulating translation in the mitochondria is still a debated issue. Overexpression of the endogenous mitochondrial deadenylase PDE12 or mitochondrially targeted cytosolic deadenylase (PARN) caused strong inhibition of protein synthesis [27, 67]. This is further supported by the observation that targeting the cytosolic poly(A)-binding protein (PABPC1) into the mitochondria did not lead to the changes in steady-state levels of the transcripts but caused inhibition of translation. Interestingly, when mitochondrial PPR protein LRPPRC was inactivated, mitochondrial transcripts lost their poly(A) tails with no universal effect on translation, suggesting that translation of some transcripts can be effective even in the absence of mRNA polyadenylation [36].

4 Conclusions and Future Prospects

As outlined in this chapter, the complex and diverse roles of polyadenylation in prokaryotic and organellar RNA metabolism have begun to emerge and with that an importance of developing appropriate methodology. Unlike those found in the eukaryotic cytosol, bacterial and organellar poly(A) tails are heterogeneous in length, many are relatively short and unstable, and some of the 3′ additions are heteropolymeric in nature [44, 73] (*see* Table 1 for summary).

Table 1
Comparison of polyadenylation mechanisms in bacteria, mitochondria, and chloroplasts

Species		Reference
Enzymes involved in polyadenylation/3′ extension		
Bacteria	PAP I (bacterial-type poly(A) polymerase 1, *E. coli*)	[2, 3, 5, 6]
	PNPase	[7, 9, 17, 19]
Mitochondria		
Human	MTPAP (PAPD1 or TUTase1, Cid1-like mitochondrial poly(A) polymerase)	[12–14]
Protists	KPAP1 (TUT5, Cid1-like kinetoplast poly(A) polymerase 1, *Trypanosoma*)	[16]
	RET1 (*Trypanosoma*)	[99]
Plants/algae	PAP I-like enzyme (*Arabidopsis*), mtPNPase	[9]
Yeast	No poly(A) polymerase identified (*S. cerevisiae*)	[10]
Chloroplasts	AtNtr1 (PAP I homologue, flowering plants [*Arabidopsis*])	[8]
	CrPAP4 (PAP I homologue, green algae [*Chlamydomonas*])	[8]
	PNPase	[100]
Enzymes involved in degradation/regulation		
Bacteria	Degradosome: PNPase, RNase E, RhlB, enolase	[25, 101–103]
	RNAse II, RNAse R, RNase J	[26, 104, 105]
	Deadenylase: None identified	
	Other: Hfq (RNA-binding protein)	[29]
Mitochondria		
Human	Ribonuclease: mtPNPase–hSuv3 degradosome complex?	[106]
	Deadenylase: PDE12	[27, 28]
	Other: PABPC5 (poly(A)-binding protein?)	[33]
	LRPPRC, SLIRP	[35, 36, 107, 108]
Protists	Degradosome-like SUV3/DSS-1 complex	[109]
	Deadenylase: None identified	
	Other: KPAF1/2 (kinetoplast PPR proteins)	[16]
Plants/algae	Ribonuclease: mtPNPase, mtRNAse II	[8]
	Deadenylase: mtPNPase	[56]
	Other: PPR proteins	[34]
Yeast	Degradosome (mtEXO): SUV3 helicase, DSS1 exoribonuclease	[60]
Chloroplasts	Ribonuclease: RNase E, RNase J, cpPNPase	[56, 110]
	Deadenylase: cpPNPase	[56, 100]
	Other: CSP41 a/b (RNA-binding/cleavage)	[111, 112]
	PPR proteins	[34]

(continued)

Table 1 (continued)

Species		Reference
Length of 3′ tails		
Bacteria	~40–50 nt (*B. subtilis*, *E. coli*)	[44, 80]
Mitochondria		
Human	~45 nt on 12 of 13 transcripts	[66]
Protists	20–50 nt and 200–300 nt	[61–64]
Plants/algae	~50 nt (14–52 nt *cox2*, maize; 31–55 nt *atp9*, *Arabidopsis*)	[113, 114]
Yeast	AU-rich dodecamer sequence	[59]
	No poly(A) tails detected (*S. cerevisiae*)	[10]
	Oligo(A) up to ~8 nt	[58]
Chloroplasts	Up to 300 nt (spinach)	[54, 55]
Nature of 3′ extensions		
Bacteria	Homopolymeric poly(A), heteropolymeric poly(A) rich	[49–51]
Mitochondria		
Human	Homopolymeric poly(A)	[66, 67]
Protists	Homopolymeric oligo (A), heteropolymeric A/U	[16, 61–64]
Plants/algae	Homopolymeric poly(A) (*Arabidopsis*, *Chlamydomonas*)	[8]
	Heteropolymeric poly(U)-rich (*Chlamydomonas*)	
Yeast	Homopolymeric oligo(A)	[58]
Chloroplasts	Homopolymeric poly(A)	[8, 9]
	Heteropolymeric poly(A) rich	[54, 55]
Role of 3′ extensions in RNA stability		
Bacteria	Destabilizing	[44, 47, 70–74]
Mitochondria		
Human	Unclear; possibly stabilizing on some mRNAs and destabilizing on others	[14, 23, 27, 36, 67, 68, 115–119]
	Possibly destabilizing on truncated RNAs/degradation intermediates	[68, 69]
Protists	Short tails: Stabilizing on partially/fully/never-edited mRNAs; destabilizing on pre-edited mRNAs	[12, 16, 63, 64]
	Long tails: Stabilizing on fully edited mRNAs	[12]
Plants/algae	Destabilizing	[9, 56, 57]
Yeast	Role of oligo(A) unknown; dodecamer sequence required for mRNA stability	[59]
Chloroplasts	Destabilizing	[53–55]

This has presented a challenge as very short and quickly degraded 3′ tails have been difficult to analyze using existing techniques developed for studying polyadenylation in the eukaryotic cytosol.

Novel techniques that aim to overcome these difficulties are being developed or adapted for the use in prokaryotes and organelles, and several are presented in the following chapters of this section (part II, Chapters 2–5 [76–79]). Among these are methods to analyze the overall polyadenylation status of the organism or an organelle in vivo by RNA isolation and a subsequent poly(A) sizing assay or the RNAseH/Northern blot analyses that examine the presence of long versus short poly(A) extensions (part II, Chapter 2 [76]). The procedure to study poly(A) tails of specific transcripts by cloning the 5′–3′ end-ligated RNA, a technique which avoids relying on oligo(dT)-dependent cDNA amplification (part II, Chapter 2 [76]), allows characterization of either homo- or heteropolymeric 3′ extensions. The use of described methods has already helped to study the distribution and length of poly(A) tails of many pre-tRNAs, mRNAs, and rRNAs in *E. coli* [47, 73, 80, 81], regulation of pre-tRNA polyadenylation [73], and details of the role and regulation of *E. coli* PAP I [75]. Moreover, a modified version of the 5′–3′ end RNA ligation and cloning assay has been used to characterize the diverse poly(A) tails of human mitochondrial transcripts [27, 67, 82, 83]. It is worth mentioning that although the roles of the long, stable poly(A) tails of eukaryotic mRNAs have been studied extensively, a much more complex picture now emerges as short, transient, and destabilizing poly(A) extensions are observed on a variety of eukaryotic RNA substrates (for a review *see* [84]). Thus, the techniques developed for the analysis of short, unstable prokaryotic and organellar 3′ tails may also prove useful in the study of the interplay between polyadenylation and deadenylation in the eukaryotic cytosol.

Enzymatic in vitro assays for studying polyadenylase activity have been developed using fractionated *E. coli* proteins obtained after affinity chromatography on immobilized DNA (part II, Chapter 3 [77]). The simplified approach has previously been used to identify new transcription factors [85–88] and to describe the oligomeric forms of RNA polymerase in *E. coli* [89]. The modern adaptation of *E. coli* protein extract fractionation coupled with different methods of monitoring the synthesis of nascent polypeptide described in this volume (part II, Chapter 3 [77]) can be used to identify unknown protein factors associated with the polyadenylation and deadenylation machinery in bacteria and to characterize mechanisms and interactions that are essential to these processes.

Deep sequencing approaches allow global analyses of changes in the RNA processing and metabolism in different conditions or, in the case of the organelles such as mitochondria, in various tissues and cell types. The recently published comprehensive map of the human mitochondrial transcriptome is an example of the vast amount of information about the complexity of organellar gene expression that such approach can provide [90]. As the majority of prokaryotic and organellar transcriptomes are generated from relatively small genomes, the next-generation RNA sequencing methods may prove to be very convenient research tools, as they offer much more detailed analysis compared to conventional transcriptomic studies [91, 92]. Two methods for preparation of libraries for deep-sequencing analysis of mitochondrial transcriptome have been described in part II, Chapter 4 [78]: the directional RNA sequencing, which allows to study the stability and abundance of RNA produced from each strand of the mitochondrial DNA, and the parallel analysis of RNA (PARE), which can be used to identify new transcripts and to investigate mitochondrial RNA processing.

Polyadenylation in prokaryotes and organelles is an exciting subject, which is extensively studied at the moment, not least due to the recent developments in suitable research methodology. However, the process and its regulation are still far from being fully understood. Future studies could employ global analyses of prokaryotic and organellar transcriptomes to understand the signals determining the presence of 3′ extensions and to answer the questions of the abundance and length of poly(A) tails of various transcripts and the effect they have on transcript translation and stability. An optimized protocol for studying mitochondrial RNA stability without affecting the transcription process has been described in part II, Chapter 5 [79]. A range of in vitro and in vivo studies can be applied to identify and characterize all of the components of the polyadenylation/deadenylation machinery and potential complexes (e.g., the PPR proteins [93, 94], ribonucleases [27, 67], or any potential poly(A)- and RNA-binding proteins [23, 67, 95–98]) and to understand the interplay between polyadenylation and other processes in bacterial/organellar gene expression. Polyadenylation-specific techniques, some of which are presented in this volume, and assisting methods such as siRNA silencing, overexpression, knockout, or mutations of studied proteins or other changes to prokaryotic/organellar physiology can be combined to help understand the biological significance of poly(A) and polynucleotide tails in bacteria and eukaryotic organelles.

References

1. Dyall SD, Brown MT, Johnson PJ (2004) Ancient invasions: from endosymbionts to organelles. Science 304:253–257
2. Just A, Butter F, Trenkmann M et al (2008) A comparative analysis of two conserved motifs in bacterial poly(A) polymerase and CCA-adding enzyme. Nucleic Acids Res 36:5212–5220
3. Cao GJ, Sarkar N (1992) Identification of the gene for an *Escherichia coli* poly(A) polymerase. Proc Natl Acad Sci U S A 89: 10380–10384
4. Sarkar D, Park ES, Emdad L et al (2005) Defining the domains of human polynucleotide phosphorylase (hPNPaseOLD-35) mediating cellular senescence. Mol Cell Biol 25:7333–7343
5. Raynal LC, Carpousis AJ (1999) Poly(A) polymerase I of *Escherichia coli*: characterization of the catalytic domain, an RNA binding site and regions for the interaction with proteins involved in mRNA degradation. Mol Microbiol 32:765–775
6. Yehudai-Resheff S, Schuster G (2000) Characterization of the *E. coli* poly(A) polymerase: nucleotide specificity, RNA-binding affinities and RNA structure dependence. Nucleic Acids Res 28:1139–1144
7. Mohanty BK, Kushner SR (2000) Polynucleotide phosphorylase functions both as a 3′ right-arrow 5′ exonuclease and a poly(A) polymerase in *Escherichia coli*. Proc Natl Acad Sci U S A 97:11966–11971
8. Zimmer SL, Schein A, Zipor G et al (2009) Polyadenylation in *Arabidopsis* and *Chlamydomonas* organelles: the input of nucleotidyltransferases, poly(A) polymerases and polynucleotide phosphorylase. Plant J 59:88–99
9. Schuster G, Stern D (2009) RNA polyadenylation and decay in mitochondria and chloroplasts. Prog Mol Biol Transl Sci 85: 393–422
10. Groot GS, Flavell RA, Van Ommen GJ et al (1974) Yeast mitochondrial RNA does not contain poly(A). Nature 252:167–169
11. Wang SW, Toda T, MacCallum R et al (2000) Cid1, a fission yeast protein required for S-M checkpoint control when DNA polymerase delta or epsilon is inactivated. Mol Cell Biol 20:3234–3244

12. Etheridge RD, Aphasizheva I, Gershon PD et al (2008) 3′ adenylation determines mRNA abundance and monitors completion of RNA editing in *T. brucei* mitochondria. EMBO J 27:1596–1608
13. Tomecki R, Dmochowska A, Gewartowski K et al (2004) Identification of a novel human nuclear-encoded mitochondrial poly(A) polymerase. Nucleic Acids Res 32:6001–6014
14. Nagaike T, Suzuki T, Katoh T et al (2005) Human mitochondrial mRNAs are stabilized with polyadenylation regulated by mitochondria-specific poly(A) polymerase and polynucleotide phosphorylase. J Biol Chem 280:19721–19727
15. Bai Y, Srivastava SK, Chang JH et al (2011) Structural basis for dimerization and activity of human PAPD1, a noncanonical poly(A) polymerase. Mol Cell 41:311–320
16. Aphasizheva I, Maslov D, Wang X et al (2011) Pentatricopeptide repeat proteins stimulate mRNA adenylation/uridylation to activate mitochondrial translation in trypanosomes. Mol Cell 42:106–117
17. Grunberg-Manago M, Oritz PJ, Ochoa S (1955) Enzymatic synthesis of nucleic acid-like polynucleotides. Science 122:907–910
18. Mohanty BK, Kushner SR (2011) Bacterial/archaeal/organellar polyadenylation. Wiley Interdiscip Rev RNA 2:256–276
19. Symmons MF, Jones GH, Luisi BF (2000) A duplicated fold is the structural basis for polynucleotide phosphorylase catalytic activity, processivity, and regulation. Structure 8: 1215–1226
20. Liu Q, Greimann JC, Lima CD (2006) Reconstitution, activities, and structure of the eukaryotic RNA exosome. Cell 127: 1223–1237
21. Lorentzen E, Dziembowski A, Lindner D et al (2007) RNA channelling by the archaeal exosome. EMBO Rep 8:470–476
22. Rossmanith W (2012) Of P and Z: mitochondrial tRNA processing enzymes. Biochim Biophys Acta 1819:1017–1026
23. Rorbach J, Minczuk M (2012) The post-transcriptional life of mammalian mitochondrial RNA. Biochem J 444:357–373
24. Borowski LS, Szczesny RJ, Brzezniak LK et al (2010) RNA turnover in human mitochondria: more questions than answers? Biochim Biophys Acta 1797:1066–1070
25. Mackie GA (2013) RNase E: at the interface of bacterial RNA processing and decay. Nat Rev Microbiol 11:45–57
26. Arraiano CM, Matos RG, Barbas A (2010) RNase II: the finer details of the modus operandi of a molecular killer. RNA Biol 7:276–281
27. Rorbach J, Nicholls TJ, Minczuk M (2011) PDE12 removes mitochondrial RNA poly(A) tails and controls translation in human mitochondria. Nucleic Acids Res 39:7750–7763
28. Poulsen JB, Andersen KR, Kjaer KH et al (2011) Human 2′-phosphodiesterase localizes to the mitochondrial matrix with a putative function in mitochondrial RNA turnover. Nucleic Acids Res 39:3754
29. Vogel J, Luisi BF (2011) Hfq and its constellation of RNA. Nat Rev Microbiol 9: 578–589
30. Hankins JS, Denroche H, Mackie GA (2010) Interactions of the RNA-binding protein Hfq with cspA mRNA, encoding the major cold shock protein. J Bacteriol 192:2482–2490
31. Feng Y, Huang H, Liao J et al (2001) *Escherichia coli* poly(A)-binding proteins that interact with components of degradosomes or impede RNA decay mediated by polynucleotide phosphorylase and RNase E. J Biol Chem 276:31651–31656
32. Yohn CB, Cohen A, Danon A et al (1998) A poly(A) binding protein functions in the chloroplast as a message-specific translation factor. Proc Natl Acad Sci U S A 95:2238–2243
33. Kazak L, Reyes A, Duncan AL et al (2013) Alternative translation initiation augments the human mitochondrial proteome. Nucleic Acids Res 41:2354–2369
34. Schmitz-Linneweber C, Small I (2008) Pentatricopeptide repeat proteins: a socket set for organelle gene expression. Trends Plant Sci 13:663–670
35. Chujo T, Ohira T, Sakaguchi Y et al (2012) LRPPRC/SLIRP suppresses PNPase-mediated mRNA decay and promotes polyadenylation in human mitochondria. Nucleic Acids Res 40:8033–8047
36. Ruzzenente B, Metodiev MD, Wredenberg A et al (2011) LRPPRC is necessary for polyadenylation and coordination of translation of mitochondrial mRNAs. EMBO J 31:443
37. August JT, Ortiz PJ, Hurwitz J (1962) Ribonucleic acid-dependent ribonucleotide incorporation. I. Purification and properties of the enzyme. J Biol Chem 237:3786–3793
38. Xu F, Lin-Chao S, Cohen SN (1993) The *Escherichia coli* pcnB gene promotes adenylylation of antisense RNAI of ColE1-type plasmids in vivo and degradation of RNAI decay intermediates. Proc Natl Acad Sci U S A 90:6756–6760
39. He L, Soderbom F, Wagner EG et al (1993) PcnB is required for the rapid degradation of RNAI, the antisense RNA that controls the copy number of ColE1-related plasmids. Mol Microbiol 9:1131–1142

40. Coburn GA, Miao X, Briant DJ et al (1999) Reconstitution of a minimal RNA degradosome demonstrates functional coordination between a 3′ exonuclease and a DEAD-box RNA helicase. Genes Dev 13:2594–2603
41. Cohen SN (1995) Surprises at the 3′ end of prokaryotic RNA. Cell 80:829–832
42. Carpousis AJ, Vanzo NF, Raynal LC (1999) mRNA degradation. A tale of poly(A) and multiprotein machines. Trends Genet 15:24–28
43. Sarkar N (1997) Polyadenylation of mRNA in prokaryotes. Annu Rev Biochem 66:173–197
44. O'Hara EB, Chekanova JA, Ingle CA et al (1995) Polyadenylylation helps regulate mRNA decay in *Escherichia coli*. Proc Natl Acad Sci U S A 92:1807–1811
45. Coburn GA, Mackie GA (1998) Reconstitution of the degradation of the mRNA for ribosomal protein S20 with purified enzymes. J Mol Biol 279:1061–1074
46. Goodrich AF, Steege DA (1999) Roles of polyadenylation and nucleolytic cleavage in the filamentous phage mRNA processing and decay pathways in *Escherichia coli*. RNA 5:972–985
47. Mohanty BK, Kushner SR (1999) Analysis of the function of *Escherichia coli* poly(A) polymerase I in RNA metabolism. Mol Microbiol 34:1094–1108
48. Haugel-Nielsen J, Hajnsdorf E, Regnier P (1996) The rpsO mRNA of *Escherichia coli* is polyadenylated at multiple sites resulting from endonucleolytic processing and exonucleolytic degradation. EMBO J 15:3144–3152
49. Bralley P, Jones GH (2002) cDNA cloning confirms the polyadenylation of RNA decay intermediates in *Streptomyces coelicolor*. Microbiology 148:1421–1425
50. Campos-Guillen J, Bralley P, Jones GH et al (2005) Addition of poly(A) and heteropolymeric 3′ ends in *Bacillus subtilis* wild-type and polynucleotide phosphorylase-deficient strains. J Bacteriol 187:4698–4706
51. Rott R, Zipor G, Portnoy V et al (2003) RNA polyadenylation and degradation in cyanobacteria are similar to the chloroplast but different from *Escherichia coli*. J Biol Chem 278:15771–15777
52. Gould SB, Waller RF, McFadden GI (2008) Plastid evolution. Annu Rev Plant Biol 59: 491–517
53. Kudla J, Hayes R, Gruissem W (1996) Polyadenylation accelerates degradation of chloroplast mRNA. EMBO J 15:7137–7146
54. Lisitsky I, Klaff P, Schuster G (1996) Addition of destabilizing poly (A)-rich sequences to endonuclease cleavage sites during the degradation of chloroplast mRNA. Proc Natl Acad Sci U S A 93:13398–13403
55. Yehudai-Resheff S, Hirsh M, Schuster G (2001) Polynucleotide phosphorylase functions as both an exonuclease and a poly(A) polymerase in spinach chloroplasts. Mol Cell Biol 21:5408–5416
56. Lange H, Sement FM, Canaday J et al (2009) Polyadenylation-assisted RNA degradation processes in plants. Trends Plant Sci 14:497–504
57. Gagliardi D, Stepien PP, Temperley RJ et al (2004) Messenger RNA stability in mitochondria: different means to an end. Trends Genet 20:260–267
58. Yuckenberg PD, Phillips SL (1982) Oligoadenylate is present in the mitochondrial RNA of *Saccharomyces cerevisiae*. Mol Cell Biol 2:450–456
59. Butow RA, Zhu H, Perlman P et al (1989) The role of a conserved dodecamer sequence in yeast mitochondrial gene expression. Genome 31:757–760
60. Dziembowski A, Piwowarski J, Hoser R et al (2003) The yeast mitochondrial degradosome. Its composition, interplay between RNA helicase and RNase activities and the role in mitochondrial RNA metabolism. J Biol Chem 278:1603–1611
61. Militello KT, Read LK (2000) UTP-dependent and -independent pathways of mRNA turnover in *Trypanosoma brucei* mitochondria. Mol Cell Biol 20:2308–2316
62. Ryan CM, Read LK (2005) UTP-dependent turnover of *Trypanosoma brucei* mitochondrial mRNA requires UTP polymerization and involves the RET1 TUTase. RNA 11:763–773
63. Ryan CM, Militello KT, Read LK (2003) Polyadenylation regulates the stability of *Trypanosoma brucei* mitochondrial RNAs. J Biol Chem 278:32753–32762
64. Kao CY, Read LK (2005) Opposing effects of polyadenylation on the stability of edited and unedited mitochondrial RNAs in *Trypanosoma brucei*. Mol Cell Biol 25:1634–1644
65. Aphasizhev R, Aphasizheva I (2011) Mitochondrial RNA processing in trypanosomes. Res Microbiol 162:655–663
66. Temperley RJ, Wydro M, Lightowlers RN et al (2010) Human mitochondrial mRNAs-like members of all families, similar but different. Biochim Biophys Acta 1797 1081–1085
67. Wydro M, Bobrowicz A, Temperley RJ et al (2010) Targeting of the cytosolic poly(A) binding protein PABPC1 to mitochondria causes mitochondrial translation inhibition. Nucleic Acids Res 38:3732–3742
68. Slomovic S, Schuster G (2008) Stable PNPase RNAi silencing: its effect on the processing and adenylation of human mitochondrial RNA. RNA 14:310–323

69. Shepard PJ, Choi EA, Lu J et al (2011) Complex and dynamic landscape of RNA polyadenylation revealed by PAS-Seq. RNA 17:761–772
70. Hajnsdorf E, Braun F, Haugel-Nielsen J et al (1995) Polyadenylylation destabilizes the rpsO mRNA of *Escherichia coli*. Proc Natl Acad Sci U S A 92:3973–3977
71. Li Z, Pandit S, Deutscher MP (1998) 3′ exoribonucleolytic trimming is a common feature of the maturation of small, stable RNAs in *Escherichia coli*. Proc Natl Acad Sci U S A 95:2856–2861
72. Li Z, Reimers S, Pandit S et al (2002) RNA quality control: degradation of defective transfer RNA. EMBO J 21:1132–1138
73. Mohanty BK, Maples VF, Kushner SR (2012) Polyadenylation helps regulate functional tRNA levels in *Escherichia coli*. Nucleic Acids Res 40:4589–4603
74. Joanny G, Le Derout J, Brechemier-Baey D et al (2007) Polyadenylation of a functional mRNA controls gene expression in *Escherichia coli*. Nucleic Acids Res 35:2494–2502
75. Mohanty BK, Kushner SR (2013) Deregulation of poly(A) polymerase I in *Escherichia coli* inhibits protein synthesis and leads to cell death. Nucleic Acids Res 41:1757–1766
76. Mohanty BK, Kushner SR (2013) *In vivo* analysis of polyadenylation in prokaryotes. In: Rorbach J, Bobrowicz AJ (eds) Polyadenylation, Methods in molecular biology. Humana, New York
77. Encalade GA, Sukhodolets MV (2013) Polyadenylation of RNA in *E. coli*: RNA polymerase-associated (rA)n-synthetic activities. In: Rorbach J, Bobrowicz AJ (eds) Polyadenylation, Methods in molecular biology. Humana, New York
78. Rackham O, Filipovska A (2013) Analysis of the human mitochondrial transcriptome using directional deep sequencing and parallel analysis of RNA ends. In: Rorbach J, Bobrowicz AJ (eds) Polyadenylation, Methods in molecular biology. Humana, New York
79. Borowski LS, Szczesny RJ (2013) Measurement of mitochondrial RNA stability by metabolic labelling of transcripts with 4-thiouridine. In: Rorbach J, Bobrowicz AJ (eds) Polyadenylation, Methods in molecular biology. Humana, New York
80. Mohanty BK, Maples VF, Kushner SR (2004) The Sm-like protein Hfq regulates polyadenylation dependent mRNA decay in *Escherichia coli*. Mol Microbiol 54:905–920
81. Mohanty BK, Kushner SR (2006) The majority of *Escherichia coli* mRNAs undergo post-transcriptional modification in exponentially growing cells. Nucleic Acids Res 34: 5695–5704
82. Temperley RJ, Seneca SH, Tonska K et al (2003) Investigation of a pathogenic mtDNA microdeletion reveals a translation-dependent deadenylation decay pathway in human mitochondria. Hum Mol Genet 12:2341–2348
83. Chrzanowska-Lightowlers ZM, Temperley RJ, Smith PM et al (2004) Functional polypeptides can be synthesized from human mitochondrial transcripts lacking termination codons. Biochem J 377:725–731
84. Eckmann CR, Rammelt C, Wahle E (2011) Control of poly(A) tail length. Wiley Interdiscip Rev RNA 2:348–361
85. Sukhodolets MV, Jin DJ (1998) RapA, a novel RNA polymerase-associated protein, is a bacterial homolog of SWI2/SNF2. J Biol Chem 273:7018–7023
86. Sukhodolets MV, Garges S (2003) Interaction of *Escherichia coli* RNA polymerase with the ribosomal protein S1 and the Sm-like ATPase Hfq. Biochemistry 42:8022–8034
87. Sukhodolets MV, Garges S, Jin DJ (2003) Purification and activity assays of RapA, the RNA polymerase-associated homolog of the SWI/SNF protein superfamily. Methods Enzymol 370:283–290
88. Zhi H, Yang W, Jin DJ (2003) *Escherichia coli* proteins eluted from mono Q chromatography, a final step during RNA polymerase purification procedure. Methods Enzymol 370:291–300
89. Kansara SG, Sukhodolets MV (2011) Oligomerization of the *E. coli* core RNA polymerase: formation of (alpha2betabeta′omega)2-DNA complexes and regulation of the oligomerization by auxiliary subunits. PLoS One 6:e18990
90. Mercer TR, Neph S, Dinger ME et al (2011) The human mitochondrial transcriptome. Cell 146:645–658
91. German MA, Luo S, Schroth G et al (2009) Construction of parallel analysis of RNA ends (PARE) libraries for the study of cleaved miRNA targets and the RNA degradome. Nat Protoc 4:356–362
92. German MA, Pillay M, Jeong DH et al (2008) Global identification of microRNA-target RNA pairs by parallel analysis of RNA ends. Nat Biotechnol 26:941–946
93. Rackham O, Filipovska A (2011) The role of mammalian PPR domain proteins in the regulation of mitochondrial gene expression. Biochim Biophys Acta 1819:1008
94. Davies SM, Lopez Sanchez MI, Narsai R et al (2012) MRPS27 is a pentatricopeptide repeat domain protein required for the translation of

mitochondrially encoded proteins. FEBS Lett 586:3555–3561

95. Rackham O, Mercer TR, Filipovska A (2012) The human mitochondrial transcriptome and the RNA-binding proteins that regulate its expression. Wiley Interdiscip Rev RNA 3: 675–695
96. Shutt TE, Shadel GS (2010) A compendium of human mitochondrial gene expression machinery with links to disease. Environ Mol Mutagen 51:360–379
97. Temperley R, Richter R, Dennerlein S et al (2010) Hungry codons promote frameshifting in human mitochondrial ribosomes. Science 327:301
98. Christian BE, Spremulli LL (2012) Mechanism of protein biosynthesis in mammalian mitochondria. Biochim Biophys Acta 1819:1035–1054
99. Aphasizheva I, Aphasizhev R (2010) RET1-catalyzed uridylylation shapes the mitochondrial transcriptome in *Trypanosoma brucei*. Mol Cell Biol 30:1555–1567
100. Germain A, Herlich S, Larom S et al (2011) Mutational analysis of *Arabidopsis* chloroplast polynucleotide phosphorylase reveals roles for both RNase PH core domains in polyadenylation, RNA 3′-end maturation and intron degradation. Plant J 67:381–394
101. Mohanty BK, Kushner SR (2000) Polynucleotide phosphorylase, RNase II and RNase E play different roles in the in vivo modulation of polyadenylation in *Escherichia coli*. Mol Microbiol 36:982–994
102. Carpousis AJ (2002) The *Escherichia coli* RNA degradosome: structure, function and relationship in other ribonucleolytic multienzyme complexes. Biochem Soc Trans 30: 150–155
103. Carpousis AJ (2007) The RNA degradosome of *Escherichia coli*: an mRNA-degrading machine assembled on RNase E. Annu Rev Microbiol 61:71–87
104. Matos RG, Barbas A, Gomez-Puertas P et al (2011) Swapping the domains of exoribonucleases RNase II and RNase R: conferring upon RNase II the ability to degrade ds RNA. Proteins 79:1853–1867
105. Li de la Sierra-Gallay I, Zig L, Jamalli A et al (2008) Structural insights into the dual activity of RNase J. Nat Struct Mol Biol 15:206–212
106. Borowski LS, Dziembowski A, Hejnowicz MS et al (2013) Human mitochondrial RNA decay mediated by PNPase-hSuv3 complex takes place in distinct foci. Nucleic Acids Res 41:1223–1240
107. Baughman JM, Nilsson R, Gohil VM et al (2009) A computational screen for regulators of oxidative phosphorylation implicates SLIRP in mitochondrial RNA homeostasis. PLoS Genet 5:e1000590
108. Sasarman F, Brunel-Guitton C, Antonicka H et al (2010) LRPPRC and SLIRP interact in a ribonucleoprotein complex that regulates posttranscriptional gene expression in mitochondria. Mol Biol Cell 21:1315–1323
109. Mattiacio JL, Read LK (2009) Evidence for a degradosome-like complex in the mitochondria of *Trypanosoma brucei*. FEBS Lett 583:2333–2338
110. Bollenbach TJ, Schuster G, Stern DB (2004) Cooperation of endo- and exoribonucleases in chloroplast mRNA turnover. Prog Nucleic Acid Res Mol Biol 78:305–337
111. Yang J, Stern DB (1997) The spinach chloroplast endoribonuclease CSP41 cleaves the 3′-untranslated region of petD mRNA primarily within its terminal stem-loop structure. J Biol Chem 272:12874–12880
112. Bollenbach TJ, Stern DB (2003) Divalent metal-dependent catalysis and cleavage specificity of CSP41, a chloroplast endoribonuclease belonging to the short chain dehydrogenase/reductase superfamily. Nucleic Acids Res 31:4317–4325
113. Lupold DS, Caoile AG, Stern DB (1999) Polyadenylation occurs at multiple sites in maize mitochondrial cox2 mRNA and is independent of editing status. Plant Cell 11: 1565–1578
114. Gagliardi D, Perrin R, Marechal-Drouard L et al (2001) Plant mitochondrial polyadenylated mRNAs are degraded by a 3′- to 5′-exoribonuclease activity, which proceeds unimpeded by stable secondary structures. J Biol Chem 276:43541–43547
115. Nagaike T, Suzuki T, Ueda T (2008) Polyadenylation in mammalian mitochondria: insights from recent studies. Biochim Biophys Acta 1779:266–269
116. Piechota J, Tomecki R, Gewartowski K et al (2006) Differential stability of mitochondrial mRNA in HeLa cells. Acta Biochim Pol 53: 157–168
117. Bobrowicz AJ, Lightowlers RN, Chrzanowska-Lightowlers Z (2008) Polyadenylation and degradation of mRNA in mammalian mitochondria: a missing link? Biochem Soc Trans 36:517–519
118. Slomovic S, Laufer D, Geiger D et al (2005) Polyadenylation and degradation of human mitochondrial RNA: the prokaryotic past leaves its mark. Mol Cell Biol 25:6427–6435
119. Slomovic S, Laufer D, Geiger D et al (2006) Polyadenylation of ribosomal RNA in human cells. Nucleic Acids Res 34:2966–2975

Chapter 19

In Vivo Analysis of Polyadenylation in Prokaryotes

Bijoy K. Mohanty and Sidney R. Kushner

Abstract

Polyadenylation at the 3′ ends of mRNAs, tRNAs, rRNAs, and sRNAs plays important roles in RNA metabolism in both prokaryotes and eukaryotes. However, the nature of poly(A) tails in prokaryotes is distinct compared to their eukaryotic counterparts. Specifically, depending on the organism, eukaryotic poly(A) tails average between 50 and >200 nt and can easily be isolated by several techniques involving oligo(dT)-dependent cDNA amplification. In contrast, the bulk of the poly(A) tails present on prokaryotic transcripts is relatively short (<10 nt) and is difficult to characterize using similar techniques. This chapter describes methods that can circumvent these problems. For example, we discuss how to isolate total RNA and characterize its overall polyadenylation status employing a poly(A) sizing assay. Furthermore, we describe a technique involving RNase H treatment of total RNA followed by northern analysis in order to distinguish length of poly(A) tails on various types of transcripts. Finally, we outline a useful procedure to clone the poly(A) tails of specific transcripts using 5′–3′ end-ligated RNA, which is independent of oligo(dT)-dependent cDNA amplification. These approaches are particularly helpful in analyzing transcripts with either short or long poly(A) tails both in prokaryotes and eukaryotes.

Key words Poly(A) tails, Poly(A) sizing assay, RNase H, RNA ligation, Northern analysis, Catrimide, RNA*snap*™

1 Introduction

For decades, *Escherichia coli* has been the primary model system for studying posttranscriptional regulation of gene expression in prokaryotes. However, despite significant progress in our understanding of RNA metabolism, many aspects of posttranscriptional regulation have been overlooked primarily because of numerous technical difficulties. For example, the first poly(A) polymerase (PAP I), the protein responsible for adding 3′ poly(A) tails to RNA molecules, was actually identified in *E. coli* more than 60 years ago [1]. Subsequently, the presence of poly(A) tails in *E. coli* and

Joanna Rorbach and Agnieszka J. Bobrowicz (eds.), *Polyadenylation: Methods and Protocols*, Methods in Molecular Biology, vol. 1125, DOI 10.1007/978-1-62703-971-0_19, © Springer Science+Business Media New York 2014

Bacillus subtilis was demonstrated [2–5], but the work was largely ignored, in part because of the low overall abundance of polyadenylated transcripts in both organisms and the lack of reliable methods to isolate and characterize poly(A) tails. In fact, systematic studies of polyadenylation in prokaryotic RNA metabolism really only started in 1992 when the structural gene for *E. coli* PAP I (*pcnB*) was identified [6].

The challenge in detection and characterization of the polyadenylation profiles in prokaryotes arises from several distinct features of prokaryotic polyadenylation. For example, eukaryotic poly(A) tails are relatively long and are found on almost all mRNAs. Tail lengths are specific to the species, such as, 200–300 nt in mammals and 70–90 nt in yeast [7, 8]. In contrast, poly(A) tails in *E. coli* are heterogeneous in length ranging from 1 to more than 50 nt [9, 10] and are present on a limited number of mRNAs, rRNAs, and pre-tRNAs. In addition, in wild-type *E. coli* ~10 % of the poly(A) tails are added by the biosynthetic activity of polynucleotide phosphorylase (PNPase) [11], a major $3' \rightarrow 5'$ exoribonuclease [12]. The tails synthesized by PNPase are heteropolymeric in nature containing all four nucleotides with ~50 % As [11, 13]. Interestingly, based on bioinformatic analysis many more prokaryotes contain PNPase orthologs rather than PAP I type enzymes. Thus, many of the methods employed for poly(A) tail detection in eukaryotes are simply not suitable for use in prokaryotes.

Over the years, our laboratory has either developed or adapted novel methods for the isolation and characterization of prokaryotic RNA as a means of studying various aspects of RNA metabolism, including the role of polyadenylation in those pathways [10–17]. Using these approaches we have recently shown that many pre-tRNAs in *E. coli* contain short poly(A) tails which are less than 10 nt in length with majority being less than 5 nt [9]. In contrast, the poly(A) tails found on mRNAs and rRNAs can be longer than 15 nt [13, 14, 18]. The percentage of pre-tRNA polyadenylation increases significantly in the absence of RNase T and RNase PH [9], the two major tRNA 3′ end processing enzymes [19]. In addition, the synthesis of PAP I in *E. coli* is tightly controlled, since its deregulation leads to the polyadenylation of mature tRNAs, inhibition of protein synthesis and cell death [20].

In this chapter we describe a series of procedures that can be used to characterize the polyadenylation of transcripts in any prokaryote. These methods include total RNA isolation procedures, a poly(A) sizing assay as well as methods to distinguish the presence of short versus long poly(A) tails using oligo(dT_{20}) and RNase H (*see* **Note 1**). Furthermore, we also describe how to characterize transcript-specific poly(A) tails by cloning and sequencing of 5′–3′ end ligated transcripts.

2 Materials

2.1 Bacterial Strains

1. For the experiments described here we employed the *E. coli* strains SK9124 (*rph-1 thyA715*/pBMK11) and SK10574 (Δ*rnt rph-1 thyA715*/pBMK11), both derivatives of MG1693 (*rph-1 thyA715*). pBMK11 (*pcnB*/Cm[R]) is an IPTG-controlled expression plasmid containing *pcnB* under *lac* promoter control [18] (*see* **Note 2**). However, any bacterial strain can be used. For gram-negative bacteria, the Catrimide procedure described below for total RNA isolation works well. The RNA*snap*™ RNA isolation procedure [17] developed recently in our laboratory works for both gram-negative and gram-positive bacteria (*see* **Note 3**).

2.2 Isolation of Steady-State RNA Using Catrimide

1. Luria broth (LB) supplemented with 50 μg/ml thymine and 20 μg/ml chloramphenicol.
2. 100 mM Isopropyl-β-D-thiogalactopyranoside (IPTG, made fresh).
3. Klett-Summerson colorimeter (No. 42 green filter) or a spectrophotometer.
4. Centrifuge tubes and screw-cap tubes (13 × 100 mm, Pyrex).
5. Microcentrifuge tubes (1.5 ml).
6. Refrigerated centrifuge and tabletop microcentrifuge with at least 16,000 × *g* rating.
7. RNase-free water (*see* **Note 4**).
8. 5× Tris–$MgCl_2$ (TM) buffer: 50 mM Tris base, 25 mM $MgCl_2$, pH 7.2 (with 1 N NaOH) (*see* **Note 5**).
9. Stop buffer: 50 mg chloramphenicol (dissolved in 12.5 ml of 95 % ethanol), 20 ml 5× TM buffer, 2.5 ml 1 M sodium azide (*see* **Note 6**), 80 ml RNase-free water.
10. Lysis buffer (prepared fresh): 10 ml 1× TM buffer, 303 μl freshly prepared lysozyme (Sigma®, 10 mg/ml in 1× TM buffer), 33 μl DNase I (DNase I recombinant, RNase-free, 10 U/μl, Roche) (*see* **Note 7**).
11. 10 % Catrimide solution (*see* **Note 8**).
12. 20 mM acetic acid.
13. 2 M LiCl (in 35 % ethanol and in water).
14. Dry ice and ethanol (70 %, 95 % and 100 %).
15. NanoDrop (Model, 2000C, Thermoscientific) (*see* **Note 9**).
16. 10 mg/ml ethidium bromide (Sigma®) (*see* **Note 10**).
17. Transilluminator.

2.3 Poly(A) Sizing Assay

1. 3′-end-labeling master mix (per 30 μl): 11 μl RNase-free water, 3 μl T4 RNA ligase buffer (10×), 3 μl ATP (10 mM), 3 μl DMSO (dimethyl sulfoxide), 1 μl RNasin® Plus RNase Inhibitor (40 U/μl, Promega), 5 μl ^{32}pCp [Cytidine 3′,5′-bis(phosphate)] (3,000 Ci/mmol, NEN, Perkin Elmer), 4 μl T4 RNA ligase (10 U/μl, NEB).
2. Stop solution: 500 mM NaCl, 10 mM EDTA (pH 8).
3. DE81 filter disc (Millipore™).
4. Ribonuclease digestion buffer (2×): 20 mM Tris–HCl (pH 8), 600 mM NaCl, 2 mM $MgCl_2$, RNase A (0.5 μg/μl), RNase T1 (8 U/μl, Ambion®).
5. Phenol–chloroform mix (5:1, pH 4.5, Ambion®)
6. GlycoBlue™ (Ambion®).
7. Ethanol (100 %).
8. Ambion® Gel Loading Buffer II.
9. Water bath (37 °C).
10. 5× Tris-Borate/EDTA (TBE) buffer, pH 8.8: 81 g Tris, 13.7 g boric acid, and 4.75 g EDTA. Make up the volume to 1,000 ml. pH should be ~8.8.
11. Polyacrylamide gel solution (14 %) (*see* **Note 11**): 43 g urea, 20 ml 5× TBE (pH 8.8), 35 ml 40 % AccuGel™ 19:1 (19:1 acrylamide–bis-acrylamide solution, National Diagnostics), 20 ml water. Three hundred seventy microliters 10 % ammonium persulfate (APS, made fresh) and 37 μl TEMED are added to 50 ml of polyacrylamide gel solution before pouring the gel (*see* **Note 12**).
12. 5′-end labeled molecular size standards (*see* **Note 13**).
13. Power supply with output range of 3,000 V DC/400 mA/400 W (such as Bio-Rad, PowerPac 3000).
14. Fixing solution: 20 % ethanol, 10 % acetic Acid.
15. Gel dryer.

2.4 RNase H Treatment

1. 1 M KCl.
2. 1 mM EDTA.
3. Oligo$(dT)_{20}$ (100 pmol/μl).
4. RNase H (USB).
5. Phenol–chloroform mix (5:1, pH 4.5, Ambion®).
6. GlycoBlue™ (Ambion®).
7. Ambion® Gel Loading Buffer II.
8. Water bath (65 and 37 °C).

2.5 Northern Analysis

1. Northern Gel apparatus (*see* **Note 14**).
2. 5× TBE buffer (pH 8): 54 g Tris, 27.5 g boric acid, and 4.65 g EDTA. Make up the volume to 1,000 ml. pH should be ~8.
3. Polyacrylamide gel solution (6 %): 17.4 g urea, 7 ml 5× TBE (pH 8), 7 ml 30 % AcrylaGel (National Diagnostics), 5.6 ml 2 % Bis-AcrylaGel (National Diagnostics), 4.3 ml water. Two hundred eighty microliters of 10 % APS (made fresh) and 28 μl TEMED are added to 35 ml of polyacrylamide gel solution before pouring the gel.
4. Positively charged nylon transfer membrane (0.45 μm, Nytran® SPC, Whatman Inc).
5. Power supply with output range of 1,000 VDC/400 mA/250 W (such as Bio-Rad, PowerPac 1000).
6. RNaseZap® (Ambion®).
7. Electroblotting apparatus (*see* **Note 15**).
8. Tris-acetate-EDTA (TAE) transfer buffer (10×): 100 mM Tris, 50 mM sodium acetate, 5 mM EDTA. Adjust the pH to 7.8 with glacial acetic acid.
9. Baking oven.
10. 10 mg/ml ethidium bromide (Sigma®) (*see* **Note 10**).
11. PerfectHyb™ Plus Hybridization Buffer (Sigma®).
12. Probes (*lpp* and *cysT*) (*see* **Note 16**).
13. T4 polynucleotide kinase (PNK) (NEB or other supplier).
14. γ-(^{32}P)-ATP, 6,000 Ci/mmol (Perkin Elmer).
15. 37 °C water bath.
16. DE81 filter disc (Millipore™).
17. Hybridization bottles and hybridization oven.
18. 20× SSC (Saline-sodium citrate: 3 M NaCl, 0.3 M sodium citrate).
19. 20× SDS.
20. Washing Buffer (0.3 % SDS and 2× SSC).
21. Fluorescence Imaging System (Storm, Amersham Biosciences) (*see* **Note 17**).

2.6 Self Ligation of RNA and RT-PCR

1. Tobacco acid pyrophosphatase (TAP) (Epicenter®).
2. T4 RNA ligase (NEB).
3. Two gene specific primers (P1 and P2, Fig. 4) (*see* **Note 18**).
4. SuperScript® III Reverse Transcriptase (Invitrogen™).
5. RNasin® Plus RNase Inhibitor (Promega).

6. 10 mM dNTP mix (dATP, dTTP, dCTP, dGTP).
7. RNase H (5 U/μl, USB®).
8. GoTaq® Green Master Mix (Promega).
9. PCR purification kit.

3 Methods

3.1 General Preparation for Working with RNA Samples

1. Due to the ubiquitous nature of RNases care must be taken to avoid RNase contamination. Always use powder-free gloves.
2. All glassware used for RNA procedures should be baked at 180 °C for 8 h, at least 1 day ahead of time. All microcentrifuge tubes and tips are used new from the package.
3. It is critical that all reagents are prepared with RNase-free water.
4. It is essential that all gel apparati and combs are made RNase free by treating with RNaseZap® (Ambion®).
5. Dilute 5× and 10× solutions (i.e., TBE and TAE) with RNase-free water.
6. Many of the procedures described here require the use of radioactive chemicals. Follow appropriate safety guidelines for use of such chemicals and disposal of radioactive solid and liquid wastes.

3.2 Growth and Harvesting of Cells for Steady-State RNA Isolation

1. Prepare all the solutions except the Stop and Lysis buffers at least 1 day in advance. Inoculate 5 ml of LB supplemented with 50 μg/ml thymine and 20 μg/ml chloramphenicol with a single colony of each bacterial strain. Incubate the culture overnight at 37 °C as a standing culture.
2. The next day, prepare the Stop buffer and dispense 5 ml each into centrifuge tubes. Allow Stop buffer to freeze in a −20 °C freezer.
3. Grow required amount of culture by adding the overnight inoculums (10 %) to LB supplemented with 50 μg/ml thymine and 20 μg/ml chloramphenicol. Shake each culture at 37 °C and periodically monitor the OD (optical density) either using a Klett meter or a spectrophotometer. When the OD reaches 50 Klett units above background (OD_{600} of 0.4), add 100 mM IPTG to a final concentration of 350 μM. Use an identical culture without IPTG as a control.
4. Just before the cultures are ready to be sampled remove the frozen Stop buffer from the freezer and crush the frozen buffer using a glass rod (*see* **Note 19**). Remove 3.5 ml of culture at various times after induction and immediately add it to the crushed Stop buffer while slowly vortexing to mix the sample.

The Stop buffer helps to reduce all further metabolic activities including minimizing RNA degradation. Keep cells on ice until ready for RNA isolation.

3.3 Isolation of Total RNA

1. At this time prepare the Lysis buffer and store on ice. Also prepare dry-ice/ethanol slurry in an ice bucket by adding 100 % ethanol to 1–2 lbs of crushed dry ice. Label the screw cap tubes (one for each sample) and put in dry ice/ethanol slurry.
2. Centrifuge the cell suspensions from **step 4** (Subheading 3.2) at 3,000 × *g* for 5 min at 4 °C, discard the supernatants, and invert each tube on a double layer of Kimwipes* to remove any residual liquid. Keep the cell pellets on ice.
3. Add 200 μl of Lysis buffer to each cell pellet and mix thoroughly. Transfer the cell suspension to a corresponding prechilled screw cap tube and dip the tube into the dry-ice slurry. The suspension should immediately freeze.
4. After 1 min transfer the frozen cells to a 37 °C water bath. As soon as the cells are thawed, shift the tubes back into the dry-ice slurry. Repeat this procedure four times to insure complete cell lysis (*see* **Note 20**).
5. Add 34 μl 20 mM acetic acid to the frozen pellet after the last freezing cycle and transfer the tubes to the 37 °C water bath.
6. Add 200 μl 10 % Catrimide solution to each melted cell suspension while still in the water bath and incubate 2–5 min while periodically mixing the samples.
7. Transfer each sample to a 1.5 ml microcentrifuge tube. The cell suspensions will remain cloudy. Centrifuge at 2,500 × *g* for 1 min in the cold room (*see* **Note 21**).
8. Remove the supernatants with a suction apparatus (*see* **Note 22**) and resuspend the pellets in 500 μl of 2 M LiCl (in 35 % ethanol).
9. Incubate at room temperature for 5 min and centrifuge in the cold room at 16,000 × *g* for 6 min. Remove the supernatants as in **step 8**.
10. Resuspend each pellet in 500 μl 2 M LiCl (in water), incubate at room temperature for 2 min and centrifuge in the cold room at 16,000 × *g* for 6 min. Remove the supernatants as in **step 8**.
11. Add 200 μl 70 % ethanol to rinse the pellet. Immediately remove the ethanol as in **step 8**. Dry the pellet in a SpeedVac concentrator for 3–5 min (*see* **Note 23**) or leave the tube open for 15 min at room temperature for air drying.
12. Add 125 μl RNase-free water and incubate at room temperature for 15–30 min to dissolve the RNA pellets (*see* **Note 24**). Gently tap the tube to mix the sample. Centrifuge the sample

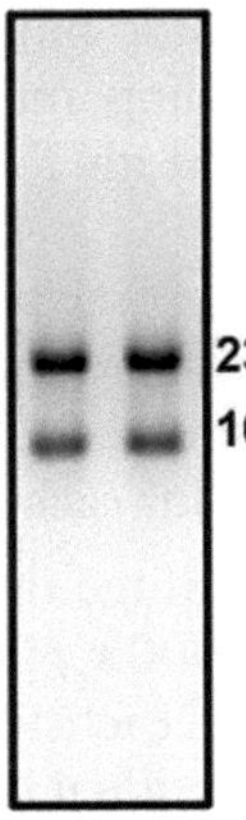

Fig. 1 Determination of RNA integrity and equal loading. Total RNA (250 ng) was separated on a 1 % agarose gel and stained with ethidium bromide (0.5 μg/ml)

for 1 min at full speed in a microcentrifuge. The supernatant containing the RNA should be transferred to a new tube. On average 100–125 μg of RNA are obtained from 3.5 ml of cultures from most strains. Slower growing cells tend to produce lower RNA yields. Store the RNA at 4 °C if it is going to be used immediately. For long term storage (>1 month) store the RNA at −20 °C.

3.4 Quantification and Quality Assessment of Total RNA

1. Use 1 μl RNA to quantify the RNA concentration using a NanoDrop (*see* **Note 9**). An OD 260/280 ratio of 1.8–2.0 indicates a good quality RNA preparation.
2. The integrity of the RNA should be verified by separating 250–500 ng of RNA on an agarose mini-gel. The gel is visualized under UV light using a transilluminator. The ribosomal RNAs (16S and 23S) will be seen as distinct bands with a intensity ratio of ~1:2 (16S:23S) (Fig. 1).

3.5 Poly(A) Tail Sizing Assay

1. The poly(A) tail sizing assay provides a global overview of the in vivo size distribution of poly(A) tails.
2. Pipette 1–20 μg (*see* **Note 25**) of each RNA sample into a microcentrifuge tube and dry in a SpeedVac (*see* **Note 23**). Resuspend the dried RNA in 2 μl of RNase-free water.
3. Add 6 μl of the 3′-end-labeling master mix to the sample. Incubate overnight (~14 h) at RT.
4. Add 50 μl Stop solution. Determine the labeling efficiency of each sample as follows:
   (a) Spot 1 μl sample at the center of a DE81 filter disc.
   (b) Dry the filter disc under a halogen lamp (or room temperature) for 10 min.

(c) Measure the total radioactivity (A) as CPM (counts per minute) using a liquid scintillation system by Cherenkov counting (i.e., without scintillation fluid).

(d) Wash the filter: 2 × 5 min each with 0.5 M Na_2HPO_4, 1 × 1 min in distilled water and finally rinse with 100 % ethanol.

(e) Dry the filter disc under a halogen lamp (or room temperature) for 10 min.

(f) Measure the remaining radioactivity (B) as CPM. Calculate the percentage of labeling efficiency using the formula (B/A) × 100.

5. Proceed to the next step if the labeling efficiencies of all the samples are within ±5 %. However, if the labeling efficiency is much greater among samples, pipette portion (μl) of samples to have identical level of end-labeled RNA to new microcentrifuge tubes based on the percentage of labeling efficiency.
6. Add an equal volume of freshly made 2× ribonuclease digestion buffer. Incubate at 37 °C in a water bath for 30 min (*see* **Note 26**). Raise the temperature to 50 °C for additional 30 min in order to facilitate the digestion of highly structured RNAs.
7. Purify the reaction mixtures by extracting once with equal volume of acid phenol–chloroform (5:1) and transferring the aqueous layers to new microcentrifuge tubes.
8. Add 1 μl GlycoBlue™ and 2.5 volume of 100 % ethanol. Precipitate the RNA overnight at −20 °C.
9. Centrifuge at full speed for 30 min in the cold room (4 °C). Remove the supernatants. Make sure that the blue RNA pellets remain stuck to the walls of the tubes.
10. Wash the pellets with 70 % ethanol and centrifuge for 5 min in the cold room. Remove the ethanol and dry the pellets in a SpeedVac (*see* **Note 23**).
11. Resuspend the dried pellets in 2.5 μl of 2× Gel Loading Buffer II.
12. Pour a suitable polyacrylamide gel (*see* **Note 11**). Allow a minimum of 2 h of polymerization time.
13. Assemble the gel apparatus with 1× TBE buffer (pH 8.8). Flush all the loading wells with a syringe to remove any unpolymerized acrylamide and urea. Run the gel at 70 W (constant power) until the buffer temperature reaches ~55 °C.
14. Dilute the labeled molecular size standards 500–1,000-fold in water. Mix equal volume of diluted labeled size standards and 2× Gel Loading Buffer II.
15. Heat samples (from **step 11**) and labeled size standards (2.5 μl) in a sand bath (95–100 °C) for 2 min. Chill all the samples on ice for 1 min and centrifuge briefly.

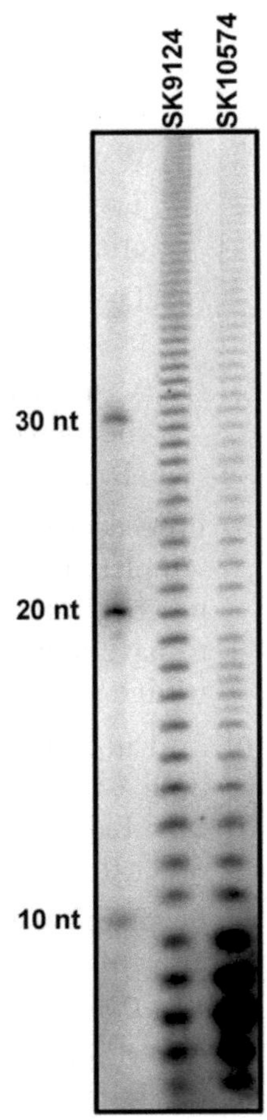

Fig. 2 Poly(A) sizing assay. One microgram of total RNA isolated from SK9124 (*rph-1*/pBMK11) and SK10574 (Δ*rnt rph-1*/pBMK11) after IPTG (350 μmol) induction was digested with RNase A and RNase T1 and separated on 14 % polyacrylamide gel. Note the presence of longer poly(A) tails in SK9124 and shorter poly(A) tails in SK10574. The *left most lane* is the dA size standards as noted

16. Turn off the power supply and flush all the wells again to remove unpolymerized acrylamide and urea before loading the samples onto the gel. Run the gel at 50 W (constant power) until the bromophenol blue dye travels ~30 cm from the wells.
17. Fix the gel by submerging the gel in the fixing solution for 3 min. Wash the gel with RNase-free water for two to three times to remove residual acetic acid. Transfer the gel to a filter paper (3 mm, Whatman®) and dry the gel in a gel dryer at 70 °C for 30 min.
18. The gel is imaged by exposing either to a PhosphorImager screen or an X-ray film.
19. A typical poly(A) sizing assay is shown in Fig. 2.

3.6 RNase H Treatment

1. Pipette 5–10 μg of total RNA in duplicate for each sample to be analyzed in separate microcentrifuge tubes.
2. Add 2 μl 1 M KCl and 2 μl 1 mM EDTA to all the samples. Add 1 μl oligo(dT)$_{20}$ (100 pmol/μl) to one set of RNA samples only. Finally add RNase-free water to all the samples to make the total volume to 20 μl.
3. Heat at 65 °C for 5 min and let the samples cool down slowly at room temperature for 30 min.
4. Add 5 μl 5× RNase H buffer and 1 μl RNase H (5 U/μl) to all samples.
5. Incubate at 37 °C for 30 min.
6. Add 80 μl water followed by 105 μl phenol–chloroform mix (5:1, pH 4.5) to the reaction.
7. Vortex for ~30 s and centrifuge at full speed for 3 min in a microcentrifuge.
8. Remove 100 μl from the top aqueous layer and place in a new microcentrifuge tube. Add 1 μl GlycoBlue™ and 1/11 volume of 3 M sodium acetate (pH 5.2) (11 μl) followed by 2.5 volume of 100 % ethanol (280 μl). Precipitate overnight at −20 °C or 1–2 h at −80 °C.
9. Centrifuge at full speed for 15 min at 4 °C. Remove the supernatant. Watch for the blue RNA pellet, which usually remains stuck at the bottom to the wall of the tube.
10. Wash the pellet with 70 % ethanol and centrifuge for 5 min in the cold room. Remove the ethanol and dry the pellets in a SpeedVac (*see* **Note 23**).
11. Resuspend the dried pellets in 5 μl of 2× Gel Loading Buffer II (Ambion®).

3.7 Northern Analysis

1. Polyacrylamide gels are ideal for separation of smaller transcripts (*see* **Note 27**). We routinely use 6 % polyacrylamide gels for separation of tRNAs with and without poly(A) tails [9, 15, 21].
2. Treat the gel apparatus and comb with RNaseZap® to remove any RNase contamination.
3. Pour the appropriate percentage of acrylamide gel, assemble the gel apparatus and fill the gel apparatus with 1× TBE (pH 8) (*see* **Note 28**). Flush all the loading wells with a syringe to remove any unpolymerized acrylamide and urea.
4. Pre-run the gel at a constant current of 60 mA until the buffer temperature reaches ~55 °C (*see* **Note 29**).
5. Turn off the power supply and flush all the wells again with 1× TBE using the syringe.

6. Heat the RNA samples at 90–95 °C in a sand bath for 3 min. Briefly centrifuge at room temperature to collect all the samples and load them while they are still warm (*see* **Note 30**).
7. Run the gel at a constant current of 50 mA for an appropriate length of time.

3.8 Transferring RNA to a Membrane

1. RNA in gel is transferred to a membrane either by electroblotting or capillary blotting using SSC (*see* **Note 31**).
2. For electroblotting of RNA, prepare 1× TAE transfer buffer from the 10× concentrate and keep in the cold room for at least 2–3 h before transfer.
3. Assemble the electroblotter according to the manufacturer's instruction and transfer the RNA to a charged nylon membrane (e.g., Nytran® SPC, Whatman Inc) in the cold room.
4. Carry out the transfer for 1 h at 15 V followed by 2 h at 40 V. Alternatively, the RNA can also be transferred overnight at 15 V.
5. Fix the RNA by placing the membrane in a baking oven at 80 °C for 30 min followed by a 10 s exposure to an UV lamp.

3.9 Probe Preparation

1. Probes are labeled by 5′ end-labeling using T4 polynucleotide kinase (PNK) in the presence of γ-(^{32}P)-ATP.
2. Add the following materials in the order listed to a 0.5 ml microcentrifuge tube: 2 μl water, 1 μl oligonucleotide (20–30 pmol/μl), 1 μl 10× T4 PNK buffer, 5 μl γ-(^{32}P)-ATP, and 1 μl T4 PNK (10 U).
3. Incubate at 37 °C for 30 min. Inactivate the PNK by heating at 90 °C for 3 min.
4. Determine the labeling efficiency of the probe as described previously (Subheading 3.5, **step 4**) (*see* **Note 32**).

3.10 Hybridization and Washing of Membrane

1. Soak the membrane in 2× SSC for 5 min and pre-hybridize in 5–7 ml of PerfectHyb™ Plus Hybridization Buffer (Sigma®) for at least 1 h using a hybridization oven set at a temperature appropriate for your probe (*see* **Note 33**).
2. Add the labeled probe directly to the hybridization buffer (*see* **Note 34**) and hybridize the membrane overnight.
3. Wash the membrane three times (1×5 and 2×15 min) at room temperature in 100–200 ml of washing buffer (*see* **Note 35**) with gentle shaking.
4. Briefly dry the membrane between two blotting papers (*see* **Note 36**), wrap in Saran wrap and expose on a PhosphorImager screen.

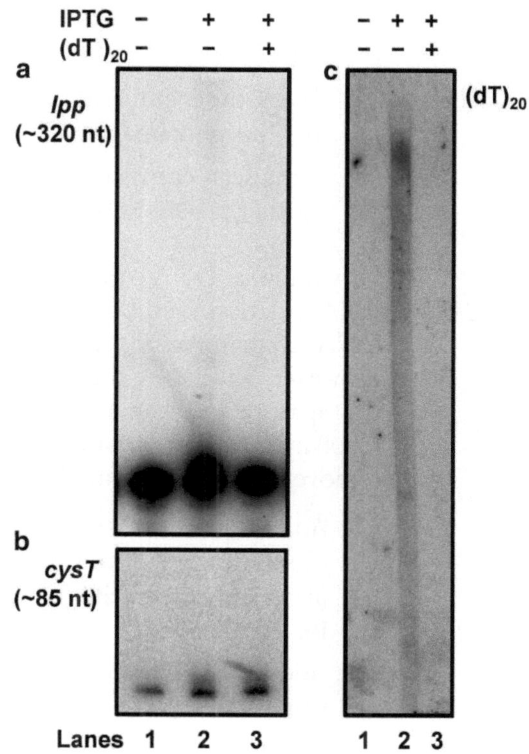

Fig. 3 Northern analysis of polyadenylation of *lpp* mRNA (**a**) and *cysT* tRNA (**b**) after IPTG induction of poly(A) tails in SK9124 (*rph-1*/pBMK11) followed by RNase H treatment. Note the smearing of both the *lpp* mRNA and *cysT* tRNA after IPTG induction of PAP I in *lane 2* compared to *lane 1* indicating addition of poly(A) tails (panels **a** and **b**). However, poly(A) tails were removed in presence of oligo(dT_{20}) by RNase H only for the *lpp* mRNA in *lane 3* (identical to *lane 1*, panel **a**), but not in *cysT* tRNA in *lane 3* (identical to *lane 2*, panel **b**) demonstrating long poly(A) tails on the *lpp* mRNA. (**c**) The same blot was stripped (Subheading 3.12) and reprobed with oligo$(dT)_{20}$ to show poly(A) tails present in *lane 2* after IPTG addition and were removed by RNase H treatment in *lane 3*. +: added and –: not added

3.11 Data Analysis

1. After appropriate time scan the PhosphorImager screen.
2. An example of northern analysis showing the nature of poly(A) tails on the *lpp* mRNA and *cysT* tRNA after RNase H treatment is shown in Fig. 3.

3.12 Stripping and Reprobing Membranes

1. Prepare 500–600 ml of 0.5× SDS from the 20× stock.
2. Heat the SDS solution in a microwave to boiling.
3. Place the membrane in a rectangular Tupperware container and pour 200–300 ml of the boiling 0.5× SDS over the membrane. Make sure that the membrane is completely immersed in the solution (*see* **Note 37**).
4. Microwave the membrane at full power for 1 min in the defrost mode for 3 min.
5. Discard the solution as radioactive waste (Subheading 3.1, **step 6**).

6. Repeat the **steps 3–5** one to two times for complete removal of the probe.
7. Check the membrane for residual radioactivity by scanning in the PhosphorImager.
8. If the scan comes out clean, the membrane can be used with another probe. Follow the steps in Subheadings 3.9 and 3.10.

3.13 Removal of Contaminating DNA from Total RNA

Very minimal DNA contamination is observed using the Catrimide RNA isolation method described here. In addition, low levels of DNA contamination are not a problem for northern analysis. However, any DNA contamination, even if it is not detected on the agarose gel, can lead to erroneous results for experiments involving reverse transcription. The DNA contamination is removed by following method (*see* **Note 38**).

1. Add 53 μl water, 20 μl 10× DNase I buffer to 125 μl of RNA (Subheading 3.3, **step 12**) followed by 2 μl of DNase I (10 U/μl) in a microcentrifuge tube (*see* **Note 39**). Incubate at 37 °C for 30 min.
2. Add 200 μl phenol–chloroform mixture (5:1, pH 4.5) to the reaction.
3. Vortex for ~30 s and centrifuge at full speed for 3 min in a microcentrifuge.
4. Remove 195 μl from the top aqueous layer and place in a new microcentrifuge tube. Add 20 μl 3 M sodium acetate (pH 5.2) followed by 540 μl 100 % ethanol. Precipitate overnight at −20 °C or 1–2 h at −80 °C.
5. Centrifuge at full speed for 15 min in the cold room (4 °C). Remove the supernatant and wash the pellet with 70 % ethanol.
6. Dry the pellet in a SpeedVac (*see* **Note 23**).
7. Resuspend the pellet in 125 μl of water. Follow the steps in Subheading 3.4 to determine the quantity and quality of the RNA.

3.14 Self-Ligation of RNA (Fig. 4)

1. The 5′ triphosphate of primary transcripts is converted in vitro to a 5′ monophosphate by TAP to facilitate RNA self-ligation.
2. Add the following materials in the order listed to a 1.5 ml microcentrifuge tube: 3 μg total RNA, 2.5 μl 10× TAP buffer, water to total volume of 24.5 μl and 0.5 μl TAP (5 U). Incubate at 37 °C for 90 min.
3. Follow the steps in Subheading 3.6 (**steps 6–10**) to purify and precipitate RNA.
4. Resuspend the RNA in 4.5 μl water. Incubate at 65 °C for 5 min (*see* **Note 40**). Quickly chill on ice for 1 min and centrifuge briefly.

5. Add 1 μl 10× RNA ligase buffer, 1 μl 10× ATP, 1 μl 10× BSA, 0.5 μl RNasin® RNase inhibitor (40 U/μl), and 2 μl RNA ligase (NEB, 10 U/μl). Incubate at 37 °C for 2 h.
6. Add 95 μl water and 105 μl phenol–chloroform mixture (5:1, pH 4.5,) to the reaction.
7. Follow the **steps 7–10** in Subheading 3.6 to purify and precipitate the RNA.
8. Resuspend the RNA in 10.5 μl water.

3.15 Reverse Transcription of Self-Ligated RNA

1. Add 2 μl Primer 1 (20 pmol/μl) (P1, Fig. 4) to the RNA from previous step (Subheading 3.14, **step 8**).
2. Heat the mixture at 65 °C for 5 min, quickly chill on ice for 1 min, briefly centrifuge in the cold room, and hold it on ice.
3. Add the following components to the RNA: 4 μl 5× first-strand buffer, 1 μl 100 mM DTT, 1 μl 10 mM dNTP mix, 0.5 μl (20 U) RNasin® RNase inhibitor, and 1 μl (200 U) SuperScript® III reverse transcriptase.
4. Initiate the reverse transcription by incubating the reaction at suitable temperature (*see* **Note 41**) and continue for 60 min.
5. Terminate the reaction by incubating at 70 °C for 15 min. Quickly chill on ice and centrifuge briefly.
6. Add 0.5 μl (2.5 U) RNase H. Incubate 20 min at 37 °C. Store at −20 °C.

3.16 PCR Amplification of cDNA

1. Amplify 2–5 μl cDNA from previous step (Subheading 3.15, **step 6**) in a 50 μl PCR reaction mixture containing GoTaq® Green Master Mix and Primer 1 and Primer 2 (P1 and P2, Fig. 4).
2. Analyze 3–5 μl of the PCR reaction on a 1–1.5 % agarose gel to determine if the amplification of the anticipated sized products occurred (*see* **Note 42**)
3. Purify the PCR product using a commercially available PCR purification kit.

3.17 Cloning of cDNA Employing Blue and White Screening

1. Use 3–5 μl purified PCR DNA fragments to clone into a low copy vector (*see* **Note 43**) that supports blue and white screening (*see* **Note 44**) using the restriction sites that were incorporated into the PCR primers (*see* **Note 18**).
2. Isolate plasmid DNA from at least 24 independent white colonies. Confirm the presence of cDNA in each plasmid by restriction digestion with the enzymes employed for the cloning step.
3. Sequence the plasmids using a commercial vendor. Using this technique we have successfully sequenced both the 5′ and 3′ ends of various tRNAs and 5S rRNA showing addition of as few as one nontemplated As at the 3′ ends [9, 16, 21].

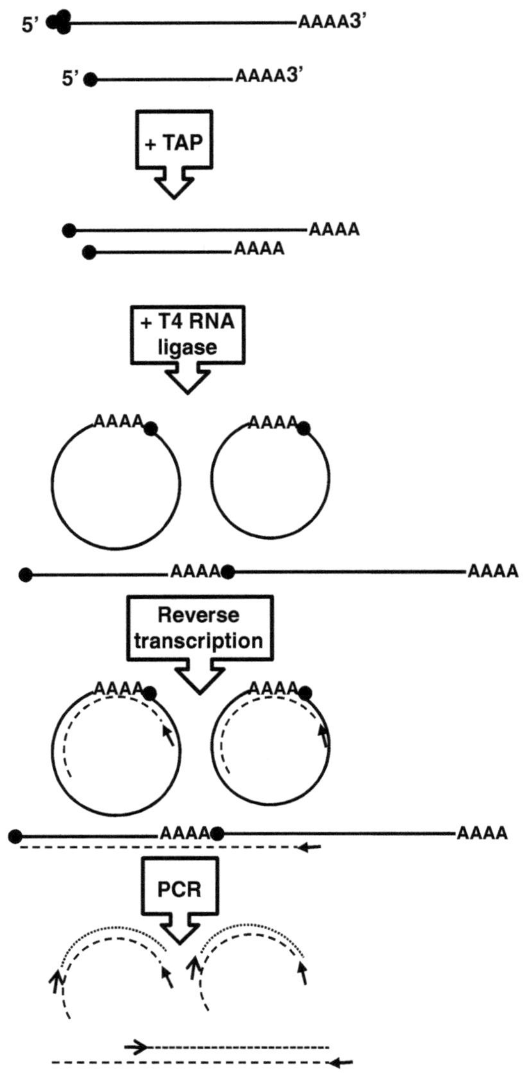

Fig. 4 RNA self-ligation, reverse transcription, and cloning of cDNAs. A graphical presentation of overall procedure. Primer P1 (→) and Primer P2 (→). single closed circle: 5′ monophosphate, three *closed circle*: 5′ triphosphate, TAP: tobacco acid pyrophosphatase

4 Notes

1. The method takes advantage of the fact that long poly(A) tails capable of hybridizing to oligo$(dT)_{20}$ primers are susceptible to RNase H degradation. In contrast, short poly(A) tails will not hybridize to oligo$(dT)_{20}$ under identical conditions and thus will be resistant to RNase H degradation. Consequently, transcripts with long poly(A) tails will have reduced size and transcripts with short poly(A) tails will remain unchanged following RNase H treatment in a northern analysis.

2. Since only ~2 % of the transcripts in a wild-type *E. coli* are polyadenylated based on oligo-dT hybridization method [22], an IPTG controlled *pcnB* expression plasmid is used to facilitate the detection of poly(A) tails by significantly raising the level of polyadenylated transcripts [14, 18].
3. The RNA*snap*™ RNA isolation procedure works for both gram-negative and gram-positive bacteria and has been described in detail [17]. It yields a more accurate representation of the in vivo levels of all the various RNA species.
4. Successful RNA isolation depends on the use of RNase-free water. We use water from a Millipore water filtration system (Model Synergy® UV), which is sterile and RNase free. We never sterilize solutions using an autoclave. Alternatively, a different commercial water filtration system or DEPC (diethyl pyrocarbonate) treated water also can be used.
5. TM buffer is stable at room temperature for 3 months and at 4 °C for 6 months.
6. Sodium azide is highly toxic. Wear a mask when weighing. Store the solution at 4 °C.
7. Mix lysozyme by pipetting up and down slowly to prevent making any bubbles. Store the lysis buffer on ice before use.
8. A cationic surfactant, trimethyl(tetradecyl)ammonium bromide (Sigma®), has been called as Catrimide for simplicity. The solution is stable for 1 year at room temperature.
9. In the absence of a NanoDrop apparatus, a spectrophotometer can be used.
10. Ethidium bromide is a known carcinogen and mutagenic agent. Employ appropriate protection in using and disposing solutions containing the chemical.
11. We routinely use 14 % polyacrylamide gel for detecting a broad range of poly(A) tails (Fig. 2). A 20 % polyacrylamide gel is ideal for detecting poly(A) tails as small as 1 nt [9, 11]. However, poly(A) tails longer than 30 nt will not separate clearly under these conditions. In that case a lower percentage (12–16 %) polyacrylamide gel can be used.
12. The unpolymerized polyacrylamide gel solution can be stored at 4 °C for later use. However, the gel solution should be brought to room temperature in order to dissolve the precipitated urea.
13. $(dA)_{10}$, $(dA)_{20}$, and $(dA)_{30}$ oligonucleotides are used as molecular size standards. Mix 5 pmol of each oligonucleotide and end-label the 5′ ends as described in Subheading 3.8.
14. We use a custom-made vertical gel apparatus, which accommodates an 18 × 18.5 cm (length × width) gel. However, a commercially available gel apparatus should work equally well.

15. We use a custom-made electroblotting apparatus that can hold two 18×18 cm gels). However, any commercially available transfer apparatus should work equally well.
16. *lpp*probe:LPP397(TACCCTCTAGATTGAGTTAATCTCC), *cysT* probe: CYST267 (CAATCCGCTACATAACCGCT).
17. Any suitable PhosphorImaging system can be used to visualize the bands. Alternatively, the bands can also be visualized by exposing the radioactive membrane to X-ray film, but the PhosphorImager scanning permits better quantification of the data.
18. Designing a primer to introduce a unique restriction site towards the 5′ end of each primer will facilitate directional cloning, which makes subsequent DNA sequencing more straightforward. The restriction sites should only be present in the cloning vector but not in part of the cDNA to be cloned.
19. If the frozen buffer is too hard to crush, warm the tubes briefly in a 37 °C water bath.
20. After the last freezing step, the cells can be stored at −70 °C indefinitely.
21. Centrifugation at room temperature at **steps 7–10** also works fine.
22. A single tip in the suction apparatus can be used for all samples. Alternatively, removing the supernatant with a pipette in **steps 8–11** is also acceptable. Only inverting the tube will leave residual liquid. Be careful not to touch or suck the pellet with tips.
23. Be careful not to overdry the pellets since it will become more difficult to dissolve the RNA. Alternatively, the pellets can be air-dried by keeping the tubes upside down on the bench.
24. An overnight incubation at 4 °C may increase the RNA yield by 10–25 %.
25. The amount of total RNA required for this assay depends on total in vivo poly(A) levels. Up to 20 μg of total RNA can be used for wild-type *E. coli* strain. However, less RNA is required for strains with abundant poly(A) levels such as will occur after IPTG induction of PAP I (Fig. 2).
26. At this step RNase A cleaves after C and U residues and RNase T1 cleaves after G residues leaving poly(A) tails labeled at their 3′ ends.
27. A 5–10 % acrylamide gel with 8.3 M urea is used to separate transcripts from <100 to ~900 nt in length.
28. 5× TBE (pH 8) very often precipitates after long term (>1 month) storage at room temperature. Discard the buffer if any visible precipitation is observed and make fresh.
29. Pre-heating the 1× TBE buffer in a microwave to ~50 °C will help to reduce the pre-run time.

30. To avoid renaturation of RNA during loading, samples can be heated and loaded in batches for a large number of samples.
31. Follow manufacturer's (such as Schleicher & Schuell) instructions for capillary blotting of RNA.
32. The labeling efficiency for each probe will vary. However, a minimum of 25 % labeling efficiency is required for detection of transcripts. A higher labeling efficiency is necessary for detection of low abundance transcripts.
33. The hybridization temperature is typically T_m (melting temperature) of the oligonucleotide minus 10 °C. The T_m of the oligonucleotide is usually provided by the manufacturer. Alternatively, the T_m of the oligonucleotide can be roughly calculated as T_m (°C) = 4 (G + C) + 2 (A + T).
34. Avoid adding labeled probes directly to the membrane to prevent localized background.
35. Suitable size sealed Tupperware (or comparable) containers are used for all washing purposes.
36. Brief drying between two blotting papers helps removing excess washing solution which sometimes may cause radioactive contamination of the PhosphorImager screen. However, care should be taken not to overdry the membrane if you want to reprobe with additional probes (*see* Subheading 3.12).
37. Adding the boiling SDS solution will immediately release the labeled probe from the membrane. Follow all precautions for handling radioactive materials.
38. Alternatively, DNA-*free*™ kit (Ambion, Austin, Texas) can also be used at this step.
39. The reaction volume can be scaled up or down depending on your needs.
40. Heating at 65 °C helps ligation of some transcripts with strong secondary structures.
41. Although the recommended maximum incubation temperature for the SuperScript® III reverse transcriptase is 50 °C, we have gone up to 52 °C for templates with extensive secondary structure. The optimum incubation temperature for a particular primer is its T_m (*see* **Note 33**) −2 to 3 °C as long as it does not exceed 52 °C.
42. The PCR products are very often heterogeneous in nature due to heterogeneous size of poly(A) tails.
43. We routinely use pWSK29, a 6–8 copy vector [23]. Any low (6–8 copies) or medium (15–20 copies) copy vector generally works well for cloning cDNA fragments containing poly(A) sequences. However, do not use ColE1 based vectors for analyzing cloned poly(A) tails.

44. Employing a blue and white selection will eliminate isolation of plasmid DNAs not containing cDNA inserts, since 99 % of the white colonies in our experience contain plasmid with cDNA insert.

Acknowledgments

This work was supported in part by grants from the National Institute of General Medical Sciences (GM57220) to SRK.

References

1. August J, Ortiz PJ, Hurwitz J (1962) Ribonucleic acid-dependent ribonucleotide incorporation. I. Purification and properties of the enzyme. J Biol Chem 237:3786–3793
2. Gopalakrishna Y, Langley D, Sarkar N (1981) Detection of high levels of polyadenylate-containing RNA in bacteria by the use of a single-step RNA isolation procedure. Nucleic Acids Res 9:3545–3554
3. Nakazato H, Venkatesan S, Edmonds M (1975) Polyadenylic acid sequences in *E. coli* messenger RNA. Nature 256:144–146
4. Sarkar N, Langley D, Paulus H (1978) Isolation and characterization of polyadenylate-containing RNA. Biochemistry 17:3468–3474
5. Srinivasan PR, Ramanarayanan M, Rabbani E (1975) Presence of polyriboadenylate sequences in pulse-labeled RNA of *Escherichia coli*. Proc Natl Acad Sci U S A 72:2910–2914
6. Cao GJ, Sarkar N (1992) Identification of the gene for an *Escherichia coli* poly(A) polymerase. Proc Natl Acad Sci U S A 89:10380–10384
7. Soucek S, Corbett AH, Fasken MB (2012) The long and the short of it: the role of the zinc finger polyadenosine RNA binding protein, Nab2, in control of poly(A) tail length. Biochim Biophys Acta 1819:546–554
8. Zhang X, Virtanen A, Kleiman FE (2010) To polyadenylate or to deadenylate: that is the question. Cell Cycle 9:4437–4449
9. Mohanty BK, Maples VF, Kushner SR (2012) Polyadenylation helps regulate functional tRNA levels in *Escherichia coli*. Nucleic Acids Res 40:4589–4603
10. O'Hara EB, Chekanova JA, Ingle CA et al (1995) Polyadenylylation helps regulate mRNA decay in *Escherichia coli*. Proc Natl Acad Sci U S A 92:1807–1811
11. Mohanty BK, Kushner SR (2000) Polynucleotide phosphorylase functions both as a 3′ → 5′ exonuclease and a poly(A) polymerase in *Escherichia coli*. Proc Natl Acad Sci U S A 97:11966–11971
12. Mohanty BK, Kushner SR (2003) Genomic analysis in *Escherichia coli* demonstrates differential roles for polynucleotide phosphorylase and RNase II in mRNA abundance and decay. Mol Microbiol 50:645–658
13. Mohanty BK, Maples VF, Kushner SR (2004) The Sm-like protein Hfq regulates polyadenylation dependent mRNA decay in *Escherichia coli*. Mol Microbiol 54:905–920
14. Mohanty BK, Kushner SR (2006) The majority of *Escherichia coli* mRNAs undergo post-transcriptional modification in exponentially growing cells. Nucleic Acids Res 34:5695–5704
15. Mohanty BK, Kushner SR (2007) Ribonuclease P processes polycistronic tRNA transcripts in *Escherichia coli* independent of ribonuclease E. Nucleic Acids Res 35:7614–7625
16. Mohanty BK, Kushner SR (2010) Processing of the *Escherichia coli leuX* tRNA transcript, encoding tRNALeu5, requires either the 3′→ 5′ exoribonuclease polynucleotide phosphorylase or RNase P to remove the Rho-independent transcription terminator. Nucleic Acids Res 38:597–607
17. Stead MB, Agrawal A, Bowden KE et al (2012) RNAsnap™: a rapid, quantitative and inexpensive, method for isolating total RNA from bacteria. Nucleic Acids Res 40:e156
18. Mohanty BK, Kushner SR (1999) Analysis of the function of *Escherichia coli* poly(A) polymerase I in RNA metabolism. Mol Microbiol 34:1094–1108
19. Li Z, Deutscher MP (1996) Maturation pathways for *E. coli* tRNA precursors: a random multienzyme process *in vivo*. Cell 86:503–512
20. Mohanty BK, Kushner SR (2013) Deregulation of poly(A) polymerase I in *Escherichia coli* inhibits protein synthesis and leads to cell death. Nucleic Acids Res 41:1757–1766

21. Mohanty BK, Kushner SR (2008) Rho-independent transcription terminators inhibit RNase P processing of the *secG leuU* and *metT* tRNA polycistronic transcripts in *Escherichia coli*. Nucleic Acids Res 36:364–375
22. Cao GJ, Sarkar N (1992) Poly(A) RNA in *Escherichia coli*: nucleotide sequence at the junction of the *lpp* transcript and the polyadenylate moiety. Proc Natl Acad Sci U S A 89:7546–7550
23. Wang RF, Kushner SR (1991) Construction of versatile low-copy-number vectors for cloning, sequencing and expression in *Escherichia coli*. Gene 100:195–199

Chapter 20

Polyadenylation of RNA in *E. coli*: RNA Polymerase-Associated $(rA)_n$-Synthetic Activities

Grant A. Encalade and Maxim V. Sukhodolets

Abstract

In *Escherichia coli*, Poly(A) polymerase (PAP) and polynucleotide phosphorylase (PNP) are key enzymes thought to be responsible for polyadenylation of the bulk of cellular RNA. In this chapter we describe enzymatic in vitro assays for monitoring $(rA)_n$-synthetic activity among fractionated *E. coli* proteins obtained after affinity chromatography on immobilized DNA. The enzymatic activities of PAP and PNP can be independently monitored among fractionated proteins due to the utilization of different nucleoside substrates (respectively, ATP and ADP) by the two enzymes. We describe two different methods for monitoring the synthesis of polyadenylate: a method based on utilization of a nucleic acid-specific fluorescent dye (RiboGreen®) and an alternative method based on utilization of P^{32}-labeled nucleoside phosphates.

Key words Polyadenylation, Poly(A) polymerase, Polynucleotide phosphorylase, RapA, RNA polymerase, RiboGreen®

1 Introduction

In our early studies, we introduced a simplified method for the purification of RNA polymerase [1, 2]; its most basic version included affinity chromatography on immobilized DNA followed by ion-exchange chromatography on a Mono Q column ([2, 3], and below, Fig. 1; utilization of a Mono Q column for the fractionation of RNA polymerase complexes was originally introduced by the Burgess group [4]). During the Mono Q step, complexes of the core RNA polymerase with different auxiliary subunits can be separated by shallow NaCl gradients [1–3, 5]. This basic protocol was instrumental in the identification of new transcription factors [1–3, 5] and in characterization of the oligomeric forms of RNA polymerase [6]. In our more recent studies, we combined this technique with mass spectrometry analyses and subsequently extended these trials to include fluorescent dye/reporter and enzymatic assays [7] (the latter assays typically being carried out in a 96-well plate format).

Joanna Rorbach and Agnieszka J. Bobrowicz (eds.), *Polyadenylation: Methods and Protocols*, Methods in Molecular Biology, vol. 1125, DOI 10.1007/978-1-62703-971-0_20, © Springer Science+Business Media New York 2014

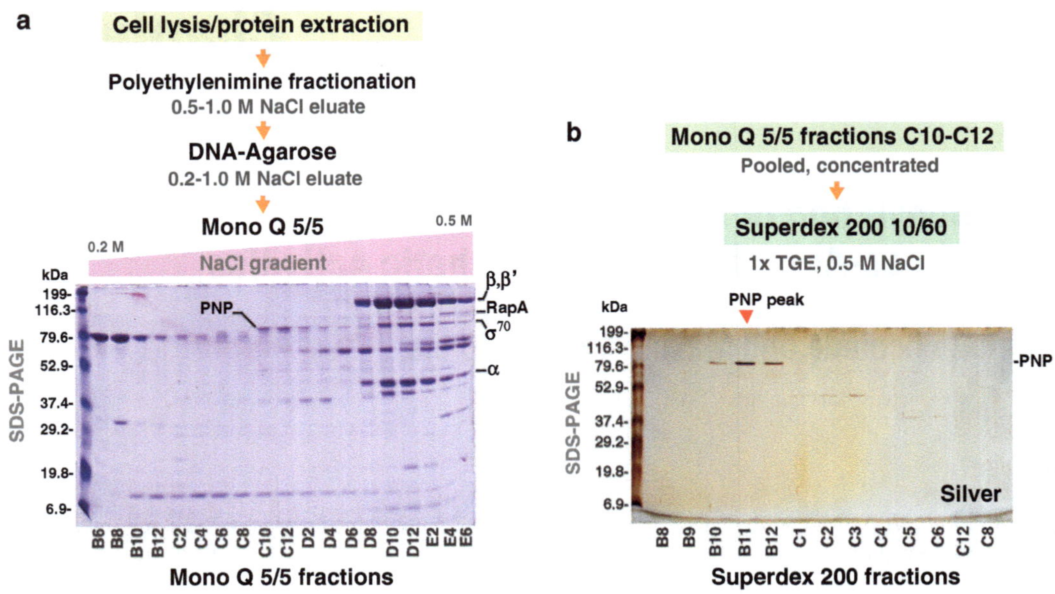

Fig. 1 Purification of native RNA polymerase complexes and PNP. (**a**) Purification of native RNA polymerase complexes containing endogenous RNA oligonucleotides and accessory enzymes involved in RNA modification [7]. *Top*: schematic for the purification procedure described in Subheading 3.1. *Bottom*: 10 μl aliquots from the fractions eluted from a Mono Q column are shown on an SDS gel (12 % ProtoGel™, National Diagnostics) stained with Coomassie Brilliant Blue. Positions of the RNA polymerase subunits and PNP are indicated. Protein size markers (Bio-Rad, broad range) are in the first lane of the gel. Note that the identity of PNP obtained in a similar purification procedure was independently verified by N-terminal protein sequencing [3]. (**b**) Purification of polynucleotide phosphorylase (PNP). *Top*: schematic for the purification procedure described in Subheading 3.2. *Bottom*: 10 μl aliquots from fractions eluted from a Superdex 200 column are shown on a silver-stained SDS gel; the position of PNP is indicated. Protein size markers (Bio-Rad, broad range) are in the first lane of the gel

In this chapter we describe enzymatic in vitro assays for monitoring Poly(A)-synthetic activity among fractionated *E. coli* proteins obtained after affinity chromatography on immobilized DNA. We describe two different methods for monitoring the synthesis of nascent polyadenylate: a method based on utilization of a nucleic acid-specific fluorescent reporter/dye (RiboGreen®) and an alternative method based on utilization of P^{32}-labeled nucleoside phosphates. Both Poly(A) polymerase (PAP or PAP I, product of the *pcnB* gene) and polynucleotide phosphorylase (PNP, product of the *pnp* gene)—the two key enzymes thought to carry out RNA polyadenylation in *E. coli* [8–10]—co-purify with RNA polymerase in this purification procedure [3, 7]. Typically, the bulk of PNP is eluted as a distinct peak in Mono Q fractions [3]; therefore, this technique can provide purified PNP after an additional gel-filtration step (this purification protocol is also described below). In contrast, PAP—present in substoichiometric amounts, possibly due to piggyback interaction with RNA polymerase—co-elutes with RNA polymerase complexes [7]. At this time, the functional significance of this interaction is not fully understood. Our current studies suggest that polyadenylation of RNA polymerase-associated RNA by an accessory enzyme (either PAP or PNP) could aid in its extraction and/or de-solubilization. The mechanism of this reaction (which involves conversion of the extruded 5′ ends of RNAs into 3′ ends through steps reminiscent to those proposed in [7]) has only recently been clarified and will be described in detail in our upcoming studies.

2 Materials

All solutions should be prepared using RNase-free, ultrapure water and only Molecular Biology-grade (or above) reagents. It is essential to follow carefully local regulations for the disposal of radioactive and chemical waste.

2.1 Enzyme Purification

1. 10× TGE (*T*ris-*G*lycerol-*E*DTA) buffer (1 L): 1 M Tris–HCl, pH 7.5 (Amresco), 100 ml; glycerol (Sigma, G5516), 500 ml; 0.5 M EDTA (MP Biomedicals), 2 ml; dithiothreitol (Sigma), 0.16 g. Add water to a volume of 1 L.
2. Extraction buffer (1 L): 1 M Tris–HCl, pH 7.5 (Amresco), 50 ml; glycerol (Sigma), 50 ml; 0.5 M EDTA (MP Biomedicals), 4 ml; 5 M NaCl (KD Medical) 47 ml; 2-mercaptoethanol (Sigma), 0.08 ml; PMSF (Sigma) 25 mg; dithiothreitol (Sigma), 16 mg. Add water to a volume of 1 L. Immediately before the extraction, add lysozyme (Sigma) to the final concentration of 0.3 mg/ml.
3. Sodium chloride, 5 M solution (KD Medical), 1 L.
4. Deoxycholate, sodium salt (Midwest Scientific, high purity grade), 10 % aqueous solution, 3 ml.
5. Polyethylenimine (Sigma), 10 % aqueous solution (pH adjusted to 7.8), 10 ml.
6. Ammonium sulfate (Sigma), 500 g.
7. Denatured (calf thymus) DNA-Agarose (GE Healthcare), 5 ml.
8. Centrifugal Filters (Amicon® Ultra, 0.5 ml/10 K).
9. Waring Blender with a variable-speed control.
10. Conductivity meter.
11. Centrifuge (Sorval RC 6 PLUS or similar).
12. AKTA FPLC (GE Healthcare) with Mono Q 5/5 and Superdex 200 gel-filtration columns (GE Healthcare).

2.2 In Vitro Polyadenylation Assays with RiboGreen®

1. RiboGreen® solution in DMSO (Invitrogen/Molecular Probes, Quant-iT™ RNA reagent, R11491).
2. 2× Reaction buffer D: 100 mM potassium acetate, 20 mM magnesium acetate, 1 mM dithiothreitol, 40 mM Tris–acetate, pH 7.8.
3. 2× Reaction buffer C: 200 mM sodium chloride, 20 mM magnesium chloride, 1 mM dithiothreitol, 100 mM Tris–chloride, pH 7.8.
4. Ultrapure adenosine-5′-triphosphate (ATP), 100 mM aqueous solution (GE Healthcare).
5. Black, flat bottom (96-well, 0.3 ml) polystyrene plates (Whatman Uniplate® or similar).
6. Fluorimeter/plate reader capable of reading fluorescence of RiboGreen® (excitation, 485 nm; emission, 538 nm) in 96-well plates, such as Fluoroskan Ascent (Thermo Scientific) or similar.

2.3 Polyadenylation Assays Utilizing ^{32}P-Labeled Nucleotides

1. ATP [Alpha-^{32}P] (MP Biomedicals, 32007U), 6,000 Ci/mmol.
2. 10× Reaction buffer D: 500 mM potassium acetate, 100 mM magnesium acetate, 5 mM dithiothreitol, 200 mM Tris–acetate, pH 7.8.
3. Purified native RNA polymerase complexes containing accessory PAP and endogenous, protein-associated RNAs (Mono Q fractions D10-E4 obtained as described above).
4. Purified RapA (isolated as described in refs. 1, 11).
5. RNase I_f (New England Biolabs), 50 U/μl.
6. Ultrapure ATP, 100 mM aqueous solution (GE Healthcare).
7. SequaGel™ reagents (National Diagnostics): UreaGel™ system Concentrate (EC-830), Diluent (EC-840), and Buffer (EC-835).
8. Ammonium persulfate, 10 % solution in water (*see* **Note 1**).
9. Loading buffer for denaturing RNA electrophoresis (3× Formamide Loading buffer). To prepare 1 ml of 3× Formamide Loading buffer, combine 0.95 ml of 99 % formamide (SigmaUltra), 0.05 ml of 0.5 M EDTA (MP Biomedicals), and 5 μl of 10 % SDS (Sigma) and 2 mg of Bromophenol Blue (Sigma); mix well by rotating the tube.
10. Flat-end, 0.37 mm Gel Loader pipette tips (Phenix Research Products, T-204R).
11. Apparatus for denaturing nucleic acid electrophoresis (such as Gibco SA 32, LABRepco), complete with a gel casting boot (LABRepco), borosilicate glass plates, 0.4 mm spacers, and combs.

12. High-voltage power supply, such as Power Supply 4000P (Thermo Scientific), or similar.
13. Electrode buffer (0.5× TBE). To make 1 L of electrode buffer, add 50 ml of 10× TBE, pH 8 (KD Medical), to 950 ml of ultrapure water.
14. Cassettes, intensifying screens (Kodak BioMax™ MS), 8 × 10 in. X-ray film (*see* **Note 2**), clear plastic wrap, film processor.
15. Ultralow (−72 °C) freezer.
16. Benchtop beta shield, shielded containers for ^{32}P-labeled samples, beta counter (*see* **Note 3**).

3 Methods

3.1 Purification of Native RNA Polymerase Complexes

In the protocol described below—which is derived from the purification procedures described in [2] and [4]—after the lysozyme treatment and the cell lysis in the presence of deoxycholate, nucleic acids and associated proteins are precipitated with polyethylenimine. The resulting pellets are subjected to a batch salt wash, which is followed by a batch salt elution of nucleic acid-associated proteins. This extraction procedure eliminates a need for DNase treatment (and thus, it is uniquely suited for affinity chromatography on immobilized DNA). This extraction procedure is followed by the affinity chromatography on immobilized denatured DNA and Mono Q chromatography steps described below:

1. Place 20 g of wild-type prototype *E. coli* cells (grown to late-exponential phase in either LB or Super Broth (*see* **Note 4**)) in a variable-speed blender with 150–200 ml of ice-cold Extraction buffer containing 0.3 mg/ml lysozyme and immediately homogenize at low to medium speed.
2. Following a 20-min incubation at room temperature, blend the mixture at high speed to shear the DNA and gradually add 3 ml of 10 % solution of deoxycholate (sodium salt) to the extract. Following an additional 20-min incubation at room temperature, blend the mixture again at high speed for 1 min to shear the DNA.
3. Spin the cell extract for 1 h at 10,000 × *g*, at 4 °C. Collect clear yellow supernatants, place in the blender, and, during continuous, slow stirring, gradually add 10 ml of 10 % polyethylenimine (pH adjusted to 7.8) to the extract. Spin the suspension for 15 min at 6,000–8,000 × *g*, at 4 °C; discard the supernatants.
4. Place the obtained Polymin P pellets in the blender with 200–300 ml of TGE buffer containing 0.5 M NaCl and homogenize at low/medium speed to avoid excessive foaming. Spin the suspension for 15 min at 6,000–8,000 × *g*, at 4 °C; discard the supernatants (the 0.5 M NaCl wash).

5. Place the pellets in the blender, resuspend in 100–150 ml of TGE buffer containing 1 M NaCl and homogenize. Spin the suspension for 15 min at 6,000–8,000 × *g*, at 4 °C; collect the supernatants (the 1 M NaCl eluate).
6. Combine supernatants in a beaker containing a magnetic bar, and, during continuous slow stirring, gradually add dry ammonium sulfate to the extract (40–41 g/100 ml) to precipitate proteins.
7. Pack 2–3 ml of denatured (calf thymus) DNA-Agarose into a cartridge or a column (operated using a standalone pump) (*see* **Note 5**). Spin ammonium sulfate suspension from the preceding purification step for 15 min at 6,000–8,000 × *g*, at 4 °C; fully discard the supernatant. Dissolve the pellet in 1× TGE buffer to a conductivity value corresponding to 1× TGE containing 0.2 M NaCl (19–20 mS/cm at 12 °C) and load onto a DNA-Agarose column at a flow rate of 0.5–2 ml/min. Wash the column with 40–50 ml of 1× TGE containing 0.2 M NaCl and elute the bound proteins with 20–30 ml of 1× TGE containing 1 M NaCl.
8. Dilute the eluate fivefold with 1× TGE, place in Superloop, and load on a Mono Q 5/5 column pre-equilibrated with 1× TGE containing 0.2 M NaCl (*see* **Note 6**). Typically, we fractionate bound proteins with linear NaCl gradients (0.2–0.6 M NaCl in 40 ml of 1× TGE) at a flow rate of 0.1–0.5 ml/min. We collect either 0.25 or 0.5 ml fractions in 96-well plates; the sealed plates can be stored at either 4 °C (for up to several weeks) or −72 °C.

3.2 Purification of Polynucleotide Phosphorylase (PNP)

1. Pool the Mono Q fractions containing PNP (identified through its Poly(A)-synthetic activity in enzymatic assays with ADP) and concentrate to 80–100 μl using a centrifugal filter (Amicon® Ultra) with a 10 kDa cutoff.
2. The resulting protein sample can then be subjected to gel filtration. We typically use a custom-packed 10 × 60 Superdex® 200 (Prep Grade, GE Healthcare) column run at 0.1–0.2 ml/min with 1× TGE containing 0.5 M NaCl as a running buffer. We collect 1 ml fractions into tall 96-well plates.
3. Protein composition of individual Superdex® 200 fractions can be evaluated from colloidal Coomassie- or silver-stained SDS gels (Fig. 1b); PNP's enzymatic activity can be measured as described below.
4. Pool the fractions containing purified PNP and concentrate in 1× TGE using a centrifugal filter (Amicon® Ultra, 0.5 ml/10 K). Add equal volume of 100 % Molecular Biology-grade glycerol to the concentrated protein sample (*see* **Note 7**); store the enzyme stock at −20 °C.

3.3 In Vitro Polyadenylation Assays with RiboGreen®

Both PNP and RNA polymerase-associated, accessory PAP obtained in the purification procedure described above carry endogenous, protein-associated RNA oligonucleotides. The presence of such oligonucleotides in RNA polymerase–PAP mixtures and attribution of the observed $(rA)_n$-synthetic activity to PAP is supported by multiple, independent lines of evidence described in our recent work [7]. In the presence of nucleoside substrates, these protein-associated RNA primers can be polyadenylated by either PAP or PNP. In vitro, this RNA synthesis can be monitored using RNA-specific fluorescent dyes. The RNA- and DNA-specific fluorescent dye RiboGreen® (Invitrogen/Molecular Probes) dramatically increases its fluorescence at 538 nm in the presence of RNA. Therefore, this fluorescent reporter-based method for the detection of nascent polyadenylate represents an attractive alternative to methods relying on radiolabeled nucleosides (*see* **Note 8**). In the protocol described below, we monitor $(rA)_n$-synthetic activity among *E. coli* proteins fractionated on a Mono Q column (Fig. 1a). Note that this method also works well for purified PNP and PAP; its application can significantly reduce the costs of enzymatic analyses in either case.

1. To prepare 2× Reaction buffer with RiboGreen®, add 3 μl of RiboGreen® stock (Invitrogen/Molecular Probes, R11491) to 3 ml of 2× Reaction buffer D (*see* above). For enzymatic assays with PAP, add 30 μl of 100 mM ATP to prepare Master mix (this amount of Master mix will be sufficient for approximately 50 individual reactions). Substitute ATP with ADP in order to assay PNP's $(rA)_n$-synthetic activity. Add 60 μl of Master mix per well of a short black 0.3 ml/96-well plate. Add 20–60 μl aliquots from individual Mono Q fractions (or purified PNP or PAP) to each well and ultrapure water, if applicable, to a final volume of 120 μl. Monitor the kinetics of RiboGreen® fluorescence ($\lambda_{ex} = 485$ nm/$\lambda_{em} = 538$ nm); a representative set of data is shown in Fig. 2.
2. Typically, enzymatic reactions with endogenous, protein-associated RNA primers display distinct lags due to a low initial concentration of substrate RNA; lags are typically followed by a steady-state phase of polyadenylate production. You can modify the described experimental setup by (1) including synthetic RNA primers of desired length and nucleotide composition, (2) varying ionic strength and the composition of Reaction buffer (*see* **Note 9**), (3) including controls lacking nucleoside di- and triphosphate substrates (the latter controls could reveal modest increases in RiboGreen® fluorescence for certain samples due to the interaction of RiboGreen® with protein-associated RNAs), etc.—in order to tailor this enzymatic assay to your specific needs.
3. Alternatively, if you are performing Mono Q chromatography in a "scouting mode" (i.e., if you are committing to Mono Q

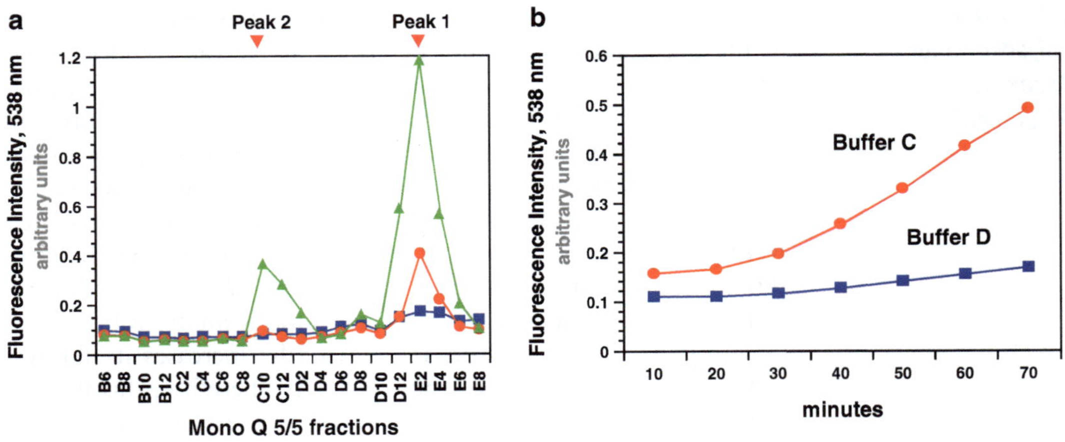

Fig. 2 Monitoring the synthesis of polyadenylate through a method based on utilization of a nucleic acid-specific fluorescent dye. (**a**) Synthesis of polyadenylate by native *E. coli* RNA polymerase–(RNA)–accessory protein(s) complexes. This panel illustrates the enzymatic assay described in Subheading 3.3. Graphs show fluorescence intensities of RiboGreen® in individual reactions corresponding to the indicated Mono Q fractions immediately after the addition of ATP (*rectangles*) or after 60 and 120 min incubations at room temperature (respectively, *circles* and *triangles*). Note the synthesis of polyadenylate by the RNA polymerase-associated PAP (*Peak 1*) and a measurable activity of PNP (*Peak 2*), which is observed at late time points due to latent ATP hydrolysis in the sample. The presence of endogenous, protein-associated RNA oligonucleotides serving as primers for the polyadenylation reaction was verified through mass spectrometry analyses of the native RNA polymerase complexes [7]; the attribution of the observed $(rA)_n$-synthetic activities to accessory PAP and PNP is based on multiple lines of biochemical and genetic evidence [7]. (**b**) Extension of endogenous, protein-associated RNA primers by purified PNP. Kinetics of polyadenylate production in 120 μl reactions containing 0.52 μg PNP (purified as described in Fig. 1b) in the presence of 1 mM ADP. Reactions were carried out as described in Subheading 3.3 using either Reaction buffer C (*circles*) or Reaction buffer D (*rectangles*)

fractionation only a relatively small portion of your protein and collecting fractions into a short 96-well plate) and your fluorimeter/plate reader is equipped with a microinjector, you can autoinject Master mix into wells of the plate. This "high-throughput" approach can considerably simplify the acquisition of data.

3.4 Polyadenylation Assays Utilizing ^{32}P-Labeled Nucleotides

In vitro, polyadenylation of endogenous RNA oligonucleotides found in native RNA polymerase complexes is enhanced in the presence of excess purified RapA, an RNA polymerase-associated SWI/SNF protein [7]. This effect is likely due to RapA-mediated remodeling of RNA polymerase–RNA complexes resulting in extrusion of the 3′ ends of RNAs [7]. Polyadenylation of RNAs with extruded 3′ ends by (accessory, RNA polymerase-associated) PAP can be monitored when this enzymatic reaction is carried out in the presence of [Alpha-^{32}P] ATP and the resulting ^{32}P-labeled RNA products are fractionated on urea gels. Below, we describe such an enzymatic assay.

1. Prepare gel assembly (for Gibco SA 32 vertical electrophoresis module consisting of a short plate [borosilicate glass,

336 mm × 196 mm], a set of 0.4 mm thick spacers, tall plate [borosilicate glass, 360 mm × 196 mm], and a Gibco SA 32 gel casting boot). We use custom-manufactured borosilicate glass plates, which, typically, we do not siliconize. If you are using Gibco SA 32 (or Gibco SA) vertical electrophoresis module, prepare 50 ml of the gel mix containing 20–25 % acrylamide/Bis (19:1) (InstaPage™-19, IBI Scientific, obtained as a 40 % stock solution), 1× TBE (KD Medical, obtained as 10× stock), and urea (Sigma, U5378, dry powder) to a final concentration of 7 M; add 60 μl of TEMED (Bio-Rad) and, immediately before gel casting, 0.3 ml of 10 % ammonium persulfate. Alternatively, if you are using a UreaGel™ kit (National Diagnostics), mix 40 ml Concentrate, 5 ml Diluent and 5 ml Buffer; before gel casting, add 0.3 ml of 10 % ammonium persulfate (Sigma). Cast gel immediately after the addition of ammonium persulfate and insert a comb after pouring in a gel mix (*see* **Note 10**). Set gel assembly aside for 2–3 h at room temperature.

2. Prepare 1 mM ATP Mix (containing [Alpha-^{32}P] ATP), which will be used to initiate the synthesis of polyadenylate. Mix the following ingredients in a 0.5 ml tube: 2 μl of 10 mM ("cold") ATP, 2 μl of [Alpha-^{32}P] ATP (approximately 25 μCi/μl), and 16 μl of ultrapure water (*see* **Note 11**).
3. Dilute purified RapA (typically stored at −20 °C as 9–20 μM glycerol stock) with ultrapure water to make 0.2 μM and 2 μM working solutions.
4. Prepare the samples by mixing 1 μl of 10× Reaction buffer D; 1–2 μl of purified native RNA polymerase complexes containing accessory PAP and endogenous, protein-associated RNAs (Mono Q fractions D10-E2); 1–2 μl of purified RapA, if present (up to a final concentration of 0.2–0.3 μM); and ultrapure water to a final volume of 8 μl. Add 2 μl of 1 mM ATP Mix (containing [Alpha-^{32}P] ATP)/tube to initiate the synthesis of polyadenylate. Incubate reactions for 30 min at 37 °C; terminate reactions by the addition of 2 volumes (20 μl) of 3× Formamide Loading buffer (*see* **Note 12**).
5. Assemble the electrophoresis module. Add Electrode buffer; load samples (4–8 μl/well) using 0.37 mm Gel Loader pipette tips. Typically, we run SA 32 urea gels at 35–45 W until Bromophenol Blue migrates approximately ¾ of the length of the gel.
6. After electrophoresis, carefully separate glass plates and transfer the gel on a flexible film support. Cover the gel with a clear plastic wrap and place the wrapped gel in a cassette. In a dark room, place an X-ray film and an intensifying screen on the gel. The optimal exposure times (at −72 °C, with Kodak BioMax MS screens) are typically between 12 and 48 h. A representative film is shown in Fig. 3.

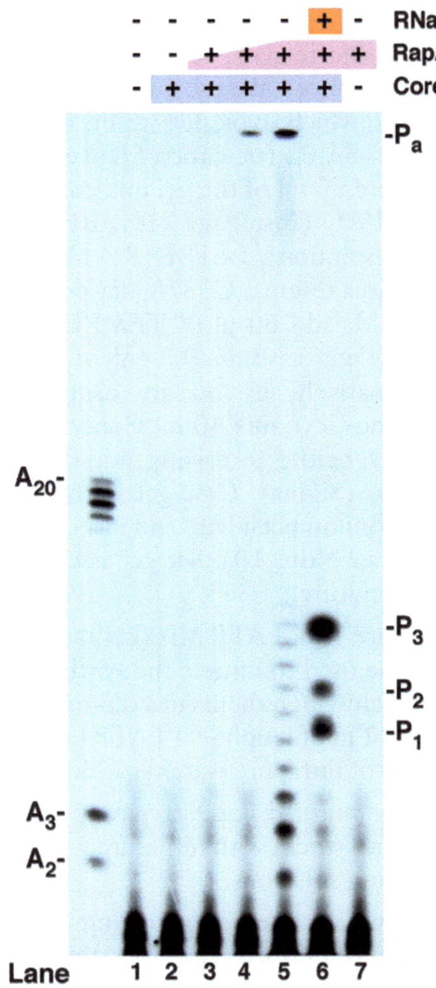

Fig. 3 Polyadenylation of endogenous RNA oligonucleotides found in native RNA polymerase–(RNA)–accessory protein(s) complexes in the presence of RNA polymerase-associated protein RapA. Reactions were carried out as described in Subheading 3.4; the composition of individual samples in lanes 1–7 is indicated in the figure. When indicated, core RNA polymerase (containing accessory proteins and RNA) (Core[a]) was present at 0.08 μg/μl; purified RapA in lanes 3, 4, and 5–7, respectively, at 33, 100, and 300 nM. Following a 30-min incubation, the reaction mixture in lane 6 was treated with 2.5 U of RNase I_f for 5 min at 37 °C. The "$A_{2\text{-}3\text{-}20}$" RNA ladder was prepared by mixing undigested, 5′ end-labeled A_{20} probe with that pretreated with purified PNP (isolated as described above) in the presence of 1 mM sodium phosphate, in Reaction buffer D (*see* **Note 9**). 5′ end labeling and gel purification of end-labeled RNAs was carried out as described [11]. P_a polyadenylate, *P1–P3* reaction products resulting from the modification of the extruded ends of protein-associated RNAs. An X-ray film exposed to a urea gel containing ^{32}P-labeled RNAs is shown

7. To expand this basic assay, following the extension reaction (but before the addition of Loading buffer), ^{32}P-labeled RNA products can be subjected to additional enzymatic (nuclease, etc.) analyses. A representative effect of 2.5 U of RNase I_f on the composition of the extension products is shown in Fig. 3, lane 6.

4 Notes

1. Aliquot 10 % ammonium persulfate aqueous stock and store aliquots frozen at −20 °C. Thaw individual aliquots weekly and store them at room temperature.
2. This film size would fit gels cast in an SA 32 module; consider 35×43 cm films if you are using an S2-type apparatus.
3. We recommend handheld counters with sample holders, such as RadEye B20-ER (Thermo Scientific), which could greatly simplify sample analyses.
4. We grow cells in 1 L batches using Super Broth (KD Medical), in which higher cell densities can be obtained.
5. This purification step has to be carried out in a cold room or a refrigerated cabinet.
6. The Mono Q chromatography can be carried out at room temperature.
7. Consider using wide-bore, sterile tips when pipetting glycerol.
8. With unrefined *E. coli* extracts—containing RNA polymerase carrying accessory DNA fragments—applicability of this method will be limited due to the contribution of RNA polymerase (specifically, core RNA polymerase) to $(rA)_n$ synthesis.
9. Note that in the presence of phosphate, PNP converts polyadenylate into 2- and 3-nucleotide-long RNA fragments, and in vitro this reaction could interfere with $(rA)_n$-synthetic activity of PNP. In vivo, contribution of PNP to RNA polyadenylation has been convincingly established [10]. The exact balance of the two activities of PNP is debatable, and it is possible that nonphysiologically low ionic strengths and artificial anion makeups of reaction buffers used in most studies (in *E. coli*, intracellular osmolarity varies widely across approximately 0.3–0.7 M range [12–14]) could account for the effects of phosphate observed in vitro.
10. Typically, we cast urea gels using 80–100 ml "gel casting" squirt bottles.
11. The unused ATP Mix can be stored at −20 °C in a shielded container.
12. To minimize RNA degradation we do not recommend boiling the samples.

Acknowledgment

This work was supported in part by NIH grant GM81803.

References

1. Sukhodolets MV, Jin DJ (1998) RapA, a novel RNA polymerase-associated protein, is a bacterial homolog of SWI2/SNF2. J Biol Chem 273:7018–7023
2. Sukhodolets MV, Garges S, Jin DJ (2003) Purification and activity assays of RapA, the RNA polymerase-associated homolog of the SWI/SNF protein superfamily. Methods Enzymol 370:283–90
3. Zhi H, Yang W, Jin DJ (2003) *Escherichia coli* proteins eluted from mono Q chromatography, a final step during RNA polymerase purification procedure. Methods Enzymol 370: 291–300
4. Hager DA, Jin DJ, Burgess RR (2003) Use of Mono Q high-resolution ion-exchange chromatography to obtain highly pure and active *Escherichia coli* RNA polymerase. Biochemistry 29: 7890–7894
5. Sukhodolets MV, Garges S (2003) Interaction of *Escherichia coli* RNA polymerase with the ribosomal protein S1 and the Sm-like hexameric ATPase Hfq. Biochemistry 42: 8022–8034
6. Kansara SG, Sukhodolets MV (2011) Oligomerization of the *E. coli* core RNA polymerase: formation of $(\alpha_2\beta\beta'\omega)_2$-DNA complexes and regulation of the oligomerization by auxiliary subunits. PLoS One 6:e18990
7. Richmond M, Pasupula RR, Kansara SG et al (2011) RapA, *Escherichia coli* RNA polymerase SWI/SNF subunit-dependent polyadenylation of RNA. Biochemistry 50:2298–2312
8. Mohanty BK, Kushner SR (1999) Analysis of the function of *Escherichia coli* poly(A) polymerase I in RNA metabolism. Mol Microbiol 34:1094–1108
9. Mohanty BK, Kushner SR (2006) The majority of *Escherichia coli* mRNAs undergo post-transcriptional modification in exponentially growing cells. Nucleic Acids Res 34:5695–5704
10. Mohanty BK, Kushner SR (2000) Polynucleotide phosphorylase functions both as a 3′–5′ exonuclease and a poly(A) polymerase in *Escherichia coli*. Proc Natl Acad Sci U S A 97:11966–11971
11. McKinley B, Sukhodolets MV (2007) *Escherichia coli* RNA polymerase-associated SWI/SNF protein RapA: evidence for RNA-directed binding and remodeling activity. Nucleic Acids Res 35:7044–7060
12. Cayley S, Lewis BA, Guttman HJ et al (1991) Characterization of the cytoplasm of *Escherichia coli* K-12 as a function of external osmolarity. J Mol Biol 222:281–300
13. Richey B, Cayley DS, Mossing MC et al (1987) Variability of the intracellular ionic environment of *Escherichia coli*. Differences between in vitro and in vivo effects of ion concentrations on protein-DNA interactions and gene expression. J Biol Chem 262:7157–7164
14. Record MT Jr, Courtenay ES, Cayley S et al (1998) Biophysical compensation mechanisms buffering *E. coli* protein-nucleic acid interactions against changing environments. Trends Biochem Sci 23:190–194

Chapter 21

Analysis of the Human Mitochondrial Transcriptome Using Directional Deep Sequencing and Parallel Analysis of RNA Ends

Oliver Rackham and Aleksandra Filipovska

Abstract

RNA sequencing using next-generation technologies provides comprehensive coverage of transcriptomes at a much greater depth than conventional transcriptomic methods. The human mitochondrial genome is relatively small, and sequencing its transcriptome provides a valuable method to investigate changes in RNA metabolism in great detail. Here we describe two methods that use next-generation technologies to investigate mitochondrial RNAs. Directional RNA sequencing enables the analyses of RNA abundance from each strand of the mitochondrial DNA. Parallel analysis of RNA ends enables the analyses of processing of mitochondrial transcripts, their termini, and annotation of any new transcripts.

Key words Next-generation sequencing, RNA, Parallel analysis of RNA ends, Mitochondria, RNA-Seq

1 Introduction

The human mitochondrial transcriptome, generated from a small and compact genome, has surprisingly complex composition and regulation. The abundance of mitochondrial RNAs, although produced from precursor polycistronic transcripts, varies between different tissues and cell types indicating that posttranscriptional processing, maturation, and degradation mechanisms play important roles in the regulation of mitochondrial gene expression (reviewed in [1–3]). There is increasing evidence that mitochondrial RNA-binding proteins control the mitochondrial transcriptome from its synthesis to its destruction and have evolved unique features to complement the unusual features of mitochondrial RNAs [1–5]. Mitochondrial RNA-binding proteins can have profound effects on cell function and energy metabolism; therefore, investigating the mechanisms by which they regulate mitochondrial gene expression is particularly important to gain insight into human health and diseases caused by mitochondrial dysfunction.

Joanna Rorbach and Agnieszka J. Bobrowicz (eds.), *Polyadenylation: Methods and Protocols*, Methods in Molecular Biology, vol. 1125, DOI 10.1007/978-1-62703-971-0_21, © Springer Science+Business Media New York 2014

Advances in next-generation technologies have provided great insight into cellular genomes and transcriptomes. Recently a comprehensive map of the human mitochondrial transcriptome revealed new and unexpected features and complexity of mitochondrial transcripts [6]. RNA sequencing and parallel analyses of RNA ends (PARE) [7, 8] can be used to investigate the human mitochondrial transcriptome, RNA abundance, RNA processing, and maturation sites and to identify new transcripts generated by the mitochondrial genome of mitochondria from different cell types or tissues [6, 9, 10] (*see* **Note 1**). Furthermore, mitochondrial RNA metabolism can be investigated using these techniques in cells where specific proteins of interest have been depleted, overexpressed, mutated, and knocked out or when mitochondrial gene expression is affected by environmental or physiological signals [11]. We have been able to analyze the abundance of mitochondrial transcripts as well as the changes in 5′ and 3′ ends of mitochondrial transcripts upon knockdown of different mitochondrial RNA-binding proteins [9]. Furthermore, an increase in the sequencing error rate at known modified nucleotides compared with non-modified nucleotides [12, 13] can be used as a marker for the presence of nucleotide modifications to suggest a role for specific proteins in mitochondrial RNA modification [9].

Here we describe two methods to prepare libraries for next-generation sequencing to analyze mitochondrial transcript abundance, processing, and stability. Directional sequencing of RNAs enables the investigation of mitochondrial RNA abundance and stability (Fig. 1), and PARE provides a global profile of mitochondrial transcript 5′ ends that precisely resolves RNA cleavage sites and also indicates unanticipated noncanonical cleavage sites (Fig. 2). These two techniques enable detailed analyses of the regulation and processing of mitochondrial RNAs.

2 Materials

All reagents for RNA work are prepared in diethylpyrocarbonate (DEPC)-treated ultrapure water, and the equipment used is treated with RNaseZap (Ambion) and rinsed in DEPC (Sigma)-treated ultrapure water before use. Sterile RNase-free filter pipette tips and other disposable plasticware should be used to prevent contamination and degradation of RNA samples. All RNA samples should be stored at −80 °C.

2.1 Materials for Isolation of Mitochondria from Cells

1. Trypsin (GIBCO, Life Technologies).
2. Swelling buffer (10 mM NaCl, 1.5 mM $MgCl_2$, and 10 mM Tris–HCl, pH 7.5).
3. Dounce tissue grinder set, 7 ml (Sigma).

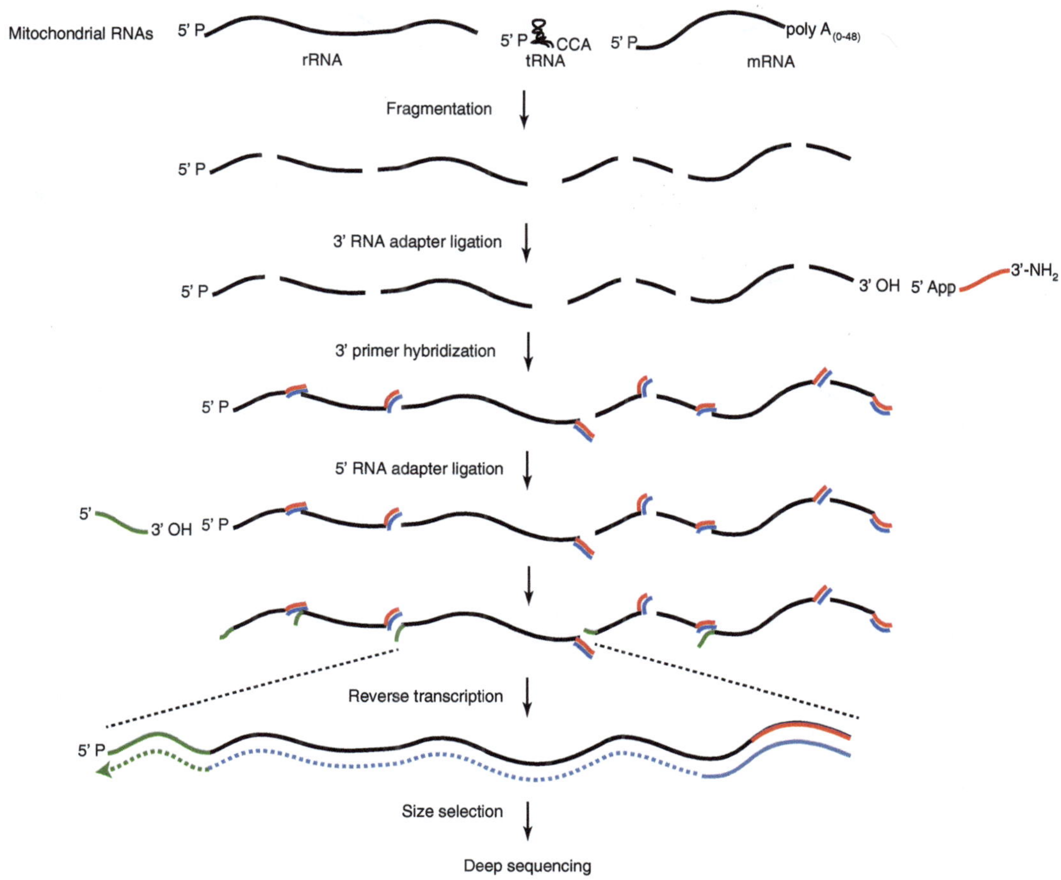

Fig. 1 Schematic representation of library preparation for directional sequencing of mitochondrial transcripts. RNA from mitochondria or mitoplasts isolated from cells or tissues is fragmented before RNA adapters are ligated at the 3′ end of the fragmented RNAs. Adapters are ligated at the 5′ end of the fragmented RNAs, and reverse transcription is initiated using primers that hybridize to the 3′ end adapters. The DNA products can be size selected using polyacrylamide gel electrophoresis and gel purification before sequencing

4. $T_{10}E_{20}$ buffer (10 mM Tris–HCl, and 1 mM EDTA, pH 7.6, with sucrose added to 250 mM final concentration by adding 1 ml of 2 M sucrose in 10 ml after adjusting the pH).
5. RNase A (Fermentas, Thermo Scientific).
6. Discontinuous sucrose gradient (sucrose added to 1.0 and 1.7 M final concentration in $T_{10}E_{20}$ buffer).
7. Pasteur pipettes.
8. miRNeasy Mini Kit (Qiagen).

2.2 Materials for Directional Sequencing of Mitochondrial RNAs

1. 5× Fragmentation Buffer (Illumina).
2. Fragmentation Stop Solution (Illumina).
3. 3 M NaOAc (pH 5.2).

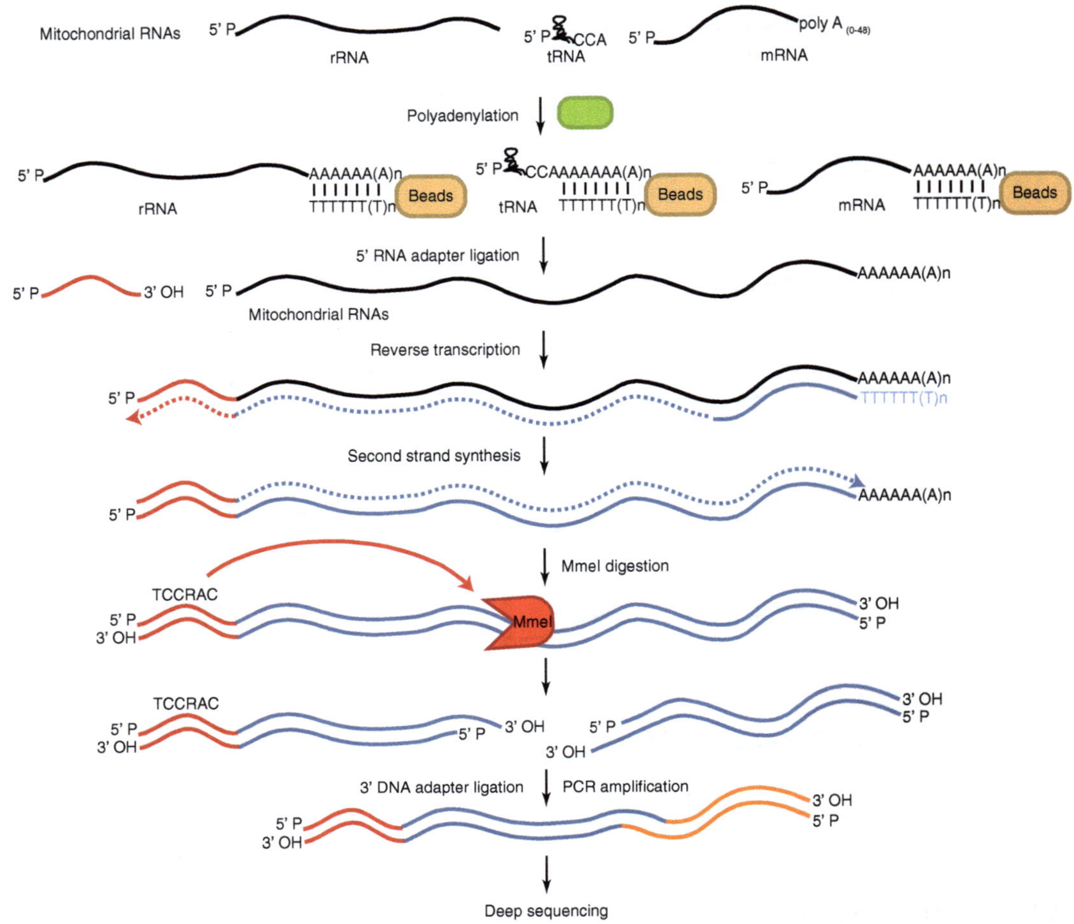

Fig. 2 Schematic representation of library preparation for parallel analysis of RNA ends of mitochondrial transcripts. Isolated mitochondrial or mitoplast RNAs are polyadenylated to ensure that all transcripts are enriched in the PARE library, including transcripts that are not generally polyadenylated such as MT-ND6 mRNA, rRNAs, and tRNAs. The polyadenylated RNAs are selected using oligo (dT) beads, and a 5′ adapter is ligated at the 5′ monophosphate of the mitochondrial RNAs that contains a unique MmeI site which cleaves 20 nt away at the 3′ recognition site of the enzyme as described in 7. An oligo (dT) primer is used for cDNA synthesis, and a primer complementary to the 5′ end adaptor is used to amplify the cDNA. The PCR product is digested with MmeI, which generates products of homogeneous size that are extracted by PAGE. Double-stranded adaptor is ligated at the 3′ end of the equal size fragments to enable their enrichment by PCR and the inclusion of barcoding sequences so that multiple samples can be sequenced on the same lane of a next-generation sequencer

4. Glycogen (Life Technologies).
5. Ethanol.
6. TruSeq RNA Small Prep Kit (Illumina).

2.3 Materials for Parallel Analysis of RNA Ends

1. Poly(A) polymerase (New England Biolabs).
2. Glycoblue (Ambion, Life Technologies).
3. Glycogen (Life Technologies).

4. RiboMinus kit (Invitrogen, Life Technologies).
5. T4 RNA Ligase (Ambion, Life Technologies).
6. Oligotex mRNA Mini Kit (Qiagen).
7. RNase OUT (Invitrogen, Life Technologies).
8. Superscript II reverse transcriptase (Invitrogen, Life Technologies).
9. MmeI restriction enzyme (New England Biolabs).
10. Phusion High-Fidelity DNA Polymerase (New England Biolabs or Thermo Scientific).
11. FastAP (Fermentas, Thermo Scientific).
12. 1-bp DNA ladder (Invitrogen, Life Technologies).
13. Phenol/chloroform/isoamyl alcohol (Roche Applied Science).
14. Chloroform:Isoamyl alcohol 24:1 (Sigma).
15. Ethidium bromide (Sigma).
16. 1× TAE buffer: 40 mM Tris base, 20 mM acetate, 2 mM EDTA, pH 8.1.
17. 1× TBE buffer: 89 mM Tris base, 89 mM boric acid, 2 mM EDTA, pH 8.0.
18. 6× loading dye: 50 % (v/v) glycerol, 0.25 m % (w/v) bromophenol blue, 0.25 % (w/v) xylene cyanol FF, and 10 mM EDTA.
19. Prepare a 12 % polyacrylamide gel by combining the following (make enough for two gels): 3.15 ml of acrylamide (30 %, 37.5:1, BioRad), 2 ml of 5× TBE (0.44 M Tris base (53.90 g/l), 0.44 M boric acid (27.50 g/l), and 10 mM EDTA, disodium, dihydrate (3.70 g/l)), 4.85 ml of DEPC water, 45 μl of 10 % APS, and 9.2 μl TEMED.
20. Synthetic RNA adaptor "5′-RNA-adaptor": 5′-GUUCAGAGUUCUACAGUCCGAC-3′ (custom synthesis by Dharmacon).
21. DNA oligonucleotides (barcode sequences are shown in lower case):

    PAREfrt5′-CGAGCACAGAATTAATACGACTTTTTTTTTTTTTTTTTV-3′.

    PAREpcr1fwd 5′-GTTCAGAGTTCTACAGTCCGAC-3′.

    PAREpcr1rev 5′-CGAGCACAGAATTAATACGACT-3′.

    PAREadapt1fwd 5′-phosphate-atcacgTCGTATGCCGTCTTC TGCTTG-3′.

    PAREadapt1rev 5′-CAAGCAGAAGACGGCATACGAcgtgatNN-3′.

    PAREadapt2fwd 5′-phosphate-cgatgtTCGTATGCCGTCTTC TGCTTG-3′.

PAREadapt2rev 5′-CAAGCAGAAGACGGCATACGAacatcg NN-3′.
PAREadapt3fwd 5′-phosphate-ttaggcTCGTATGCC-GTCTTCT GCTTG-3′.
PAREadapt3rev 5′-CAAGCAGAAGACGGCATACGAgcctaa NN-3′.
PAREadapt4fwd 5′-phosphate-tgaccaTCGTATGCC-GTCTTCT GCTTG-3′.
PAREadapt4rev 5′-CAAGCAGAAGACGGCATACGAtggtca NN-3′.
PAREadapt5fwd 5′-phosphate-acagtgTCGTATGCC-GTCTTC TGCTTG-3′.
PAREadapt5rev 5′-CAAGCAGAAGACGGCATACGAcactgt NN-3′.
PAREpcr2fwd 5′-AATGATACGGCGACCACCGACAGGTT CAGAGTTCTACAGTCCGA-3′.
PAREpcr2rev 5′-CAAGCAGAAGACGGCATACGA-3′.
PARE sequencing primer 5′-CCACCGACAGGTTCAG AGTTCTACAGTCCGAC-3′.

22. PCR buffer $-Mg^{2+}$ (Invitrogen, Life Technologies).

3 Methods

All procedures are carried out at room temperature unless otherwise specified.

3.1 Isolation of Mitochondria from Cells

1. Add trypsin to cells from confluent 15 cm dishes and harvest by low-speed centrifugation (300 × *g*). Suspend the cell pellet in 7 ml/15 cm dish of swelling buffer (10 mM NaCl, 1.5 mM $MgCl_2$, and 10 mM Tris–HCl, pH 7.5).
2. Swell the cells for 4–5 min on ice and homogenize five times with a loose glass homogenizer and then ten times with a tight glass homogenizer.
3. Adjust the sucrose concentration to 250 mM by adding 1 ml of 2 M sucrose in 10 mM Tris–HCl and 1 mM EDTA, pH 7.6 ($T_{10}E_{20}$ buffer).
4. Centrifuge the suspension at 1,300 × *g* for 3 min, suspend the pellet in 250 mM sucrose in $T_{10}E_{20}$ buffer, and centrifuge again at 1,300 × *g* for 3 min. Pool the two supernatants and use in **step 5**.
5. Centrifuge the supernatant at 15,000 × *g* for 15 min to sediment the mitochondria.
6. Add 1 ml of 250 mM sucrose in $T_{10}E_{20}$ buffer to the pellet and centrifuge at 15,000 × *g* for 15 min.
7. Add 100 μl of 250 mM sucrose in $T_{10}E_{20}$ buffer to the pellet.

8. Add 1 μl of RNase A to the 100 μl of mitochondrial suspension and incubate for 30 min on ice.
9. Layer the mitochondrial suspension on a discontinuous sucrose gradient (1.0 and 1.7 M) in $T_{10}E_{20}$ buffer and centrifuge at 70,000 × *g* for 40 min.
10. Recover the mitochondrial fraction from the interface between the two sucrose cushions, using a Pasteur pipette or an equivalent. Typically the mitochondrial fraction is collected into 300–400 μl.
11. Wash the recovered mitochondrial suspension in 250 mM sucrose in $T_{10}E_{20}$ buffer by centrifugation at 15,000 × *g* for 10 min.
12. Purify RNA from isolated mitochondria using the miRNeasy Mini Kit according to the manufacturer's instructions (*see* **Note 2**).

3.2 Directional Sequencing of Mitochondrial RNAs

1. To a sterile, RNase-free 1.5 ml microcentrifuge tube, add the following:

   4 μl 5× Fragmentation buffer.

   1–2 μg mitochondrial RNA.
   DEPC-treated ultrapure water to 20 μl.
2. Incubate the reaction at 94 °C for 5 min.
3. At the end of the incubation add 2 μl of fragmentation stop solution and place the tube on ice.
4. Precipitate the RNA by adding 2 μl of 3 M NaOAc (pH 5.2), 2 μl of glycogen (20 μg/μl), and 60 μl of 100 % ethanol and incubate at −80 °C for 30 min or overnight.
5. Collect the RNA by centrifugation at 20,200 × *g* for 25 min at 4 °C.
6. Carefully remove the ethanol with a pipette without disrupting the RNA pellet.
7. Wash the pellet with 300 μl of 70 % ethanol, centrifuge the pellet as in **step 6**, and remove the 70 % ethanol with a pipette.
8. Air-dry the pellet for 10 min at room temperature.
9. Resuspend the RNA in 10 μl of RNase-free water.
10. Prepare the library by following the instructions in the TruSeq RNA Small Prep Kit (*see* **Note 3**).

3.3 Parallel Analysis of RNA Ends

Prepare mitochondrial RNA as described in Subheading 3.1.

3.3.1 Polyadenylation of RNA

1. To a sterile, RNase-free 1.5 ml microcentrifuge tube, add the following:
   5 μg of mitochondrial RNA in a volume of 38 μl.
   5 μl of poly(A) polymerase reaction buffer.

5 μl of 10 mM ATP.
2 μl poly(A) polymerase (10 U).

2. Incubate the tube at 37 °C for 30 min.

3.3.2 RNA Precipitation

1. To the polyadenylation reaction add 2 μl of Glycoblue, 30 μl of 3 M NaOAc, and 900 μl of cold 100 % ethanol. Centrifuge at 4 °C for 30 min at maximum speed using a bench top microfuge.
2. Wash the pellet with 500 μl of 70 % ethanol. Use a pipette to remove the supernatant, and discard it. Do not disturb the RNA pellet.
3. Air-dry the pellet, and resuspend it in 5 μl of nuclease-free water.

3.3.3 Removal of Cytoplasmic rRNA Using the RiboMinus Kit

1. Use the RiboMinus Kit essentially as described in the manufacturer's instructions, except concentrate the final supernatant containing the rRNA-depleted RNA as outlined in the next section.

3.3.4 RNA Precipitation

1. To the rRNA-depleted RNA add 2 μl of Glycoblue, 30 μl of 3 M NaOAc, and 900 μl of cold 100 % ethanol. Centrifuge at 4 °C for 30 min at maximum speed using a bench top microfuge.
2. Wash the pellet with 500 μl of 70 % ethanol. Use a pipette to remove the supernatant, and discard it. Do not disturb the RNA pellet.
3. Air-dry the pellet, and resuspend it in 15 μl of nuclease-free water.

3.3.5 Ligation of the RNA Adapter

1. To the entire RNA sample (from above), add the following:
   1 μl of 5′-RNA adapter (200 μM).
   2 μl of 10× RNA ligase buffer.
   2 μl of T4 RNA ligase.
2. Mix gently, and incubate at 37 °C for 1 h.

3.3.6 RNA Extraction and Precipitation

1. Add 250 μl of nuclease-free water and 300 μl of phenol:chloroform:isoamyl alcohol (25:24:1) to the ligation reaction. Mix vigorously using a vortex for 10 s and centrifuge for 5 min at maximum speed using a bench top microfuge. Remove aqueous phase, and add 300 μl of chloroform:isoamyl alcohol (24:1). Mix vigorously using a vortex for 10 s, and centrifuge for 5 min at maximum speed using a bench top microfuge. Keep the aqueous phase.
2. To the retained aqueous phase add 2 μl of Glycoblue, 30 μl of 3 M NaOAc, and 900 μl of cold 100 % ethanol. Centrifuge at 4 °C for 30 min at top speed using a bench top microfuge.

3. Wash the pellet with 500 µl of 70 % ethanol. Use a pipette to remove the supernatant, and discard it. Do not disturb the RNA pellet.
4. Air-dry the pellet, and resuspend it in 250 µl of nuclease-free water.

3.3.7 Purifying the Poly(A) RNA with Oligotex Mini Kit

1. Use the Oligotex Mini Kit essentially as described in the manufacturer's instructions for isolation of poly A+mRNA from total RNA using the Oligotex mRNA Spin-Column Protocol.
2. At the elution step use 35 µl of hot (70 °C) buffer OEB. Pellet the Oligotex beads in a bench top centrifuge for 1 min at maximum speed. Reheat the eluate to 70 °C, elute in the same microcentrifuge tube by pipetting up and down three or four times to resuspend the resin, and centrifuge again for 1 min at maximum speed.

3.3.8 Reverse Transcription

1. To 29 µl of the purified ligated RNA from above add 1 µl of 100 µM PARErt and heat to 65 °C for 5 min. Cool to 4 °C (it is most convenient to use 100 µl thin-wall PCR tubes in a PCR machine to perform this and the following steps as a single program).
2. Add the following to the RNA/primer mix:
   10 µl of 5× first-strand buffer.
   2.5 µl of 10 mM dNTP mix.
   2.5 µl of 0.1 M DTT.
   2 µl of RNaseOUT (40 U/µl).
   3 µl of Superscript II reverse transcriptase (200 U/µl).
3. Mix gently by pipetting up and down, and incubate at 42 °C and gradually raise the temperature to 45 °C for 3 h (increasing 1 °C every 45 min). Inactivate the reverse transcriptase at 70 °C for 10 min.

3.3.9 cDNA Precipitation

1. Transfer the reverse transcriptase reaction mix to a 1.5 ml microcentrifuge tube and add 2 µl of Glycoblue, 30 µl of 3 M NaOAc, and 900 µl of cold 100 % ethanol. Centrifuge at 4 °C for 30 min at top speed using a bench top microfuge.
2. Wash the pellet with 500 µl of 70 % ethanol. Use a pipette to remove the supernatant, and discard it. Do not disturb the RNA pellet.
3. Air-dry the pellet, and resuspend it in 7 µl of nuclease-free water.

3.3.10 Short PCR Amplification

1. Add the following to two 100 µl thin-wall PCR tubes to make two identical reaction mixes:
   3 µl of cDNA (from above).
   33.5 µl of nuclease-free water.

10 μl of 5× HF buffer.
1 μl of dNTP mix.
1 μl of PAREpcr1fwd (20 μM).
1 μl of PAREpcr1rev (20 μM).
0.5 μl of Phusion High-Fidelity DNA Polymerase.

2. Cycle the reactions as follows: (a) 98 °C for 30 s; (b) 98 °C for 20 s, 60 °C for 30 s, and 72 °C for 3 min, repeat five times; and (c) 72 °C for 7 min.
3. Perform a control PCR reaction in parallel with those above. In this reaction use the remaining 1 μl of cDNA and repeat the cycles outlined in (b) 35 times. This reaction should generate a smear when examined on a 1 % agarose gel.

3.3.11 DNA Precipitation

1. Combine the two identical PCR reactions in a 1.5 ml microcentrifuge tube and add 2 μl of Glycoblue, 30 μl of 3 M NaOAc, and 900 μl of cold 100 % ethanol. Centrifuge at 4 °C for 30 min at top speed using a bench top microfuge.
2. Wash the pellet with 500 μl of 70 % ethanol. Use a pipette to remove the supernatant, and discard it. Do not disturb the DNA pellet.
3. Air-dry the pellet, and resuspend it in 22 μl of nuclease-free water.

3.3.12 MmeI Digestion

1. Add the following to the purified PCR product, from above:

3 μl of 10× NEB buffer 4.
3 μl of 10× SAM (500 μM).
2 μl of MmeI (2 U/μl).

2. Incubate at 37 °C for 2 h.
3. Dephosphorylate the cut PCR fragments with 2 μl of FastAP. Mix gently by pipetting up and down, and incubate for 1 h at 37 °C.
4. Inactivate the FastAP at 75 °C for 5 min, and then place the reaction mix on ice.

3.3.13 First PAGE Purification

1. Prepare a 12 % polyacrylamide gel, as described above.
2. Add 6× DNA loading buffer to the dephosphorylated DNA from above. Mix and load the DNA between three lanes of the polyacrylamide gel (~14 μl of DNA in each lane). Load the 10-bp DNA ladder as a size standard in one lane (use 2 μl of a 1 μg/μl stock plus 6× DNA loading buffer).
3. Run the gel at 180 V in TBE running buffer until bromophenol blue and xylene cyanol dye fronts are well separated and stain in 50 ml of 1× TAE containing ethidium bromide for 5 min.
4. Use a razor blade to excise gel bands containing the MmeI-cleaved fragments (corresponding to 42 bp, these will not be

visible in the PARE samples but the DNA ladder should be used as a guide) (*see* **Note 4**). Combine the gel pieces in one 2 ml tube and finely crush with a sterile plunger from a 1 ml syringe.

5. Add 2 volumes of 0.3 M NaCl (typically ~350 μl) to the crushed gel pieces and elute overnight at room temperature on a rotating mixer.
6. Pellet the crushed gel pieces for 10 min at maximum speed in a bench top centrifuge.
7. Remove supernatant (~200 μl), add another 200 μl 0.3 M NaCl to the gel pieces, vortex, centrifuge again, and combine supernatants. Centrifuge combined supernatants at maximum speed in a bench top centrifuge and transfer to a new tube; repeat until no traces of gel are carried over. Add 2.5 volumes of 100 % ethanol (1 ml) and 2 μl of Glycoblue and incubate at −80 °C overnight.
8. Centrifuge the samples at maximum speed at 4 °C for 1 h. Wash the pellet with 70 % ethanol, and air-dry it for 10 min to eliminate as much ethanol as possible.

3.3.14 Ligation of 3′ DNA Adapters

1. Anneal PAREadapt1fwd and PAREadapt1rev to create the 3′ DNA adapter (or pairs of PAREadapt2fwd and PAREadapt2rev, or PAREadapt3fwd and PAREadapt3rev, to create samples with alternative barcodes). Add 5 μl PAREadapt1fwd (100 μM), 5 μl PAREadapt1rev (100 μM), 2 μl PCR buffer − Mg^{2+}, and 8 μl nuclease-free water in a thin-wall 100 μl PCR tube. Heat to 100 °C for 5 min in a PCR machine, then turn off the PCR machine, and allow to cool to room temperature. Place tubes on ice. Transfer to 1.3 ml of nuclease-free water in a 1.5 ml microcentrifuge tube and mix to make a 1.5 μM stock.
2. Take the gel-purified DNA from above and add:

   16 μl nuclease-free water.
   4 μl of the 3′ DNA adapter (1.5 μM).
   2.5 μl ligation buffer.
   2.5 μl of T4 DNA ligase.
3. Incubate at room temperature for 2 h.

3.3.15 Second PAGE Purification

1. Add 6× DNA loading buffer to the ligation mix from above and load between two lanes of a polyacrylamide gel (~15 μl each lane) and one lane with the 10-bp DNA ladder.
2. Run the gel at 180 V in TBE running buffer until bromophenol blue and xylene cyanol dye fronts are well separated and stain in 50 ml of 1× TAE containing ethidium bromide for 5 min.
3. Excise the regions of the gel corresponding to ~69 nt, transfer them to a 2 ml tube, crush, and precipitate as outlined for the first PAGE purification.

3.3.16 Long PCR Amplification

1. Resuspend the precipitated DNA from above in 36.5 μl of nuclease-free water; transfer to a 100 μl thin-wall PCR tube, and add the following:

   10 μl of 5× HF buffer.
   1 μl of dNTP mix.
   1 μl of PAREpcr2fwd (P5, 20 μM).
   1 μl of PAREpcr2rev (P7, 20 μM).
   0.5 μl of Hot Start Phusion DNA polymerase.

2. Cycle the reactions as follows: (a) 98 °C for 30 s; (b) 98 °C for 20 s, 60 °C for 30 s, and 72 °C for 15 s, repeat 21 times; and (c) 72 °C for 3 min.

3.3.17 Third PAGE Purification

1. Add 6× DNA loading buffer to the PCR from above and load between three lanes of a polyacrylamide gel (~20 μl per lane) and one lane with 10-bp DNA ladder.
2. Run the gel at 180 V in TBE running buffer until bromophenol blue and xylene cyanol dye fronts are well separated and stain in 50 ml of 1× TAE containing ethidium bromide for 5 min.
3. Excise the regions of the gel corresponding to ~69 nt, transfer them to a 2 ml tube, crush, and precipitate as outlined for the first PAGE purification, except use 2 μl glycogen (20 μg/ml) in place of Glycoblue.
4. Resuspend the pellet in 10 μl of nuclease-free water.
5. Sequence using Illumina or analogous next-generation sequencing platform.

4 Notes

1. Although optimized for the human mitochondrial transcriptome these methods could be used for analyses of RNAs from diverse organelles.
2. To increase the proportion of genuine mitochondrial RNAs, mitochondria can be further processed to generate mitoplasts [6].
3. The TruSeq Small RNA Prep Kit alone can be used to prepare libraries that are enriched for small mitochondrial RNAs such as tRNAs and sRNAs. Use gel electrophoresis to size select the RNAs of interest, a 55–90 nt cutoff would enrich for mitochondrial tRNAs specifically, and a 15–50 nt cutoff would enrich for mitochondrial sRNAs.
4. It is important to cut thin bands that are around the expected size for the PCR products from the gels for the PARE libraries. This ensures that the library will be specific for the products of interest and reduce adaptor contamination.

Acknowledgements

We thank the Australian Research Council and the National Health and Medical Research Council for the award of our fellowships (FT0991008 and FT0991113) and grants (APP1058442, APP1045677, APP1041582, APP1023460, APP1005030 and DP140104111). We thank the members of our groups and colleagues that have contributed to our research and GeneWorks for directional RNA-Seq library preparation and sequencing services.

References

1. Rackham O, Mercer RT, Filipovska A (2012) The human mitochondrial transcriptome and the RNA-binding proteins that regulate its expression. Wiley Interdiscip Rev RNA 3:675–695
2. Rorbach J, Minczuk M (2012) The post-transcriptional life of mammalian mitochondrial RNA. Biochem J 444:357–373
3. Shutt TE, Shadel GS (2010) A compendium of human mitochondrial gene expression machinery with links to disease. Environ Mol Mutagen 51(5):360–379
4. Temperley RJ, Wydro M, Lightowlers RN et al (2010) Human mitochondrial mRNAs—like members of all families, similar but different. Biochim Biophysica Acta 1797: 1081–1085
5. Christian EB, Spremulli LL (2012) Mechanism of protein biosynthesis in mammalian mitochondria. Biochim Biophys Acta 1819: 1035–1054
6. Mercer TR, Neph S, Dinger ME et al (2011) The human mitochondrial transcriptome. Cell 146:645–658
7. German MA, Luo S, Schroth G et al (2009) Construction of parallel analysis of RNA ends (PARE) libraries for the study of cleaved miRNA targets and the RNA degradome. Nat Protoc 4:356–362
8. German MA, Pillay M, Jeong D-H et al (2008) Global identification of microRNA-target RNA pairs by parallel analysis of RNA ends. Nat Biotechnol 26:941–946
9. Lopez Sanchez MI, Mercer TR, Davies SM et al (2011) RNA processing in human mitochondria. Cell Cycle 10:2904–2916
10. Rackham O, Shearwood A, Mercer TR et al (2011) Long noncoding RNAs are generated from the mitochondrial genome and regulated by nuclear-encoded proteins. RNA 17: 2085–2093
11. Rackham O, Filipovska A (2012) The role of mammalian PPR domain proteins in the regulation of mitochondrial gene expression. Biochim Biophys Acta 1819:1008–1016
12. Ebhardt H, Tsang H, Dai D et al (2009) Meta-analysis of small RNA-sequencing errors reveals ubiquitous post-transcriptional RNA modifications. Nucleic Acids Res 37:2461–2470
13. Findeiss S, Langenberger D, Stadler P et al (2011) Traces of post-transcriptional RNA modifications in deep sequencing data. Biol Chem 392:305–313

Chapter 22

Measurement of Mitochondrial RNA Stability by Metabolic Labeling of Transcripts with 4-Thiouridine

Lukasz S. Borowski and Roman J. Szczesny

Abstract

Determination of RNA stability is one of the basic issues addressed in studies of RNA metabolism. In a standard approach used for RNA half-life measurement synthesis of RNA is inhibited and then the steady-state level of RNA is quantified and used for calculations. Here, we present an optimized protocol for mitochondrial RNA stability studies without perturbation of transcription and present results produced for the mitochondrial CytB messenger RNA. This method was originally described for nuclear transcripts and involves metabolic labeling of RNA with 4-thiouridine.

Key words Mitochondrial RNA, 4-Thiouridine (4sU), RNA half-life, Metabolic labeling, RNA stability

1 Introduction

The steady-state level of RNA is determined by two opposite processes of its creation and destruction. Their efficiency can be regulated by different means, like protein–RNA interaction or RNA folding. In mammals, RNA stability factors that regulate transcripts synthesized in the nucleus have been identified [1]. Much less is known about their counterparts acting in mitochondria [2]. A typical RNA stability measurement encompasses inhibition of transcription and tracking of RNA depletion, which enables the RNA's half-life calculation. This, however, requires specific inhibition of RNA polymerase, which can be difficult to achieve, especially for mitochondrial RNA polymerase, for which specific inhibitors are unknown. So far stability of human mitochondrial RNA (mtRNA) has been studied by inhibition of mtRNA synthesis with actinomycin D [3] or high concentration of ethidium bromide [4]. Both of them inhibit transcription indirectly, which can hamper data interpretation, for example by inhibition of mitochondrial DNA replication. Here, we describe a detailed protocol for mtRNA stability measurement without perturbation of transcription that was

Joanna Rorbach and Agnieszka J. Bobrowicz (eds.), *Polyadenylation: Methods and Protocols*, Methods in Molecular Biology, vol. 1125, DOI 10.1007/978-1-62703-971-0_22,

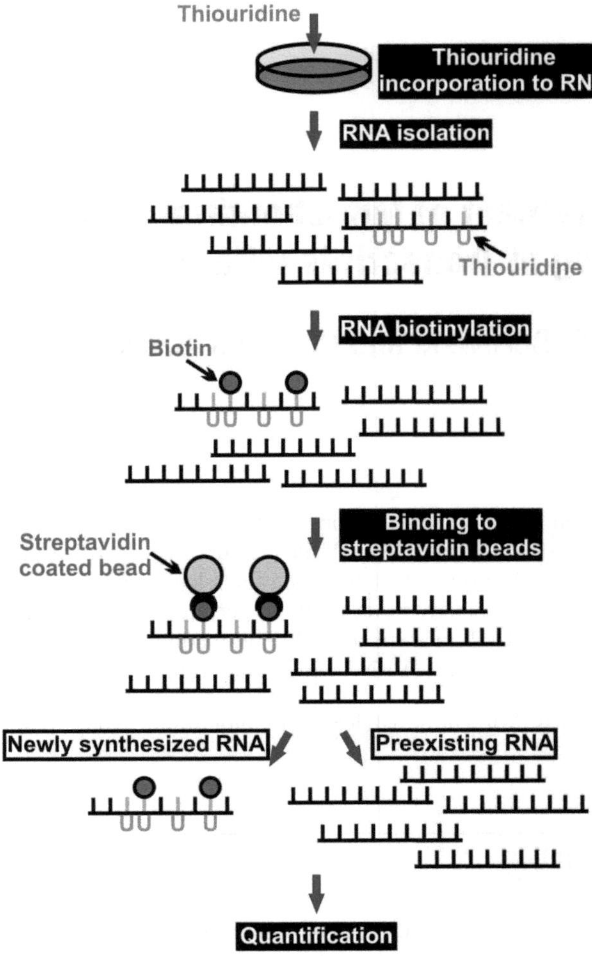

Fig. 1 *Scheme of the procedure.* Cells are cultured in the presence of thiouridine, then total RNA is isolated, and a fraction of newly synthesized thiouridine-labeled RNA is biotinylated and separated from preexisting RNA. Amount of a given transcript in particular fractions and/or total RNA is quantified, and half-life is calculated

originally applied to measurement of nuclear transcripts [5]. We use metabolic labeling of RNA with a naturally occurring nucleoside analog, 4-thiouridine (4sU), which is not toxic and does not affect transcription; this is considered to be the least invasive technique of RNA stability measurement currently available [6]. The procedure consists of the following steps: (1) transcriptional labeling of RNA with 4sU and isolation of RNA, (2) biotinylation of RNA, (3) separation of RNA into preexisting and newly synthesized fractions, and (4) quantification (Fig. 1). The half-life of a transcript is calculated based on the ratio of its preexisting fraction to its total amount. The results for mitochondrial CytB mRNA are presented (Fig. 2).

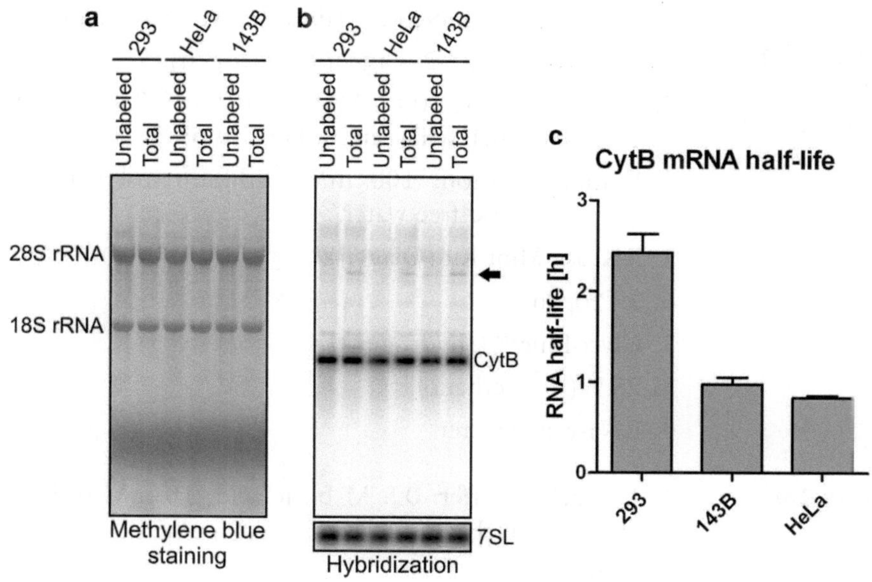

Fig. 2 *Measurement of half-life of mitochondrial CytB mRNA.* Indicated cells were incubated with thiouridine and subjected to the procedure described above. (**a**) Methylene blue staining of the membrane; cytoplasmic rRNAs are marked. (**b**) An autoradiogram of hybridization with riboprobe complementary to CytB mRNA, encompassing 15088–15499 nucleotides of mtDNA. An *arrow* indicates unprocessed nascent transcript present mainly in total RNA but not in the preexisting RNA fraction (*unlabeled*). 7SL RNA was detected using probe encompassing 103–259 nucleotides of sequence NR_002715. (**c**) Calculated CytB mRNA half-life. Mean values of four measurements are shown. Standard deviation is marked

2 Materials

2.1 Labeling of RNA with 4sU and Isolation of RNA

1. 50 mM 4sU (Sigma) dissolved in sterile phosphate-buffered saline (PBS) (*see* **Note 1**). Store at −20 °C.
2. TRI Reagent (Sigma).
3. Chloroform.
4. 2-Propanol.
5. 75 % (v/v) ethanol.
6. RNase-free water.

2.2 Biotinylation of RNA

1. 10× Biotinylation buffer: 100 mM Tris–HCl, pH 7.4, 10 mM EDTA (*see* **Note 2**). Store at 4 °C.
2. 1 mg/mL EZ-Link HPDP-Biotin (Pierce) in dimethylformamide (DMF) (*see* **Note 3**). Store at 4 °C.
3. Chloroform/isoamyl alcohol (24:1). Prepare fresh.
4. 5 M NaCl.
5. 2-Propanol.
6. 75 % (v/v) ethanol.
7. RNase-free water.

2.3 Separation of RNA into Preexisting and Newly Synthesized Fractions

1. Streptavidin-coupled Dynabeads™ M-280 (Invitrogen).
2. The DynaMag™-2 magnet (Invitrogen).
3. Washing buffer: 100 mM Tris–HCl, pH 7.4, 10 mM EDTA, 1 M NaCl, 0.1 % Tween20. Heat to 65 °C.
4. Elution solution: 100 mM Dithiothreitol (DTT). Prepare fresh in RNase-free water.
5. RNeasy Mini Kit (Qiagen).
6. 2-Propanol.
7. GlycoBlue™ (Invitrogen).
8. 75 % (v/v) ethanol.
9. RNase-free water.

2.4 Northern Blot Analysis

1. 10× NBC buffer: 0.5 M boric acid, 10 mM sodium acetate, 50 mM NaOH. Autoclave.
2. Agarose.
3. 37 % formaldehyde.
4. Formamide.
5. 10× Loading dye: 15 % Ficoll, 0.25 % bromophenol blue, 0.25 % xylene cyanol in 0.1 M EDTA, pH 8.0.
6. 10× SSC: 1.5 M sodium chloride, 150 mM sodium citrate. Autoclave.
7. Nytran N, 0.45 μm nylon blotting membranes (Whatman, part of GE Healthcare).
8. Methylene blue solution: 0.02 % methylene blue in 0.3 M sodium acetate, pH 5.2.
9. T7 transcription kit (Fermentas).
10. [α-^{32}P] UTP (10 mCi)/mL (Hartmann Analytic).
11. DNase I recombinant, RNase-free (Roche).
12. Ultrafree-MC Centrifugal Filter Units with Microporous Membrane (Millipore).
13. PerfectHyb™ Plus Hybridization Buffer (Sigma).
14. Washing buffer 1: 2× SSC, 0.1 % sodium dodecyl sulfate (SDS). Heat to 65 °C.
15. Washing buffer 2: 0.2× SSC, 0.1 % SDS. Heat to 65 °C.

3 Methods

3.1 Labeling of RNA with 4sU and Isolation of RNA

1. Add 4sU to the cell culture medium to a final concentration of 250 μM (*see* **Note 4**). Take some cell culture medium off the dish, and add it to the falcon tube, to which required amount of

4sU was added (*see* **Note 5**). Mix gently, and put back it in the dish. This step must be done as quickly as possible (*see* **Note 6**).

2. Put the plates back into the incubator and incubate for 60 min.
3. Terminate labeling by quick removal of the cell culture medium and addition of TRI reagent (under the hood). Use 1, 3, and 5 mL of TRI reagent for 60, 100, and 145 mm dishes, respectively. Pipette thoroughly to ensure complete cell lysis.
4. Transfer the lysate into 1.5 or 15 mL tube and incubate at room temperature for 5 min. At this step samples can be frozen at −80 °C or can be processed further.
5. Add 0.2 mL of chloroform per 1 mL of lysate and shake for 2–3 min.
6. Incubate at room temperature for 10 min.
7. Centrifuge at 12,000 × *g* for 15 min at 4 °C.
8. Transfer aqueous upper phase to a new tube and add 1 volume of 2-propanol. Mix thoroughly by inverting the tube.
9. Incubate at room temperature for 10 min.
10. Centrifuge at 12,000 × *g* for 10 min at 4 °C.
11. Remove the supernatant, and wash the pellet with 0.9 mL of 75 % ethanol. Add ethanol, mix gently by inverting the tube (*see* **Note 7**), and incubate at room temperature for 5 min.
12. Centrifuge at 12,000 × *g* for 5 min at 4 °C.
13. Remove the supernatant. Spin briefly, and remove residual ethanol.
14. Leave the tube open on the bench for 5–10 min to dry the pellet. To avoid contamination cover the tube with the sterile side of parafilm.
15. Dissolve the RNA in RNase-free water. Add water, incubate for 5–10 min at room temperature, and then pipette gently. Measure the concentration and adjust to 1 μg/μL. At this step the RNA can be frozen at −80 °C or can be processed further.

3.2 Biotinylation of RNA

1. Prepare biotinylation mixture: 100 μL of 10× biotinylation buffer, 100 μL of RNA (at 1 μg/μL), 200 μL EZ-Link HPDP-Biotin, and 600 μL of RNase-free water (*see* **Note 8**).
2. Incubate for 90 min with rotation at room temperature.
3. Add 1 volume of chloroform/isoamyl alcohol (24:1), and shake for 2–3 min.
4. Incubate at room temperature for 5 min.
5. Centrifuge at 16,000 × *g* for 5 min at 4 °C.
6. Transfer aqueous upper phase to a fresh tube and add 1 volume of 2-propanol and 1/10 volume of 5 M NaCl. Mix thoroughly.

7. Incubate at room temperature for 5 min.
8. Centrifuge at 16,000 × *g* for 20 min at 4 °C.
9. Remove the supernatant, and wash the pellet with 0.9 mL of 75 % ethanol. Add ethanol, mix gently by inverting the tube (*see* **Note 7**), and incubate at room temperature for 5 min.
10. Centrifuge at 16,000 × *g* for 5 min at 4 °C.
11. Remove the supernatant. Spin briefly, and remove residual ethanol.
12. Leave the tube open on the bench for 5–10 min to dry the pellet. To avoid contamination cover the tube with the sterile side of parafilm.
13. Dissolve the RNA in RNase-free water. Add water, incubate for 5–10 min at room temperature, and then pipette gently. Measure the concentration and adjust to 1 μg/μL. At this step RNA can be frozen at −80 °C or can be processed further.

3.3 Separation of RNA into Preexisting and Newly Synthesized Fractions

1. Incubate biotinylated RNA for 10 min at 65 °C. After incubation place the samples immediately on ice and incubate for 5 min.
2. Add 100 μL of streptavidin-coupled Dynabeads to a new tube and wash with 1 volume of washing buffer. Rotate for 3 min at room temperature, place the tubes in the DynaMag™-2 magnet stand, and remove the buffer (*see* **Note 9**).
3. Add 100 μg of biotinylated RNA (*see* **Note 10**) to 100 μL of streptavidin-coupled Dynabeads (prewashed) and incubate with rotation at room temperature for 15 min.
4. Place the samples in the DynaMag™-2 magnet stand (*see* **Note 9**), and transfer the unbound RNA into new tubes (preexisting RNA fraction).
5. Shake the beads gently to disrupt the pellet, and add 800 μL of washing buffer heated to 65 °C.
6. Incubate with rotation for 5 min at room temperature.
7. Place the samples in the DynaMag™-2 magnet stand (*see* **Note 9**), and transfer the buffer after wash to tubes containing the unbound RNA fraction (*see* **Note 11**).
8. Repeat **steps 5**–**7** two more times with washing buffer heated to 65 °C and three times with the washing buffer at room temperature but discarding the washing solution.
9. To elute labeled RNA add to the beads 100 μL of elution solution. Shake gently, and incubate with rotation for 3 min at room temperature.
10. Place the samples in the DynaMag™-2 magnet stand (*see* **Note 9**), and transfer the labeled RNA (newly synthesized RNA fraction) to a new tube.

11. After 5 min repeat **steps 9–10** with the same beads and collect the eluates with those obtained in **step 10**.
12. Recover RNA from preexisting and newly synthesized fractions using RNeasy Mini Kit (Qiagen) according to the manufacturer's instructions.
13. Optionally RNA can be recovered by 2-propanol precipitation (**steps 6–13**, Subheading 3.2). In the case of newly synthesized RNA fraction precipitation must be enhanced by addition of 30 μg of GlycoBlue. During precipitation of preexisting RNA fraction addition of NaCl can be omitted as there is already 1 M NaCl present in the washing buffer.

3.4 Northern Blot Analysis (*See* Note 12)

1. Prepare 1 % agarose gel in 1× NBC buffer.
2. When the gel has cooled down to ~65 °C add 2.5 mL of 37 % formaldehyde per 100 mL of the gel (under a fume hood), and pour the gel into the gel tray (also under a fume hood). Let it cool down for 45 min.
3. Add 1.5–2 μg of RNA (max 5 μL) to the denaturing mixture composed of 2 μL of 10× NBC buffer, 3 μL of 37 % formaldehyde, and 10 μL of formamide. Prepare fresh.
4. Heat the samples to 65 °C for 5 min.
5. Add 2 μL of 10× loading dye, and spin briefly.
6. Mix once by pipetting and load the samples into the wells. Run the gel with constant voltage of 90 V.
7. Transfer RNA to a Nytran N nylon blotting membrane (Whatman) by overnight capillary transfer in 10× SSC.
8. Wash the membrane briefly with 2× SSC.
9. Immobilize RNA by ultraviolet cross-linking (120 mJ/cm^2).
10. Stain the membrane with methylene blue solution for 30–60 s until bands of ribosomal RNA are visible. Wash twice for 5 min with deionized water to remove the background staining (Fig. 2a). The membrane can be stored at −20 °C or subjected to hybridization (*see* **Note 13**).
11. Prepare radiolabeled riboprobes (*see* **Note 14**) by in vitro transcription using a T7 transcription kit and [α-^{32}P] UTP. Mix 4 μL of 5× transcription buffer, 0.8 μL Ribolock (40 U/μL), 1 μL of ATP/CTP/GTP mixture (10 mM each), 2.4 μL of 100 μM UTP, 100–200 ng of DNA template (*see* **Note 15**), 5 μL of [α-^{32}P] UTP, H_2O to 19 μL, and 1 μL of T7 or SP6 polymerase.
12. Incubate at 37 °C for 2 h.
13. Remove the template by adding of 1 U of DNaseI. Incubate at 37 °C for 15 min.

14. Add 50 μL of deionized water and transfer to the Ultrafree-MC Centrifugal Filter Unit.
15. Centrifuge at 12,000 × *g* for 1 min (*see* **Note 16**).
16. Incubate the membrane with PerfectHyb Hybridization Buffer at 65 °C with rotation for 30 min.
17. Heat the radiolabeled riboprobe at 70 °C for 90 s, and add 35 μL (*see* **Note 17**) of the probe directly to the PerfectHyb.
18. Incubate the membrane with the hybridization solution overnight at 65 °C with rotation.
19. Discard hybridization solution, and wash the membrane three times with washing buffer 1 for 8 min and once with washing buffer 2 for 5 min. Perform washings in 65 °C, and use buffers heated to 65 °C.
20. Expose the membrane to PhosphorImager screen overnight.
21. Scan the screen (Fig. 2b).

3.5 Quantification

1. Using radiograms and proper software (e.g., Image Quant, Multi Gauge), quantify the level of preexisting RNA and total RNA for each sample (*see* **Note 18**). For standardization use 7SL or rRNA labeling (or any other unrelated/unaffected transcript).
2. Calculate the ratio of the preexisting RNA to the total RNA. The ratio should be below "1"; if it is not, use the correction factor calculated by Dolken et al. [5] for the NIH/3T3 cell line (0.81)—multiply the ratios by this factor.
3. Calculate RNA half-life (Fig. 2c) from the equation

$$t_{1/2} = -\frac{t \ln 2}{\ln(1 - R)}$$

where t is time (in hours) of RNA labeling with 4sU (1 h) and R is the ratio of the amount of preexisting RNA to the amount of total RNA.

4 Notes

1. Should be stored in small aliquots (200–300 μL) at −20 °C and only thawed once.
2. Prepare with RNase-free water, and store at 4 °C in aliquots of 2 mL.
3. Heat gently during preparation to ensure complete solubilization, and store at 4 °C.
4. The influence of 4sU treatment on cells can be assessed by testing their survival/growth rate after treatment. Incubate cells

with 4sU for the intended time, wash cells, add culture medium without 4sU, and on the following days examine the cell numbers by chamber counting or any commercially available viability test, such as Alamar Blue (Invitrogen) and compare to untreated cells.

5. Depending on how much RNA is needed, labeling of RNA can be performed on different size dishes (60, 100, or 145 mm diameter). During labeling the volume of the cell culture medium can be reduced to 2, 5, or 10 mL, respectively. Different amounts of RNA can be isolated depending on the cell line: a 60 mm dish (~90 % confluence) yields ~120 μg (293), ~60 μg (143B), and ~30 μg (HeLa, NIH/3T3). We recommend using 100 or 145 mm dishes for 143B, HeLa, and NIH/3 T3 cell lines.
6. Do not handle more than four plates at a time.
7. Invert the tube so that the cap covered with organic solvent is also washed.
8. Use 30–100 μg of total RNA and 2 μg of EZ-Link HPDP-Biotin per 1 μg of RNA.
9. It is helpful to spin the beads briefly (5 s, at 8,000 × *g*) before putting the sample in the magnet stand.
10. Use 50–100 μg of biotinylated RNA and 1 μL of streptavidin-coupled Dynabeads per 1 μg of RNA.
11. The preexisting RNA fraction is the RNA not bound to the streptavidin-coupled Dynabeads and RNA recovered from the first wash of the beads.
12. Here we use northern analysis for quantification, but quantitative PCR can also be used where applicable, for example when the mature transcript can be distinguished from its precursor or degradation intermediates.
13. Because both strands of mitochondrial DNA are entirely transcribed, detection (regardless northern or qPCR) must be strand specific. Mitochondrial antisense transcripts were reported (mirror RNAs [7] and long noncoding RNAs [8]) which preclude proper quantification of the amount of sense transcripts unless strand-specific detection is applied. We recommend testing specificity of the probes using RNA isolated from cells devoid of mitochondrial DNA (rho0; we use derivatives of 143B cells).
14. Oligonucleotides labeled with T4 polynucleotide kinase can be used as well; however, hybridization temperature must be adjusted.
15. As template, use gel-extracted PCR products, 200–400 bp long, obtained in reaction with primers containing overhangs with T7 and SP6 polymerase promoter, forward and reverse, respectively.

Sequences of promoters in the forward direction are as follows: T7—TAATACGACTCACTATAGGG, SP6—ATTTAGGTGA CACTATAGAAT.

16. **Steps 14** and **15** decrease the amount of unincorporated [α-^{32}P] UTP which offers two benefits: (a) lower radioactivity of solutions which will be handled in further steps and (b) lower background signal of the hybridization.
17. The rest of the probe can be stored at −20 °C.
18. RNA half-life can be calculated from the amount of total RNA and the amount of preexisting RNA as well as from the amount of total RNA and the amount of newly synthesized RNA. As the amount of the newly synthesized RNA fraction obtained in experiments is always very small, for northern analysis we recommend using preexisting RNA fractions.

Acknowledgements

The authors thank Aleksander Chlebowski for his help with the manuscript. Authors are supported by the Ministry of Science and Higher Education of Poland [0542/IP1/2011/71]. LSB is the recipient of the Stipend for Young Researchers from the Foundation for Polish Science. RJS is the recipient of a TEAM scholarship from the Foundation for Polish Science (TEAM/2008-2/1) and a scholarship for outstanding young scientists from the Polish Ministry of Science and Higher Education.

References

1. Chlebowski A, Lubas M, Jensen TH et al (2013) RNA decay machines: the exosome. Biochim Biophys Acta 1829:552–560
2. Rorbach J, Minczuk M (2012) The post-transcriptional life of mammalian mitochondrial RNA. Biochem J 444:357–373
3. Piechota J, Tomecki R, Gewartowski K et al (2006) Differential stability of mitochondrial mRNA in HeLa cells. Acta Biochim Pol 53:157–168
4. Nagao A, Hino-Shigi N, Suzuki T (2008) Measuring mRNA decay in human mitochondria. Methods Enzymol 447:489–499
5. Dolken L, Ruzsics Z, Radle B, Friedel CC et al (2008) High-resolution gene expression profiling for simultaneous kinetic parameter analysis of RNA synthesis and decay. RNA 14:1959–1972
6. Balagopal V, Fluch L, Nissan T (2012) Ways and means of eukaryotic mRNA decay. Biochim Biophys Acta 1819:593–603
7. Szczesny RJ, Borowski LS, Brzezniak LK et al (2010) Human mitochondrial RNA turnover caught in flagranti: involvement of hSuv3p helicase in RNA surveillance. Nucleic Acids Res 38:279–298
8. Rackham O, Shearwood AM, Mercer TR et al (2011) Long noncoding RNAs are generated from the mitochondrial genome and regulated by nuclear-encoded proteins. RNA 17:2085–2093

Part III

Deadenylation and Decay

Chapter 23

Deadenylation and Its Regulation in Eukaryotic Cells

Xiaokan Zhang, Frida E. Kleiman, and Emral Devany

Abstract

Messenger RNA deadenylation is a process that allows rapid regulation of gene expression in response to different cellular conditions. The change of the mRNA poly(A) tail length by the activation of deadenylation might regulate gene expression by affecting mRNA stability, mRNA transport, or translation initiation. Activation of deadenylation processes are highly regulated and associated with different cellular conditions such as cancer, development, mRNA surveillance, DNA damage response, and cell differentiation. In the last few years, new technologies for studying deadenylation have been developed. Here we overview concepts related to deadenylation and its regulation in eukaryotic cells. We also describe some of the most commonly used protocols to study deadenylation in eukaryotic cells.

Key words Deadenylation, Eukaryotic cells, Regulation, CCR4-NOT, PARN

1 Introduction

Gene expression is extensively regulated at various steps from transcriptional initiation to RNA processing, translation efficiency, and posttranslational modification of the protein. Gene regulation is essential for organisms by allowing precise spatial, temporal, and quantitative control of protein production and enhancing adaptability and versatility in different cellular conditions, such as development, mRNA surveillance, inflammatory response, cell differentiation, cancer, and DNA damage response.

Almost all eukaryotic mRNA precursors, with the exception of histones, undergo a co-transcriptional polyadenylation. Briefly, the 3′ end of the precursor mRNA is cleaved followed by the addition of a non-templated polyadenylated tail, which in mammalian cells is of approximately 200–300 adenosines [1, 2]. Most mRNAs undergo this first round of polyadenylation that confers stability to the transcript. However, the length of poly(A) tails changes in different conditions affecting mRNA stability, transport, or translation initiation, and hence gene expression. The mechanisms controlling deadenylation are dynamic and highly regulated. Deadenylation

Joanna Rorbach and Agnieszka J. Bobrowicz (eds.), *Polyadenylation: Methods and Protocols*, Methods in Molecular Biology, vol. 1125, DOI 10.1007/978-1-62703-971-0_23, © Springer Science+Business Media New York 2014

processes are a widespread strategy that allows rapid control of mRNA stability, and represent additional checkpoint targets in the regulation of gene expression.

2 Deadenylation and Deadenylases

In eukaryotic cells, the initiation and rate-limiting step in mRNA decay is the removal of the mRNA poly(A) tail, a process known as deadenylation [3–5]. Deadenylases progressively degrade the poly(A) tail from the 3′ end. Once the poly(A) tail is removed, the mRNAs can be degraded by the exosome in the 3′ to 5′ direction by a complex of exonucleases [6]. Alternatively, mRNA degradation can start by deadenylation followed by DCP1:DCP2-mediated decapping and 5′-3′ exonucleolytic degradation by the XRlp 5′ exonuclease [7, 8].

Deadenylases are the key catalytic exoribonucleases that are required for proper regulation of poly(A) tail length by degrading the mRNA in a 3′ to 5′ direction. In eukaryotic cells, a number of deadenylases have been identified and studied in detail. All known deadenylases belong to two superfamilies deadenylases: DEDD (Asp-Glu-Asp-Asp)-type and EEP (endonuclease-exonuclease-phosphatase)-type [9]. While POP2, poly(A)-specific ribonuclease (PARN), poly(A) nuclease (PAN2), and CAF1 deadenylase belong to DEDD superfamily, CCR4-NOT and Nocturnin deadenylases belong to EEP superfamily [10, 11]. Interestingly, some multi-subunit deadenylases are composed from subunits from both families [12].

It has been shown that deadenylases localize both in the nucleus and the cytoplasm and are involved in different cellular processes. The best characterized deadenylases so far are the CCR4-POP2-NOT multi-subunit complex, which is the predominant deadenylase in eukaryotes, PARN, a divalent metal-ion dependent, processive and cap-interacting exoribonuclease, and PAN, which is involved in early steps of poly(A) tail metabolism.

3 *cis*- and *trans*-Acting Factors Involved in Deadenylation

The role of deadenylation in different cellular processes is either to decrease the total mRNA levels or to regulate the levels of specific mRNAs through sequence elements present in their 3′ untranslated region (3′UTR). These *cis*-elements are recognized by *trans-acting* RNA-binding protein factors that recruit the deadenylases, resulting in the regulation of deadenylation and gene expression in a sequence specific manner (Fig. 1). Examples of these *cis-acting* regulatory signals include AU-rich elements (AREs) and microRNA (miRNA) targeting sites [13–15].

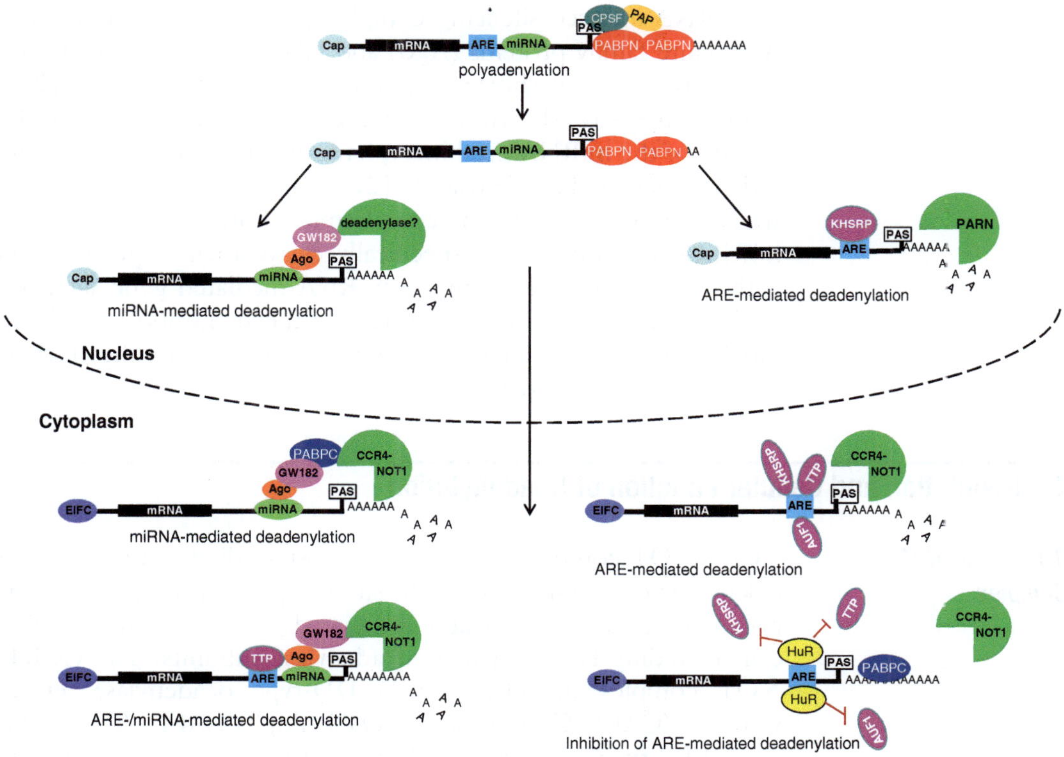

Fig. 1 Model for miRNA- and ARE-mediated deadenylation pathways in different cellular compartments. Polyadenylation signal (PAS), cleavage-polyadenylation specificity factor (CPSF), poly(A) polymerase (PAP), AU-rich element (ARE), microRNA (miRNA), Cap-binding complex (CBC), ribosome (Ribos), eukaryotic translation initiation factor complex (EIFC), poly(A) binding protein (PABP), Argonaute (Ago), tristetraprolin (TTP), butyrate response factor 1 (BRF1), AU-rich binding factor 1 (AUF1), and KH type splicing regulatory protein (KHSRP)

The ARE typically contains one or several AUUUA pentamer repeats within the 3′UTR of mRNAs coding proteins involved in cell differentiation, cell growth regulation, and responses to external stimuli. A number of *trans*-acting factors, known as ARE-binding proteins (ARE-BPs), regulate ARE-mediated decay and, subsequently, modulate the stability of ARE-containing mRNAs. ARE-BPs, such as tristetraprolin (TTP), butyrate response factor 1 (BRF1), AU-rich binding factor 1 (AUF1), and KH type splicing regulatory protein (KSRP), can recruit deadenylases to the target mRNAs and promote deadenylation. Alternatively, other ARE-BPs, such as HuR, can block the recruitment of deadenylase and exosome resulting in the inhibition of deadenylation [16–19]. For example, PARN deadenylase has been shown to be involved in ARE-mediated deadenylation: either KSRP [20] or CUG-BP [21] or TTP [22, 23] recruits PARN to ARE-containing mRNAs to initiate the poly(A) tail shortening that precedes the degradation by the exosome.

miRNAs are involved in the regulation of deadenylation by base pairing to the 3′UTR of target mRNAs. The miRNA bound to miRNA-induced silencing complex (miRISC), which contains Argonaute family proteins (Ago) and GW182, and a deadenylase induce poly(A) tail shortening and mRNA destabilization [24]. Many studies have shown that the levels of specific miRNAs, or the activity of the miRNA machinery, have direct consequences in the miRNA-mediated deadenylation [25–27]. Although most of these studies have focused on cytoplasmic deadenylation, nuclear miRISCs are also able to specifically cleave their target mRNAs with high efficiency resulting in miRNA-mediated gene silencing in the nucleus [28, 29]. However, the mechanism(s) underlying miRNA-mediated regulation of gene expression in the nucleus remain(s) to be elucidated.

4 Regulation and Cellular Function of Deadenylation

4.1 CCR4-NOT Complex

CCR4-NOT complex is one of the most studied deadenylases in eukaryotic cytoplasm. This multi-subunit protein complex is composed of at least seven stably associated subunits in mammalian cells including two enzymatic deadenylase subunits. The CCR4-NOT complex includes one DEDD-type deadenylase, either CNOT7/CAF1/CAF1a or CNOT8/Pop2/CAF1b/Calif, and one EEP-type enzyme, either CNOT6/CCR4/CCR4a or CNOT6L/CCR4-like/CCR4b [30].

In addition to its roles in ARE- and miRNA-dependent decay, the CCR4-NOT complex can interact with other group of RNA-binding proteins. One of these factors, NANOS2, is involved in germ cell development [31]. The direct interaction of NANOS2 with CNOT1 protein in the CCR4-NOT complex induces the change of their localization to P-bodies, which are known as centers of RNA degradation, resulting in the sexual development of germ cells by suppressing meiosis and promoting male-type differentiation. Other interactors of CCR4-NOT are two human PUFs: PUM1 and PUM2. PUFs are RNA-binding proteins involved in stem cell proliferation, fertility, and memory formation [32]. Human PUMs physically associate with CNOT deadenylase subunits, including CNOT6, -6L, -7, and -8, and recruit CCR4-NOT deadenylase complex to repress target mRNAs. Interestingly, the expression of CNOT6, CNOT6L, and CNOT7 deadenylases is altered in leukemia patients, suggesting a possible role for these deadenylases in cancer [33].

4.2 PARN

PARN is the major nuclear deadenylase in mammalian cells. It is reported to bind to both 5′ cap and poly(A) tail concurrently increasing its processivity and enhancing the mRNA degradation rate [34, 35]. It has been reported that cap binding complex (CBC) inhibits PARN, representing a mechanism by which PARN is recruited to the nascent pre-mRNA and ensuring that PARN does not degrade the poly(A) tail of pre-mRNAs [36]. PARN also interacts with mRNA 3′ processing factor CstF1 and the tumor suppressor BARD1 upon DNA damage reverting PARN inhibition by 5′ cap [37]. Another PARN integrator is the peripheral exosome subunit Mpp6. This interaction mediates the recruitment of the exosome to deadenylated mRNAs for subsequent 3′ → 5′ degradation [38].

PARN is involved in the degradation of several cancer-related genes [21, 23, 37] and can functionally interact with different tumor suppressors, such as BARD1 [37] and p53 [39]. Recent studies have shown that PARN expression [33] and phosphorylation state are altered in cancer, indicating that PARN deadenylase regulation is critical in tumorigenesis. PARN also participates in cytoplasmic functions as part of the dendritic cytoplasmic polyadenylation element binding protein (CPEB)-associated polyadenylation apparatus that is involved in hippocampal synaptic plasticity during learning and memory processes. Synaptic stimulation induces phosphorylation of CPEB, PARN expulsion from the ribonucleoprotein complex, and polyadenylation of mRNAs in dendrites [40].

5 Use of High-Throughput Methods to Evaluate Poly(A) Tail Length

High-throughput studies are useful to have a broad picture of targets affected by deadenylases or interacting proteins. Microarray analysis has been used successfully to determine RNAs with different poly(A) tail length. For example microarray experiments designed to measure polyadenylated RNAs revealed that U4 and U6 snRNAs as well as many different small nucleolar RNAs (snoRNAs) are overexpressed in certain mutants of the CCR4-NOT complex [41]. Pathway analysis of regulated genes indicated that the p53 signaling pathway is the most significantly affected by PARN knockdown in nonstress conditions, suggesting that PARN expression has a specific effect on the expression of genes associated with p53-mediated signaling [39]. Microarray analysis has also shown that PARN regulates the expression of genes involved in mRNA metabolism, transcription, and cell motility in mouse myoblasts [42]. After determining the mRNAs affected by high-throughput methods it is important to validate each mRNA through poly(A) tail length assays. Meijer et al. [43] described a technique that allows the isolation of mRNAs with different length of poly(A) tails.

6 In Vitro Deadenylation Assays

In vitro deadenylation assays are a very important tool to investigate properties of individual deadenylases. In fact, most of the deadenylases were identified using in vitro deadenylation assays [44–46]. Unlike polyadenylation reactions, deadenylation can be reconstituted in vitro. Through this method it is also possible to dissect components and protein factors that regulate deadenylation and deadenylases, such as activators or inhibitors, using recombinant proteins in in vitro assays. For example, the addition of GST-CstF1 and His-p53 to in vitro reactions by PARN results in the activation of its deadenylase activity [37, 39]. Importantly, those studies should test any potential deadenylase activity of the interactors. Although the in vitro assays are powerful tools, in vivo analysis are essential to validate the interactions and regulation of deadenylases in cellular systems. The protocols described in the following chapters are used to analyze the most studied eukaryotic deadenylases.

7 Deadenylation Using Eukaryotic Cell Extracts

Extracts system is a powerful approach to study deadenylation in a cell-free assay. First, the use of cell extracts allows the analysis of deadenylation in the presence of deadenylases and all their regulators, mimicking functional cellular condition. Indeed, cell extracts are particularly useful for examining deadenylase complexes with multiple subunits without affecting the assemble of the subunits. Second, extracts systems allow monitoring the regulation of deadenylation processes in response to different treatments or cellular conditions, such as cell differentiation, DNA damage response, inflammatory response and cancer. For example, the study of HeLa cells nuclear extracts has revealed that PARN-dependent deadenylation is strongly activated in response to UV-induced DNA damage [37]. Third, this approach permits to study mechanisms of mRNA deadenylation in variety of cell lines and allows assaying deadenylation process in different cell fractions and mRNA targets. So far, the deadenylation assays have been successfully performed in HeLa, RKO, HCT116, and many other cell lines, using both cytoplasmic extracts [21, 47] and nuclear extracts [37]. Besides, the effect of deadenylases on specific mRNA targets could be determined by analyzing the change in their length by using poly(A) test (PAT) assay as described in [42] and/or changes in mRNA stability by real-time reverse transcriptase PCR [37, 39, 48].

One disadvantage of using extract systems is that it does not allow the study of individual deadenylases, making difficult the characterization of their functions. However, the use of extracts from cells treated with siRNAs targeting the deadenylase of interest has addressed that problem. For example, the siRNA-mediated knockdown of deadenylases has helped to identify the role of PARN in nuclear deadenylation [37]; p53 as an activator of PARN-dependent deadenylation [39], PARN participation in wound healing [42]; the role of CAF1, but not POP2, in miRNA-mediated poly(A) removal [49]; the role of CCR4 in cell proliferation and cell survival [48]. Although the use of eukaryotic cell extracts is powerful, the reconstitution of deadenylation systems using recombinant components, both deadenylases and regulators, is still a necessary assay to confirm deadenylases function, such as mechanistic analysis, enzymatic properties, reaction kinetics and cofactor requirements.

References

1. Bentley DL (2005) Rules of engagement: co-transcriptional recruitment of pre-mRNA processing factors. Curr Opin Cell Biol 17(3): 251–256
2. Sheets MD, Wickens M (1989) Two phases in the addition of a poly(A) tail. Genes Dev 3(9):1401–1412
3. Wilusz CJ, Gao M, Jones CL et al (2001) Poly(A)-binding proteins regulate both mRNA deadenylation and decapping in yeast cytoplasmic extracts. RNA 7(10):1416–1424
4. Chen CY, Shyu AB (2003) Rapid deadenylation triggered by a nonsense codon precedes decay of the RNA body in a mammalian cytoplasmic nonsense-mediated decay pathway. Mol Cell Biol 23(14):4805–4813
5. Parker R, Song H (2004) The enzymes and control of eukaryotic mRNA turnover. Nat Struct Mol Biol 11(2):121–127
6. Beelman CA, Parker R (1995) Degradation of mRNA in eukaryotes. Cell 81(2):179–183
7. Caponigro G, Parker R (1996) mRNA turnover in yeast promoted by the MATalpha1 instability element. Nucleic Acids Res 24(21): 4304–4312
8. Mukherjee D, Gao M, O'Connor JP et al (2002) The mammalian exosome mediates the efficient degradation of mRNAs that contain AU-rich elements. EMBO J 21(1–2):165–174
9. Goldstrohm AC, Wickens M (2008) Multifunctional deadenylase complexes diversify mRNA control. Nat Rev Mol Cell Biol 9(4): 337–344
10. Thore S, Mauxion F, Seraphin B et al (2003) X-ray structure and activity of the yeast Pop2 protein: a nuclease subunit of the mRNA deadenylase complex. EMBO Rep 4(12):1150–1155
11. Zuo Y, Deutscher MP (2001) Exoribonuclease superfamilies: structural analysis and phylogenetic distribution. Nucleic Acids Res 29(5): 1017–1026
12. Doidge R, Mittal S, Aslam A et al (2012) The anti-proliferative activity of BTG/TOB proteins is mediated via the Caf1a (CNOT7) and Caf1b (CNOT8) deadenylase subunits of the Ccr4-not complex. PLoS One 7(12):e51331
13. Lagnado CA, Brown CY, Goodall GJ (1994) AUUUA is not sufficient to promote poly(A) shortening and degradation of an mRNA: the functional sequence within AU-rich elements may be UUAUUUA(U/A)(U/A). Mol Cell Biol 14(12):7984–7995
14. Chen CY, Shyu AB (1995) AU-rich elements: characterization and importance in mRNA degradation. Trends Biochem Sci 20(11):465–470
15. Zubiaga AM, Belasco JG, Greenberg ME (1995) The nonamer UUAUUUAUU is the key AU-rich sequence motif that mediates mRNA degradation. Mol Cell Biol 15(4):2219–2230
16. Ma WJ, Cheng S, Campbell C et al (1996) Cloning and characterization of HuR, a ubiquitously expressed Elav-like protein. J Biol Chem 271(14):8144–8151
17. Fan XC, Steitz JA (1998) Overexpression of HuR, a nuclear-cytoplasmic shuttling protein, increases the in vivo stability of ARE-containing mRNAs. EMBO J 17(12):3448–3460
18. Barreau C, Paillard L, Osborne HB (2005) AU-rich elements and associated factors: are there unifying principles? Nucleic Acids Res 33(22):7138–7150
19. Westmark CJ, Bartleson VB, Malter JS (2005) RhoB mRNA is stabilized by HuR after UV light. Oncogene 24(3):502–511
20. Gherzi R, Lee KY, Briata P et al (2004) A KH domain RNA binding protein, KSRP, promotes ARE-directed mRNA turnover by recruiting the degradation machinery. Mol Cell 14(5):571–583
21. Moraes KC, Wilusz CJ, Wilusz J (2006) CUG-BP binds to RNA substrates and recruits PARN deadenylase. RNA 12(6):1084–1091
22. Korner CG, Wahle E (1997) Poly(A) tail shortening by a mammalian poly(A)-specific 3′-exoribonuclease. J Biol Chem 272(16):10448–10456
23. Lai WS, Kennington EA, Blackshear PJ (2003) Tristetraprolin and its family members can promote the cell-free deadenylation of AU-rich element-containing mRNAs by poly(A) ribonuclease. Mol Cell Biol 23(11):3798–3812

24. Fabian MR, Mathonnet G, Sundermeier T et al (2009) Mammalian miRNA RISC recruits CAF1 and PABP to affect PABP-dependent deadenylation. Mol Cell 35(6):868–880
25. Johnson SM, Grosshans H, Shingara J et al (2005) RAS is regulated by the let-7 microRNA family. Cell 120(5):635–647
26. Sampson VB, Rong NH, Han J et al (2007) MicroRNA let-7a down-regulates MYC and reverts MYC-induced growth in Burkitt lymphoma cells. Cancer Res 67(20):9762–9770
27. Wakiyama M, Takimoto K, Ohara O et al (2007) Let-7 microRNA-mediated mRNA deadenylation and translational repression in a mammalian cell-free system. Genes Dev 21(15):1857–1862
28. Robb GB, Brown KM, Khurana J et al (2005) Specific and potent RNAi in the nucleus of human cells. Nat Struct Mol Biol 12(2): 133–137
29. Nishi K, Nishi A, Nagasawa T et al (2013) Human TNRC6A is an Argonaute-navigator protein for microRNA-mediated gene silencing in the nucleus. RNA 19(1):17–35
30. Doidge R, Mittal S, Aslam A et al (2012) Deadenylation of cytoplasmic mRNA by the mammalian Ccr4-Not complex. Biochem Soc Trans 40(4):896–901
31. Suzuki A, Saba R, Miyoshi K et al (2012) Interaction between NANOS2 and the CCR4-NOT deadenylation complex is essential for male germ cell development in mouse. PLoS One 7(3):e33558
32. Van Etten J, Schagat TL, Hrit J et al (2012) Human Pumilio proteins recruit multiple deadenylases to efficiently repress messenger RNAs. J Biol Chem 287(43):36370–36383
33. Maragozidis P, Karangeli M, Labrou M et al (2012) Alterations of deadenylase expression in acute leukemias: evidence for poly(a)-specific ribonuclease as a potential biomarker. Acta Haematol 128(1):39–46
34. Martinez J, Ren YG, Nilsson P et al (2001) The mRNA cap structure stimulates rate of poly(A) removal and amplifies processivity of degradation. J Biol Chem 276(30):27923–27929
35. Dehlin E, Wormington M, Korner CG, Wahle E (2000) Cap-dependent deadenylation of mRNA. EMBO J 19(5):1079–1086
36. Balatsos NA, Nilsson P, Mazza C et al (2006) Inhibition of mRNA deadenylation by the nuclear cap binding complex (CBC). J Biol Chem 281(7):4517–4522
37. Cevher MA, Zhang X, Fernandez S et al (2010) Nuclear deadenylation/polyadenylation factors regulate 3′ processing in response to DNA damage. EMBO J 29(10):1674–1687
38. Lehner B, Sanderson CM (2004) A protein interaction framework for human mRNA degradation. Genome Res 14(7):1315–1323
39. Devany E, Zhang X, Park JY et al (2013) Positive and negative feedback loops in the p53 and mRNA 3′ processing pathways. Proc Natl Acad Sci U S A 110(9):3351–3356
40. Udagawa T, Swanger SA, Takeuchi K et al (2012) Bidirectional control of mRNA translation and synaptic plasticity by the cytoplasmic polyadenylation complex. Mol Cell 47(2): 253–266
41. Azzouz N, Panasenko OO, Colau G et al (2009) The CCR4-NOT complex physically and functionally interacts with TRAMP and the nuclear exosome. PLoS One 4(8):e6760
42. Lee JE, Lee JY, Trembly J et al (2012) The PARN deadenylase targets a discrete set of mRNAs for decay and regulates cell motility in mouse myoblasts. PLoS Genet 8(8):e1002901
43. Meijer HA, Bushell M, Hill K et al (2007) A novel method for poly(A) fractionation reveals a large population of mRNAs with a short poly(A) tail in mammalian cells. Nucleic Acids Res 35(19):e132
44. Astrom J, Astrom A, Virtanen A (1991) In vitro deadenylation of mammalian mRNA by a HeLa cell 3′ exonuclease. EMBO J 10(10): 3067–3071
45. Lowell JE, Rudner DZ, Sachs AB (1992) 3′-UTR-dependent deadenylation by the yeast poly(A) nuclease. Genes Dev 6(11): 2088–2099
46. Sachs AB, Deardorff JA (1992) Translation initiation requires the PAB-dependent poly(A) ribonuclease in yeast. Cell 70(6):961–973
47. Garneau NL, Wilusz CJ, Wilusz J (2008) Chapter 5. In vivo analysis of the decay of transcripts generated by cytoplasmic RNA viruses. Methods Enzymol 449:97–123
48. Mittal S, Aslam A, Doidge R et al (2011) The Ccr4a (CNOT6) and Ccr4b (CNOT6L) deadenylase subunits of the human Ccr4-Not complex contribute to the prevention of cell death and senescence. Mol Biol Cell 22(6):748–758
49. Piao X, Zhang X, Wu L et al (2010) CCR4-NOT deadenylates mRNA associated with RNA-induced silencing complexes in human cells. Mol Cell Biol 30(6):1486–1494

Chapter 24

Assaying mRNA Deadenylation In Vitro

Mandy Jeske, Claudia Temme, and Elmar Wahle

Abstract

Deadenylation is the removal of poly(A) tails from mRNA. This chapter presents two methods to assay deadenylation in vitro. The first is a quick and quantitative assay for the degradation of radiolabeled poly(A) that can easily be adapted to be used for many different enzymes. The second method uses an extract from *Drosophila* embryos to catalyze the deadenylation of an RNA dependent on a specific sequence that also directs deadenylation in vivo.

Key words Deadenylases, PARN, CCR4–NOT complex, *nanos*

1 Introduction

Deadenylases are poly(A)-specific 3′-exonucleases. Their main biological role is the shortening and, finally, removal of the poly(A) tails of mRNAs. Deadenylation is typically the first step in the degradation of eukaryotic mRNAs. Deadenylation can also contribute to translational repression of mRNAs. In eukaryotes, three types of deadenylases are known, the PAN2/PAN3 heterodimer, homodimeric PARN, and, most importantly, the CCR4–NOT complex [1, 2]. Addition of oligo(A) tails to all kinds of RNAs can also be a means to facilitate the attack of 3′ exonucleases leading either to processing of the RNA or its complete degradation [3]. In these reactions, 3′ exonuclease degradation is not limited to the poly(A) tail. Thus the enzymes involved, mostly the exosome and its associated 3′-exonucleases, are not considered deadenylases. The distinction is not always clear, though.

This chapter describes two types of assays useful in studies of deadenylases. The first, release of acid-soluble nucleotides from homogeneously labeled poly(A), is convenient for rapidly assaying the activity of any poly(A) degrading enzyme in vitro. While this type of assay is useful, it does not reflect the biological deadenylation reaction. Rates of deadenylation vary widely between different mRNAs in vivo as they are regulated by specific RNA sequences via

Joanna Rorbach and Agnieszka J. Bobrowicz (eds.), *Polyadenylation: Methods and Protocols*, Methods in Molecular Biology, vol. 1125, DOI 10.1007/978-1-62703-971-0_24, © Springer Science+Business Media New York 2014

RNA-binding proteins or miRNAs. The second method that will be described here is a complex *Drosophila* embryo extract system that reproduces this type of sequence-dependent mRNA deadenylation observed in cells.

1.1 Measuring the Activity of the Poly(A)-Specific Ribonuclease by a TCA Precipitation Assay

The fact that biological polymers, both nucleic acids and proteins, are precipitated by trichloroacetic acid (TCA), whereas their building blocks remain soluble, is the time-honored basis of many assays either to determine the synthesis of polymers from their precursors or to measure polymer degradation. For example, the in vitro activity of poly(a)-specific ribonuclease (PARN) can be rapidly assayed by the release of radiolabeled degradation products from homogeneously labeled poly(A) [4]. For this purpose, TCA is added at the end of the reaction to precipitate the remaining polymer. The precipitate is removed by centrifugation, and the acid-soluble products in the supernatant are quantitated by scintillation counting. While we describe the specific reaction conditions for PARN, the assay can easily be adapted to any enzyme able to degrade poly(A). For example, the nonspecific Endonuclease G has been characterized with the same type of assay [5].

For substrate preparation, a short oligo(A) primer is elongated with α-^{32}P-labeled ATP in an unspecific polyadenylation reaction. In our hands, use of a truncated version of the bovine poly(A) polymerase [6] yields poly(A) with a relatively homogeneous length distribution, whereas use of the yeast enzyme Pap1p [4] produces polymers with a broader length distribution, perhaps due to a preference of this enzyme for longer RNAs. Other commercially available enzymes, like *E. coli* poly(A) polymerase or the "noncanonical" enzyme Cid1 from *S. pombe*, have not been tested for the purpose in our lab but are likely to work as well.

Solubility in TCA is not restricted to mononucleotides; short oligonucleotides are soluble as well. We have not determined the exact cutoff, but the end products of RNA degradation by Endonuclease G, which are up to five nucleotides long, are acid soluble, whereas we have found that oligomers longer than 20 nucleotides are precipitated. Obviously, poly(A) degradation can be assayed by methods with much higher resolution, like gel electrophoresis or thin-layer chromatography, but these are more time consuming and quantitation is more cumbersome.

1.2 A Drosophila Embryo Extract System to Measure Sequence-Dependent mRNA Deadenylation (Embryo Extract Deadenylation Assay)

While many examples of sequence-dependent mRNA deadenylation have been studied in vivo, few in vitro systems have been reported that would allow biochemical details of the reaction to be analyzed. Potential difficulties include the dependence of the reaction on multiple components or the prevalence of nonspecific RNases. In vitro systems reproducing sequence-dependent mRNA deadenylation have been developed from *Xenopus* eggs [7, 8] and from mammalian cells [9]. We have found that extract from 0- to

2-h-old *Drosophila* embryos reproduces deadenylation of the *nanos* mRNA [10, 11]. As initially observed in vivo, the in vitro reaction depends on two Smaug recognition elements (SREs), the binding sites for the regulatory protein Smaug. Although the same extracts also recapitulate the SRE-dependent translational repression seen in vivo [10, 12], deadenylation is independent of translation. However, SRE-dependent deadenylation does depend on a high ATP concentration maintained by the addition of an ATP-regenerating system.

To study deadenylation, in vitro-synthesized, radiolabeled mRNA substrates are used. RNA substrates must contain at least one SRE and should be short enough to allow a good separation of deadenylated product from polyadenylated substrate by denaturing polyacrylamide gel electrophoresis. For convenience, the mRNA's poly(A) tail should be encoded within the plasmid used for runoff transcription. Such constructs and alternative ways for RNA substrate preparation have been described in detail [11]. To confirm that shortening of the RNA is due to deadenylation, digestion with RNase H in the presence of oligo(dT) can be used [10].

2 Materials

Use analytical grade reagents. Prepare all solutions using double-distilled or similar high-grade water. For all solutions coming into contact with RNA, use water made RNase-free by treatment with diethylpyrocarbonate (DEPC): Supplement double-distilled water with 0.1 % (v/v) DEPC, leave on a magnetic stirrer overnight, and autoclave to inactivate non-reacted DEPC. Note that poly(A) is less sensitive to RNases, being resistant against, e.g., RNase A. Nevertheless, care should be taken in handling this type of RNA as well. Use all other precautions commonly used to prevent RNase contaminations. For example, wear disposable gloves and use disposable plasticware handled with appropriate care.

2.1 Materials for the TCA Precipitation Assay

1. Oligo(A) primer: Buy a synthetic RNA-oligonucleotide of the desired length (~15 nucleotides) (*see* **Notes 1** and **2**). Dissolve the primer in DEPC-treated water at a concentration of at least 20 μM (with respect to 3′ ends). Store at −20 °C.
2. 5× poly(A) polymerase reaction buffer for bovine poly(A) polymerase: 125 mM Tris–HCl pH 8.3, 200 mM KCl, 50 % glycerol, 0.25 mM EDTA, 0.1 % Nonidet P-40 (NP-40); 1 mg/ml RNase-free bovine serum albumin (BSA) (*see* **Note 3**); or 4× polyadenylation buffer for the yeast Pap1p enzyme: 80 mM Tris–HCl, pH 7.0, 40 % glycerol, 200 mM KCl, 0.8 mM EDTA, 0.4 mg/ml RNase-free BSA (*see* **Note 3**). Store these buffers at −20 °C.
3. ATP, 25 mM: A commercial solution may be used. If you make up your own solution from solid ATP, adjust the pH to neutral

to prevent hydrolysis. Measure the ATP concentration in a spectrophotometer (ε_{259} = 15,400 at pH 7.0) and dilute to 25 mM. Store at −20 °C.

4. 25 mM $MnCl_2$, 25 mM: Store in small aliquots at −20 °C.
5. 25 mM Dithiothreitol (DTT), 25mM: Dilute from a 1 M stock solution kept at −20 °C, and store in small aliquots at the same temperature (*see* **Note 4**).
6. α-^{32}P-labeled ATP (3,000–5,000 Ci/mmol).
7. Poly(A) polymerase: We use the mammalian variant His-PAP531 [6] expressed in *E. coli* and purified over a Ni-NTA and a MonoQ column. Yeast poly(A) polymerase is available from Affymetrix or MCLAB. *E. coli* poly(A) polymerase, which has not been tested in our lab, can be purchased from NEB or Invitrogen, and *S. pombe* Cid1 can be obtained from NEB.
8. Phenol:chloroform (1:1): Mix one volume of buffer-saturated phenol (molecular biology grade) with one volume of chloroform, and let phases separate. The lower phase contains the phenol:chloroform mixture. Store in a refrigerator.
9. 10 M Ammonium acetate, 10 M.
10. 96 % ethanol, analytical grade.
11. 70 % ethanol, analytical grade.
12. Sephadex G50 spin columns (GE Healthcare): Self-made columns can be used as well if care is taken to avoid RNase contamination.
13. Poly(A) (e.g., Sigma Aldrich or Roche Diagnostics) (*see* **Note 5**): Dissolve in DEPC-treated water and measure the concentration in a spectrophotometer (ε_{258} = 9,800 with respect to mononucleotides); concentration should be at least 5 mM with respect to mononucleotides. Store at −20 °C.
14. 4× PARN reaction buffer: 80 mM HEPES pH 7.0, 0.6 M potassium acetate, 4 mM magnesium acetate, 0.08 % NP-40, 0.8 mg/ml RNase-free BSA, 40 % (v/v) glycerol.
15. Freshly dissolved 16 % (w/v) TCA: The solution can be stored on ice for a day and presumably significantly longer.
16. Scintillation cocktail.

2.2 Materials for the Deadenylation Assay Using Drosophila Embryo Extract

2.2.1 Materials for Maintaining Flies, Collecting Embryos, and Preparing Extract

1. Refer to standard procedures for fly propagation [13]. We maintain flies at 25 °C.
2. Fly cages: We use 3–4 cages with a volume of 14 l each.
3. Embryo collection trays and yeast paste: Follow published protocols [14, 15].
4. Set of three stapled sieves with different mesh sizes: upper, middle, and lower with 710, 355, and 125 μm mesh, respectively (Retsch, Haan, Germany); or self-made sieve (*see* **Note 6**).

5. Paintbrush.
6. Dounce homogenizer with tight-fitting pistil (Wheaton): Use a 1, 2, or 7 ml homogenizer for the lysis of up to 500 mg, 1 g, or 3.5 g embryos, respectively.
7. Paper towels.
8. Sodium hypochlorite solution (Roth).
9. 1 ml pipette tips with cut ends.
10. Dewar filled with liquid nitrogen.
11. Lysis buffer (2×): 60 mM HEPES–KOH, pH 7.4, 200 mM potassium acetate, 4 mM magnesium acetate. Store at −20 °C.
12. DTT: 1 M solution in DEPC-treated water. Store at −20 °C (*see* **Note 4**).
13. Protease inhibitor Pefabloc SC (Roche Applied Science): 20 mg/ml solution in DEPC-treated water. Store in aliquots at −20 °C.

2.2.2 Materials for In Vitro Synthesis and Purification of the Labeled RNA Substrate

1. Purified and linearized DNA template. Store at −20 °C.
2. Bacteriophage RNA polymerase, 50 u/μl, enzyme type depending on the transcription vector. Store at −20 °C.
3. Transcription buffer, supplied with the RNA polymerase. Store at −20 °C.
4. RNasin, 40 u/μl (Promega). Store at −20 °C.
5. DTT: 1 M solution in DEPC-treated water; store at −20 °C (*see* **Note 4**).
6. Nucleotide (NTP) mix containing 10 mM each ATP, CTP, GTP, and 1 mM UTP. Assemble from commercial nucleotide stock solutions. Store at −20 °C.
7. [α-^{32}P]-UTP, 3,000 Ci/mmol, 10 μCi/μl (Hartmann Analytics).
8. RNase-free DNase I, 10 u/μl (Roche Applied Science). Store at −20 °C.
9. Water-saturated phenol:chloroform (1:1) (*see* Subheading 2.1, **step 8**).
10. Ammonium acetate, 10 M.
11. 96 % ethanol, analytical grade.
12. 70 % ethanol, analytical grade.
13. Scintillation cocktail.

2.2.3 Materials for the Deadenylation Assay

The following reagents are prepared with DEPC-treated water and stored at −20 °C:

1. HEPES–KOH, pH 7.4, 1 M.
2. Potassium acetate, 1.25 M.
3. Magnesium acetate, 0.5 M.

4. Yeast tRNA (Roche Applied Science): Dissolve in water, and determine the concentration by measuring the absorption at 260 nm, assuming that OD of 1 corresponds to 0.04 mg/ml (*see* **Note 5**). Adjust the concentration to 2.5 mg/ml.
5. ATP, 50 mM: Dissolve ATP in water, adjust pH to 7 with NaOH, determine the concentration in a spectrophotometer with $\varepsilon_{259} = 15{,}400$, and dilute to 50 mM. Alternatively, a commercial solution can be used.
6. Creatine phosphate (Sigma): 1 M solution in water; store in aliquots.
7. Creatine kinase, 10 mg/ml: Dissolve creatine kinase (Roche Applied Science) in a buffer consisting of 20 mM HEPES–KOH, pH 7.4 and 50 % glycerol. The enzyme remains active for at least 1 year.
8. Formamide loading buffer: Supplement deionized formamide with 0.1 % (w/v) bromophenol blue, 0.1 % (w/v) xylene cyanol, and 10 mM EDTA pH 8. Store in aliquots.

Store the following reagents at 4 °C:

9. Stop solution (25 mM EDTA, pH 8): Dissolve EDTA in water and adjust the pH with KOH.
10. Water-saturated phenol:chloroform (1:1) (*see* Subheading 2.1).
11. Ammonium persulfate (APS): 10 % solution in water.
12. *N,N,N′,N′*-tetramethyl-ethylenediamine (TEMED; Merck).

The following reagents are stored at room temperature:

13. Ammonium acetate/glycogen solution: Supplement 10 M ammonium acetate with 0.2 mg/ml glycogen (20 mg/ml glycogen stock is available from Roche Applied Science).
14. 96 % ethanol, analytical grade.
15. TBE buffer (10×): 0.9 M Tris, 0.9 M borate, 20 mM EDTA.
16. Denaturing polyacrylamide gel stock solution: 20 % acrylamide/*N,N′*-methylenebisacrylamide solution (19:1; Serva), 1× TBE, and 8.3 M urea. Keep in the dark.
17. Denaturing polyacrylamide gel dilution solution: 1× TBE and 8.3 M urea.
18. TBE running buffer (1×): Prepare from 10× TBE stock solution.

3 Methods

Obey the usual safety rules for handling of ^{32}P!

3.1 Methods for the TCA Precipitation Assay

3.1.1 Preparation of Radiolabeled Poly(A)

1. For the use of bovine poly(A) polymerase, assemble, on ice, a 50 μl polyadenylation reaction containing 150 pmol (with respect to 3′ ends) oligo(A) primer in 1× PAP buffer, 0.5 mM DTT, 0.5 mM $MnCl_2$, 0.5 mM unlabeled ATP, and 30–50 μCi [α-^{32}P]-ATP. Mix well, add the required amount of poly(A) polymerase (*see* **Note 7**), mix again gently avoiding the formation of bubbles, and transfer the tube to 37 °C to start the reaction. Incubate for 1 h at 37 °C. For yeast poly(A) polymerase, use the same conditions, except that the buffer for the yeast enzyme and 0.7 mM $MnCl_2$ are used and the incubation temperature is 30 °C (*see* **Note 8**).
2. At any time during the incubation period, remove a small aliquot (e.g., 1 μl) from the reaction mixture, mix with scintillation cocktail, and count the radioactivity (in counts per min, cpm) in a scintillation counter (*see* **Note 9**). Calculate the total cpm in the reaction mixture and divide by the total amount of ATP (25,000 pmol in this case, ignoring the chemical concentration of the labeled ATP) to obtain the specific radioactivity of ATP in the reaction. Under the conditions used, expect 2,500–5,000 cpm/pmol.
3. At the end of the incubation period, extract the reaction mixture with phenol/chloroform and precipitate RNA from the aqueous phase by adding 1/3 volume of 10 M ammonium acetate and 2.5 volumes (with respect to the total volume) of 96 % ethanol (*see* **Note 10**).
4. Centrifuge in a microcentrifuge at maximum speed for 20 min, carefully remove the supernatant, and wash the pellet with 200 μl 70 % ethanol.
5. Centrifuge again for 5 min at maximum speed, remove the supernatant, let the pellet air-dry, and dissolve the RNA in 50 μl DEPC-treated water. Both after precipitation and, more importantly, after washing, it may be useful to check with a Geiger–Müller counter that the pellet is not lost. Save supernatants until you are sure that you have not lost the pellet.
6. Apply the RNA to a Sephadex G50 spin column to remove residual non-incorporated ATP: Insert the spin column into a 2 ml microfuge tube and centrifuge for 2 min at 2,000 × *g*. Insert the column into a fresh 1.5 ml tube, apply the RNA solution gently on top of the column bed, and centrifuge again at 2,000 × *g* for 2 min. The flow-through will contain the purified RNA, whereas non-incorporated ATP will be retained on the column.
7. Take an aliquot from the flow-through of the spin column, and measure the radioactivity in a scintillation counter.

Calculate the total cpm in the sample, and, using the previously determined specific radioactivity, calculate the amount of AMP incorporated into the polymer (in pmol). If desired, the length of the poly(A) product can be determined by electrophoresis of an aliquot on a 10 % urea-polyacrylamide gel next to a radioactively labeled standard. The procedure described above yields an average length of poly(A) of 400–600 nt in our hands.

8. Store the RNA at –20 °C.

3.1.2 TCA Precipitation Assay

1. Assemble, on ice, a reaction mixture as a suitable multiple of 50 μl, and include a safety margin for pipetting inaccuracy. For example, if you plan to run ten assays plus two controls, assemble 13×50 μl. Each reaction contains non-labeled poly(A) corresponding to 1 nmol of AMP together with 50,000–100,000 cpm of labeled poly(A), prepared as above, in 1× PARN buffer plus 0.5 mM DTT (*see* **Note 11**). Mix well, and dispense aliquots of 50 μl into individual microfuge tubes on ice. Start the reactions by addition of enzyme to individual tubes and transfer to 37 °C (*see* **Note 12**). Two control reactions receive no enzyme. One is incubated with the enzyme-containing reactions and serves as the background control. Leave the other tube on ice.
2. After the desired reaction time, stop the reactions by the addition of 150 μl ice-cold 16 % TCA and keep the sample on ice. Add the same amount of TCA to both control reactions (*see* **Note 13**).
3. Centrifuge the samples including the background control in a refrigerated centrifuge at 4 °C at maximum speed for 20 min (*see* **Note 14**). Keep the other control reaction on ice.
4. Transfer 100 μl each of the supernatants to scintillation vials, add scintillation cocktail, and measure the released radioactivity in your samples. Subtract the value of the background (=no enzyme) control from the values of the enzyme-containing samples. Vortex the second control tube that has not been centrifuged and also count 100 μl. From the number obtained, calculate the amount of radioactivity (in cpm) in the complete reaction volume and divide (in this example) by 1,000 pmol (the amount of poly(A) in nucleotides present in the reaction) to obtain the specific radioactivity in your substrate (in cpm/pmol) (*see* **Note 15**). From the radioactivity counts in the supernatants of the individual reactions and the specific radioactivity, calculate the amounts (pmol) of nucleotides released.

3.2 Deadenylation Assays Using Drosophila Embryo Extract and Labeled RNA Substrates

Drosophila embryo extract is prepared from embryos collected every 2 h. The method is based on the one originally described by Tuschl and colleagues [16] and can be used for extract preparation from about 50 mg to at least 3 g embryos per collection time point. For larger preparations (>5 g embryos per 2-h collection) use the equipment described in [14] or [15].

3.2.1 Preparation of Drosophila Embryo Extract

1. Collect 0–2-h-old embryos by changing the yeast-smeared collection trays in the fly cages every 2 h. Pre-warm the new trays for about 1 h prior to use. Immediately proceed with extract preparation (*see* **Note 16**).
2. Use a soft paintbrush and running cold tap water to rinse embryos off the trays and into the stack of sieves: The upper sieve retains flies, the middle sieve retains fly body parts, and the lower sieve retains embryos. Yeast cells are washed through all sieves.
3. Dechorionate embryos at room temperature: Use distilled water to flush embryos out of the sieve into a beaker containing distilled water and mix with 1 volume of sodium hypochlorite solution (bleach). Incubate with occasional swirling until embryos have lost their dorsal appendages and tend to stick to the glass and to each other (approximately 1 min) (*see* **Notes 17** and **18**).
4. Pour dechorionated embryos back into the 125 μm mesh sieve and wash with cold tap water until the chlorine smell has disappeared (*see* **Note 19**).
5. Blot the embryos with paper towels from the bottom of the sieve to remove water.
6. With a spatula, transfer the embryos into a dounce homogenizer of known weight that is kept on ice. Weigh again to determine the mass of embryos.
7. Per 1 g of embryos add 1 ml of ice-cold 1× lysis buffer supplemented with 5 mM DTT and 1 mg/ml Pefabloc SC. Homogenize the embryos on ice: Move the pistil up and down until it slides easily within the glass tube (approximately 5–10 strokes).
8. Dispense 500 μl aliquots of the lysate into 1.5 ml Eppendorf tubes and spin in a refrigerated microcentrifuge for 20 min at 14,500 ×*g* and 4 °C. This separates the lysate into a fatty upper layer, a pellet containing cell nuclei and debris, and the soluble cytoplasmic phase in the middle.
9. Touch the fatty upper layer with a 1 ml pipette tip with the point cut off. The layer will stick to tip and can be taken off.

10. Collect the cytoplasmic phase, dispense in suitable aliquots into microfuge tubes, and drop them into liquid nitrogen (*see* **Note 20**). The total protein concentration of the extract can be determined by commercially available Bradford reagent and should be near 30 mg/ml.
11. Store the extract in liquid nitrogen (*see* **Note 21**).

3.2.2 Preparation of the RNA Substrate

The template for the SRE-containing RNA should be cloned into one of the standard transcription vectors for bacteriophage RNA polymerases (T3, T7, or SP6). A suitable poly(A) tail should also be encoded in the plasmid. We have been using a tail length of 72 nucleotides. For runoff transcription, the template should be linearized immediately downstream of the poly(A) tail. We have found that 3 non-A nucleotides downstream of the tail do not interfere with SRE-dependent deadenylation, but 40 nucleotides do [10]. The total length of the transcript should not be more than a few hundred nucleotides, since shorter RNAs are less sensitive to RNases and easier to analyze by gel electrophoresis.

1. Prepare the template DNA for runoff transcription: Digest 5–20 μg of plasmid DNA with the appropriate restriction endonuclease, and extract the linearized DNA with phenol:chloroform and then chloroform. Precipitate by the addition of 10 M ammonium acetate to a final concentration of 2.5 M and 2.5 volumes of ethanol. Dissolve the DNA in DEPC-treated water, and determine the concentration in a spectrophotometer, assuming that $A_{260} = 1$ corresponds to 0.05 mg/ml.
2. Perform runoff transcription in a 25 μl reaction containing 1 μg of purified, linearized template DNA, 75 u of the appropriate bacteriophage RNA polymerase, 1× transcription buffer, 30 mM DTT, 50 u RNasin, 2.5 μl NTP mixture, and 2 μl (20 μCi) [α-^{32}P]-UTP. Incubate for 2 h at 37 °C (*see* **Note 22**).
3. Remove the DNA template by adding 1 μl RNase-free DNase I and incubating for 20 min at 37 °C.
4. At any point up to here, remove 1 μl from the transcription mix and measure the radioactivity by scintillation counting (*see* **Note 9**). Calculate the total radioactivity in the reaction mixture (in cpm) and divide by 2,500 pmol (the total amount of UTP in the reaction) to obtain the specific activity in the UTP pool. Under the conditions used, expect about 20,000 cpm/pmol.
5. Purify the RNA by phenol:chloroform and chloroform extraction followed by precipitation with ammonium acetate and ethanol (*see* **Note 23**).
6. Wash the pellet with 70 % ethanol, and dissolve the RNA in 30–50 μl DEPC-treated water. Remove 1 μl, and measure the

radioactivity by scintillation counting. Calculate the total radioactivity (in cpm) incorporated into RNA. Calculate the RNA concentration from the total radioactivity, the specific radioactivity of the UTP calculated in step 4, and the length and UMP content of your RNA (for long RNAs, assuming a U content of 25 % should be precise enough) (*see* **Note 24**).

7. Store RNA at −20 °C.

3.2.3 Deadenylation Assay

1. Thaw aliquots of *Drosophila* embryo extract as needed (*see* **Note 25**), RNA substrate, and stock solutions for the deadenylation reaction. Keep reagents on ice.
2. Assemble the deadenylation reaction containing 5–20 nM (with respect to 3′ ends, not nucleotides) labeled poly(A)$^{+}$ RNA, 20–40 % (v/v) embryo extract, 16 mM HEPES–KOH pH 7.4, 50 mM potassium acetate, 2.5 mM magnesium acetate, 250 μg/ml tRNA, 80 μg/ml creatine kinase, 20 mM creatine phosphate, and 0.8 mM ATP. Add DEPC-treated water to the desired final volume. The concentrations indicated ignore components introduced with the extract. Mix the components gently but thoroughly; avoid the formation of air bubbles. No reaction takes place as long as the mixture is left on ice. To study the time course of sequence-dependent deadenylation, set up a reaction with a volume of 10 μl for each intended time point. If you wish to vary a component of the assay, set up the required multiple of the reaction volume, leaving out the component to be varied, mix gently, dispense aliquots into individual tubes, and add the missing component.
3. Start the reaction by transfer to 25 °C. To stop the reaction, mix 10 μl of the reaction mixture with 190 μl of pre-chilled stop solution and put on ice.
4. Add 1 volume of phenol:chloroform (200 μl) to each sample and vortex vigorously. Spin for 30 min at maximum speed and at room temperature in a microcentrifuge, remove 180 μl of the upper aqueous phase into a new 1.5 ml tube, and repeat the phenol/chloroform extraction.
5. Transfer only 120 μl of the upper phase into another 1.5 ml tube, and precipitate the RNA by addition of 40 μl of the ammonium acetate/glycogen solution and 400 μl ethanol. Mix by vortexing or inverting the tube several times, and sediment the precipitate for 30 min at maximum speed and room temperature in a microcentrifuge.
6. Completely remove the supernatant. You may wish to use a Geiger–Müller counter to make sure that you do not lose the RNA. Let the pellet air-dry, and dissolve it in 4 μl of formamide loading buffer by vortexing. Denature the RNA by heating to 90 °C for 3 min. Also mix a suitable aliquot of

unreacted substrate RNA with formamide loading buffer for comparison.

7. Prepare a denaturing polyacrylamide/urea gel according to standard protocols [17]. For routine analysis, a 20 cm gel provides sufficient resolution. The optimal acrylamide concentration depends on the length of your RNA. Load your samples together with a suitable radiolabeled size marker (restriction digest of plasmid DNA labeled with Polynucleotide Kinase), run the gel in 1× TBE, and analyze by phosphoimaging.

4 Notes

1. As the TCA precipitation assay measures the release of acid-soluble, radioactively labeled nucleotides, the labeled part of the poly(A) substrate should be long compared to the unlabeled primer. Thus, use of a short primer is advisable.
2. While oligo(A) can also be prepared by partial alkaline hydrolysis of commercially available poly(A) and subsequent size fractionation, we do not recommend this method since additional steps would be necessary to remove the phosphate group remaining at the 3′ end generated by hydrolysis.
3. RNase-free BSA can be obtained commercially. Alternatively, it can be prepared by reductive methylation, which inactivates RNases. For a protocol, see [18].
4. DTT is prone to oxidation. Store at high concentration in tightly closed containers at −20 °C, and dilute small aliquots as needed. Do not store solutions, in particular dilute solutions, for an extended period of time.
5. Commercial polynucleotides are often not entirely free of proteins. It may be useful to perform a phenol/chloroform extraction after the polynucleotide has been dissolved, followed by chloroform extraction and ethanol precipitation. As the pellet will be big, wash thoroughly with 70 % ethanol to remove traces of phenol. Also because of the size of the pellet, allow sufficient time for it to dissolve in DEPC-treated water.
6. To handle a small amount of embryos (<200 mg), use a self-made sieve [11] and remove flies and body parts from the collected embryos with forceps.
7. The amount of PAP required must be determined empirically. The goal is to incorporate a large fraction (~80 %) of the ATP in the reaction mixture into polymer. The amount of enzyme needed to achieve this obviously depends on the type of enzyme and on the quality and age of the preparation. For bovine PAP, 0.1 μg is a good starting point.

8. If you use any other type of poly(A) polymerase, use the buffer conditions recommended by the supplier.
9. If the radioactivity in the sample is too high, the aliquot from the reaction mixture should be diluted with water, and an aliquot of the dilution should be counted.
10. Use of ammonium acetate for precipitation reduces the coprecipitation of mononucleotides.
11. The mixture of unlabeled poly(A) and ^{32}P-labeled poly(A) is convenient to keep the chemical concentration of the substrate constant, and add just the amount of label required to allow quantitation of nucleotides released. In principle, different length distributions of labeled and unlabeled poly(A) might be an issue, but we have not experienced problems with this. Another potential problem is that both alkaline hydrolysis and many RNases leave phosphates at the 3′ end, which may inhibit 3′ exonucleases; thus, a partially degraded batch of poly(A) may be resistant to degradation. The easiest way to check the state of the 3′ ends is to examine the ability of the polynucleotide to be extended by poly(A) polymerase.
12. Adjust buffer conditions, reaction temperature, and concentration of poly(A) to the requirements of any enzyme that you wish to assay. Choose conditions that result in a linear response to incubation time and enzyme concentration. If possible, use a concentration of poly(A) that will saturate the enzyme.
13. If you use low concentrations of poly(A) substrate or much of the poly(A) is degraded, a carrier may be necessary to facilitate the precipitation of the remaining polymer. In this case, the reaction should be stopped without addition of TCA, e.g., by excess EDTA in the case of divalent metal ion-dependent nucleases. The stop reagent should be mixed with the carrier (e.g., 10 μg of yeast RNA per reaction) before addition to the reactions. Then TCA can be added. Final TCA concentration should be 10–12 %.
14. Once TCA has been added to stop the reaction, centrifugation must be performed within the next 30 min, as the substrate is slowly degraded by the low pH of the sample. For the same reason, refrigeration before and during centrifugation is necessary. If samples have to be stored for longer times, stop the reaction by other means (e.g., addition of EDTA) and add the TCA shortly before centrifugation.
15. This calculation ignores the chemical concentration of the radiolabeled poly(A). Use the specific radioactivity of your synthesis product (determined in Subheading 3.1.1, **step 7**) to judge if this is justified. Remember that ^{32}P has a half-life of 14.3 days, so the specific activity drops accordingly with storage.

16. It is essential to prepare extract immediately after each 2-h embryo collection to obtain high deadenylation activity.
17. If you have collected the embryos in a self-made collection sieve, submerge them in this sieve in a beaker containing 1:2 diluted sodium hypochlorite solution.
18. Dechorionation is critical: Insufficient dechorionation hampers lysis of embryos, and excessive bleaching damages them. As the efficiency of the sodium hypochlorite solution decreases upon storage, control the dechorionation process carefully by eye.
19. If the sodium hypochlorite is not removed completely, it may inactivate the extract.
20. Choose the size of the aliquot such that refreezing after use is not necessary.
21. Storage at −80 °C is possible but not recommended for long periods of time.
22. During the deadenylation reaction in *Drosophila* embryo extract, a cap structure on the 5′ end of the RNA substrate is not required for RNA stability. Thus, transcription is performed without a cap dinucleotide.
23. The quality of the transcript should be checked by running an aliquot on a denaturing polyacrylamide/urea gel alongside radioactive markers. Gel purification of the RNA is not necessary if the products are uniform and of the expected length. Removal of non-incorporated nucleotides is efficient if the RNA is ethanol-precipitated in the presence of 2.5 M ammonium acetate. The absence of free nucleotides can be checked by thin-layer chromatography [10]. If necessary, repeat the precipitation or purify the RNA over a Sephadex G50 spin column (*see* Subheading 3.1.1, **step 6**).
24. If time has passed between the determination of the specific activity of the UTP in the reaction mixture and the preparation of the RNA, take into account the half-life of ^{32}P (14.3 days).
25. We have always thawed extract by slight warming but only until it was just thawed. We have not tested whether slow thawing on ice makes any difference.

References

1. Goldstrohm AC, Wickens M (2008) Multifunctional deadenylase complexes diversify mRNA control. Nature Rev Mol Cell Biol 9:337–344
2. Harnisch C, Moritz B, Rammelt C et al (2012) Eukaryotic RNases and their partners in RNA degradation and biogenesis. In: Chanfreau G, Tamanoi F (eds) The enzymes, vol 31. Academic Press, New York, pp 181–211
3. Houseley J, Tollervey D (2009) The many pathways of RNA degradation. Cell 136: 763–776
4. Körner C, Wahle E (1997) Poly(A) tail shortening by a mammalian poly(A)-specific

3′-exoribonuclease. J Biol Chem 272: 10448–10456

5. Temme C et al (2009) The Drosophila melanogaster gene cg4930 encodes a high affinity inhibitor for endonuclease G. J Biol Chem 284:8337–8348
6. Martin G, Keller W (1996) Mutational analysis of mammalian poly(A) polymerase identifies a region for primer binding and a catalytic domain, homologous to the family X polymerases, and to other nucleotidyl transferases. EMBO J 15:2593–2603
7. Legagneux V, Omilli F, Osborne HB (1995) Substrate-specific regulation of RNA deadenylation in Xenopus embryo and activated egg extracts. RNA 1:1001–1008
8. Voeltz GK, Ongkasuwan J, Standart N et al (2001) A novel embryonic poly(A) binding protein, ePAB, regulates mRNA deadenylation in Xenopus egg extracts. Genes Dev 15:774–788
9. Marchese FP et al (2010) MAPKAP kinase 2 blocks tristetraprolin-directed mRNA decay by inhibiting CAF1 deadenylase recruitment. J Biol Chem 285:27590–27600
10. Jeske M, Meyer S, Temme C et al (2006) Rapid ATP-dependent deadenylation of nanos mRNA in a cell-free system from Drosophila embryos. J Biol Chem 281:25124–25133
11. Jeske M, Wahle E (2008) Cell-free deadenylation assays using Drosophila embryo extracts. Meth Enzymol 448:107–118
12. Jeske M, Moritz B, Anders A et al (2011) Smaug assembles an ATP-dependent stable complex repressing nanos mRNA translation at multiple levels. EMBO J 30:90–103
13. Sission JC (2000) In: Sullivan W, Ashburner M, Hawley RS (ed) Drosophila protocols. Cold Spring Harbor Laboratory Press, Cold Spring Harbor, NY, pp 540–551
14. Moritz M (2000) In: Sullivan W, Ashburner M, Hawley RS (ed) Drosophila protocols. Cold Spring Harbor Laboratory Press, Cold Spring Harbor, NY, pp 570–575.
15. Gebauer F, Hentze M (2007) Studying translational control in Drosophila cell-free systems. Meth Enzymol 429:23–33
16. Tuschl T, Zamore PD, Lehmann R et al (1999) Targeted mRNA degradation by double-stranded RNA in vitro. Genes Dev 13:3191–3197
17. Sambrook J, Russell DW (2001) Molecular cloning, vol 1. Cold Spring Harbor Laboratory Press, Cold Spring Harbor, NY
18. Wahle E (1991) A novel poly(A)-binding protein acts as a specificity factor in the second phase of messenger RNA polyadenylation. Cell 66:759–768

Chapter 25

Assaying mRNA Deadenylation In Vivo

Claudia Temme and Elmar Wahle

Abstract

Deadenylation is the removal of poly(A) tails from mRNA. Here, we present two methods for assaying deadenylation in vivo. The first is a method for measuring bulk poly(A) tail lengths. When combined with a block in transcription, the method can be used for measuring the rate of bulk poly(A) tail shortening. The second is an RT-PCR method to determine the poly(A) tail lengths of individual RNAs. Again in combination with a block of transcription, the method permits the rate of deadenylation of an individual RNA to be measured.

Key words Poly(A) tail, Polyadenylation, Deadenylation, Deadenylase, mRNA decay

1 Introduction

Non-templated adenylate residues can be added to RNA in vivo for at least two purposes. The first is to facilitate attack on the RNA by a 3′ exonuclease. In this case, oligo(A) tails are added by noncanonical poly(A) polymerases [1, 2]. The oligo(A) tails are typically short, and, as they attract 3′ exonucleases, they are also short-lived. The 3′ exonucleases acting on them usually continue into the non-poly(A) part, either processing or degrading the RNA [3]. The second purpose of adding A residues is to generate a functional mRNA. These poly(A) tails are long, e.g., ~250 nucleotides in mammalian cells, they are added by a canonical poly(A) polymerase [4, 5], and they are essential for the function of the mRNA, for example, in the initiation of translation. These tails are gradually shortened in the cytoplasm, and their shrinking below a certain threshold triggers decay of the mRNA [3, 6, 7].

Numerous methods can be used to assess the presence of poly(A) tails and their lengths. This chapter describes two such assays that are being widely used in studies of in vivo deadenylation. As an alternative to the two methods described here, high-resolution Northern blotting can be used, provided the RNA to be examined is sufficiently abundant. If one aliquot of the RNA

Joanna Rorbach and Agnieszka J. Bobrowicz (eds.), *Polyadenylation: Methods and Protocols*, Methods in Molecular Biology, vol. 1125, DOI 10.1007/978-1-62703-971-0_25, © Springer Science+Business Media New York 2014

preparation is digested with RNase H in the presence of oligo(dT) and a second aliquot is left untreated, the size difference between them corresponds to the length of the poly(A) tail. If the RNA to be examined is long so that the size difference caused by the RNase H/oligo(dT) digestion is not resolved by the gel, the RNA can be cut by RNase H in the presence of a specific antisense DNA oligonucleotide, and the Northern can then be probed for the downstream fragment, again with and without RNaseH/oligo(dT) treatment (for examples of these methods, *see* ref. 8). If still higher resolution and/or information about the exact nucleotide sequences at the 3′ end are desired, conventional or high-throughput sequencing of cDNAs may be the method of choice (for examples, *see* refs. 9, 10).

1.1 Measurement of Bulk Poly(A) Tail Length

The first method to be described here is based on the resistance of poly(A) to digestion by RNase A (which is pyrimidine-specific) and RNase T1 (which cuts after guanosine). If the digestion is carried out after 3′ end-labeling of total RNA, the analysis of the remaining nuclease-resistant fragments provides a rapid assessment of bulk poly(A) tail length. We describe a current adaptation of a method that goes back to the early days of molecular biology [11, 12]. When steady-state RNA samples are analyzed in this way, populations of poly(A) can be expected that extend from very short up to the length of newly synthesized poly(A) tails characteristic for the organism—between 200 and 250 nucleotides in mammalian cells and 70–80 nucleotides in yeast. Rates of bulk poly(A) shortening can be obtained if the method is applied to RNA isolated at various time points after inhibition of mRNA synthesis. Commonly used methods to achieve this include treatment of mammalian cells with actinomycin D or yeast cells with thiolutine. In yeast, a temperature-sensitive allele of RNA polymerase II can also be used [13–15]. We have also combined this method of poly(A) tail analysis with the isolation of newly synthesized RNA based on pulse-labeling with 4-thiouridine [16] (U. Kühn and E. Wahle, unpublished data).

We prefer the use of [α-^{32}P]-3′deoxy ATP (cordycepin triphosphate) and poly(A) polymerase for 3′ labeling of RNA [17]. Due to the absence of the 3′-hydroxyl, the reaction is restricted to the addition of a single nucleotide. Under so-called nonspecific reaction conditions [24], poly(A) polymerase extends all kinds of RNA molecules. While preferences for certain RNAs cannot be excluded, the poly(A) tails, which are of interest here, provide an efficient and uniform type of substrate so that biases in labeling are unlikely to affect the results. We are using a variant of the bovine poly(A) polymerase (PAP531) [18], which we prepare ourselves. Commercially available *S. cerevisiae* poly(A) polymerase (Pap1p) can be used as an alternative [17]. Other commercially available enzymes which include poly(A) polymerase from *E. coli* and Cid1 from *S. pombe*, are likely to work as well, but have not been tested for this purpose in our lab. As an alternative to the use of poly(A) polymerase and 3′-dATP, 3′ labeling can be performed by RNA ligase and ^{32}P-labeled pCp [19].

The RNase digestion is a critical step in this procedure, because the substrate specificity of RNase A and T1 is not absolute; nonspecific RNA degradation can occur at low salt concentration or high enzyme to substrate ratio. Therefore, an excess of unlabeled carrier RNA is added to the reaction. The use of control RNAs to ascertain the specificity of the digestion is also recommended.

1.2 PAT Assay to Determine the Poly(A) Tail Length of Specific mRNAs

The ligation-mediated poly(A) tail test (LM-PAT) method is an RT-PCR procedure that yields information about poly(A) tail lengths of specific mRNAs. In principle, the poly(A) tail length of any RNA can be determined by an RT-PCR assay in which an oligo(dT)-anchor primer is combined with a gene-specific upstream primer. However, the oligo(dT)-anchor primer can anneal anywhere on the poly(A) tail, so that the distribution of PCR products is biased towards shorter chains compared to the distribution of the RNA template. Moreover, shorter cDNAs are more efficiently amplified. The LM-PAT assay, originally developed by Sallés and Strickland [20, 21], was designed to favor priming of cDNA synthesis at the very end of the poly(A) tail. For this purpose, an array of 5′phosphorylated $(dT)_{12}$ oligonucleotides is hybridized to the poly(A) tail of RNAs in the presence of T4 DNA ligase at a relatively high temperature. This converts most of the poly(A) tail into a poly(A)-poly(dT) duplex but tends to leave short single-stranded 3′ oligo(A) overhangs that were too short to stably hybridize with oligo(dT) under the conditions used. Then the temperature is lowered, and an excess of an oligo(dT)-anchor primer is added. At the lower temperature, the primer can hybridize to the 3′overhang and be ligated to the product of the previous ligation. The resulting poly(dT) with the anchor sequence at its 5′ end is now used as the primer for reverse transcription. Subsequent PCR amplification using the anchor primer and a gene-specific primer leads to a family of products, the size distribution of which reflects the different poly(A) tail lengths of the original mRNA plus the upstream section of the polyadenylated RNA that was included in the amplification. Ethidium bromide staining can be used to detect the PCR products. However, the use of radioactively labeled PCR primers in combination with denaturing electrophoresis, as described here, provides better resolution and permits product detection over a much larger range of concentrations.

In a related assay (extension PAT, ePAT), polyadenylated RNA is first extended by DNA polymerase. A DNA oligonucleotide with a 3′ oligo(dT) sequence is hybridized to the tail and provides the template for addition of a DNA tag, which, in the end, is used for PCR amplification [22].

As with the bulk method described above, the PAT assay can be used to measure the rates of deadenylation if it is applied to samples taken at various time points after the stop of transcription. As the method is applied to specific mRNA species, regulated promoters can be used instead of the inhibition of bulk transcription. Possibilities include galactose-regulated promoters in yeast [15], tetracycline-regulated promoters, or the regulation of a promoter by heat shock [8, 15, 23]. Regulated promoters also offer the opportunity to perform pulse-chase experiments rather than measuring mRNA decay from steady-state.

As with the bulk method, control RNAs can be used. Product specificity is one potential problem of the PAT assay, since only the 5′ primer is gene-specific. Nested 5′ primers may be used. Possible specificity controls include sequencing the product or shortening it with restriction enzymes predicted to cleave in the amplified sequence upstream of the poly(A) tail. Although we have not used this method, one can also imagine blotting of the product and hybridization with specific probes. Note that product heterogeneity does not necessarily reflect the size distribution of poly(A) tails but possibly also heterogeneity in the sites of poly(A) addition.

2 Materials

Use analytical grade reagents. Prepare all solutions using double-distilled or similar high-grade water. For all solutions coming into contact with RNA, use water made RNase-free by treatment with diethylpyrocarbonate (DEPC): Supplement double-distilled water with 0.1 % (v/v) DEPC, leave on a magnetic stirrer overnight, and autoclave to inactivate non-reacted DEPC. Note that poly(A) is less sensitive to RNases, being resistant against, e.g., RNase A. Nevertheless, care should be taken in handling this type of RNA as well. Use all other precautions commonly used to prevent RNase contaminations. For example, wear disposable gloves and use disposable plastic ware handled with appropriate care.

2.1 Materials for Bulk Poly(A) Tail Measurement

1. 5× polyadenylation buffer (125 mM Tris–HCl pH 8.3, 200 mM KCl, 50 % glycerol, 0.25 mM EDTA, 2.5 mM $MnCl_2$, 0.1 % Nonidet P-40 (NP-40), 1 mg/ml RNase-free bovine serum albumin (BSA)) for bovine poly(A) polymerase or 4× polyadenylation buffer (80 mM Tris–HCl pH 7.0, 40 % glycerol, 200 mM KCl, 2.8 mM $MnCl_2$, 0.8 mM EDTA, 0.4 mg/ml RNase-free BSA) for *S. cerevisiae* poly(A) polymerase. Store at –20 °C (*see* **Note 1**).
2. Poly(A) polymerase. We use the mammalian variant His-PAP531 [18] expressed in *E. coli* and purified over a Ni-NTA and a MonoQ column. *S. cerevisiae* poly(A) polymerase is available from Affymetrix or MCLAB. *E. coli* poly(A)-polymerase, which

has not been tested in our lab, can be purchased from NEB or Invitrogen, and *S. pombe* Cid1 can be obtained from NEB (*see* **Note 2**).

3. [α-^{32}P] 3′deoxy ATP (5,000 Ci/mmol, 10 μCi/μl; Perkin Elmer).
4. 5 mM DTT (*see* **Note 3**).
5. Total cellular RNA to be analyzed (*see* **Note 4**). Determine the concentration by spectrophotometry assuming that $A_{260} = 1$ corresponds to 40 μg/ml. Store the RNA at −20 °C or, for long-term storage, at −80 °C.
6. Optional: Polyadenylated control RNA generated by in vitro transcription (*see* **Note 5**).
7. Phenol/chloroform (1:1). Mix 1 volume of buffer-saturated phenol (molecular biology grade) with 1 volume of chloroform and let the phases separate. The lower phase contains the phenol:chloroform mixture. Store in a refrigerator.
8. 10 M ammonium acetate.
9. RNase A (5 ng/μl) in water. Store at −20 °C.
10. RNase T1 (12.5 u/μl; Sigma-Aldrich). Store the stock solution in a refrigerator. Dilute the appropriate amount of the stock solution in 50 mM Tris–HCl pH 7.5, 100 mM KCl, 10 % glycerol shortly before use. Store the dilution on ice.
11. 10× RNase T1-buffer (1 M KCl, 0.5 M Tris–HCl pH 7.5, 10 mg/ml yeast total RNA) (*see* **Note 6**). Store at −20 °C.
12. 2× Proteinase K buffer (0.2 M Tris–HCl pH 7.9, 0.3 M NaCl, 25 mM EDTA, 2 % SDS). Store at room temperature.
13. Proteinase K (Merck). Dissolve at 20 mg/ml in 0.1 M Tris–HCl pH 7.9, 0.15 M NaCl, 12.5 mM EDTA, 50 % (v/v) glycerol and store at −20 °C.
14. 1 mg/ml rRNA (Roche) (*see* **Note 6**). Store at −20 °C.
15. 96 % Ethanol, analytical grade.
16. 70 % Ethanol, analytical grade.
17. Formamide loading buffer (80 % deionized formamide, 1 mM EDTA, 0.2 mg/ml each bromophenol blue and xylene cyanol). Store at −20 °C.

2.2 Materials for PAT Assay

1. 3 M sodium acetate.
2. 96 % Ethanol, analytical grade.
3. 70 % Ethanol, analytical grade.
4. 10 M ammonium acetate.
5. [γ-^{32}P]-ATP (3,000–5,000 Ci/mmol, 10 μCi/μl; Perkin Elmer). Store as recommended by the supplier.

Store the following reagents at −20 °C:

6. RNA preparation to be analyzed (*see* **Note 4**). Determine the concentration by spectrophotometry assuming that $A_{260} = 1$ corresponds to 40 μg/ml. Store the RNA at −20 °C or, for long-term storage, at −80 °C.
7. 50 ng/μl $(dT)_{12}$, 5′phosphorylated using Polynucleotide Kinase and ATP.
8. 1 μg/μl oligo(dT) anchor primer (5′gCg AgC TCC gCg gCC gCg TTT TTT TTT TTT 3′).
9. Gene-specific primers (10–100 μM in water) (*see* **Note 7**).
10. Optional: Polyadenylated control RNA generated by in vitro transcription (*see* **Note 5**).
11. M-MLV reverse transcriptase (RNaseH minus) and corresponding reaction buffer (Promega).
12. T4 DNA ligase (400 u/μl ; NEB) and 10× T4 DNA ligase buffer.
13. *Taq* DNA Polymerase and corresponding reaction buffer (e.g., GE Healthcare or other supplier).
14. T4-Polynucleotide Kinase (PNK) and corresponding reaction buffer (10 u/μl; NEB).
15. 100 mM DTT (*see* **Note 3**).
16. dNTP mix: dATP, dGTP, dCTP, dTTP at 10 mM each; assemble from commercial stock solutions. Store at −20 °C.
17. 10 mM ATP. Dilute from a commercial stock solution. If you make up your own solution from solid ATP, adjust the pH to neutral to prevent hydrolysis. Measure the ATP concentration in a spectrophotometer ($\varepsilon_{259} = 15{,}400$ at pH 7.0) and dilute to 10 mM. Store at −20 °C.
18. 10× TBE buffer: 0.9 M Tris, 0.9 M boric acid, 20 mM EDTA.
19. Denaturing polyacrylamide gel stock solution: 20 % acrylamide/*N*,*N*′-methylenebisacrylamide solution (19:1; Serva), 1× TBE and 8.3 M urea; keep in the dark.
20. Denaturing polyacrylamide gel dilution solution: 8.3 M urea in 1× TBE.

3 Methods

Obey the usual safety rules for handling of ^{32}P!

3.1 Bulk Poly(A) Tail Measurement

3.1.1 End-Labeling of Total RNA with PAP

1. Assemble, on ice, a 20 μl reaction containing 1–1.5 μg of the RNA preparation to be analyzed in 1× polyadenylation buffer plus 0.5 mM DTT, 20 μCi [α-^{32}P] 3′-deoxy ATP, and 100–500 ng poly(A) polymerase (*see* **Note 8**). Optional: Set up a reaction

with ~0.1 μg of the control RNA bearing a poly(A) tail of a defined length in the same way (*see* **Note 9**).

2. Incubate the reaction for 1 h at 37 °C (bovine poly(A) polymerase) or 30 °C (yeast poly(A) polymerase).
3. Add 130 μl of H_2O and 150 μl of phenol/chloroform to the reaction and vortex vigorously. Separate the phases in a microcentrifuge and transfer the upper aqueous phase, approximately 150 μl, to a new tube.
4. Add 1/3 volume of 10 M ammonium acetate and precipitate the RNA with 2.5 volumes (with respect to total aqueous volume) of 96 % ethanol.
5. Centrifuge for 15 min at maximum speed in a microcentrifuge.
6. Take off the supernatant, add 200 μl of 70 % ethanol to the pellet and centrifuge again for 5 min. Remove the supernatant and air-dry the pellet (*see* **Note 10**).
7. Dissolve the labeled RNA in 15 μl of H_2O.
8. Take 1 μl and determine the amount of radioactivity in a scintillation counter for the following RNase digestion reaction (*see* **Notes 11** and **12**).

3.1.2 RNase T1 and RNase A Digestion of the Labeled RNA

1. Assemble a 20 μl reaction on ice containing 2 μl of 10× RNase T1 buffer and 100,000–500,000 cpm of the end-labeled RNA. If you want to compare different RNA samples, all should contain similar amounts of radioactivity. Add 5 ng RNase A and 25 u RNase T1 and incubate for 30 min at 30 °C. If available, incubate 10,000 cpm of the labeled control RNA under the same conditions (*see* **Note 13**).
2. Prepare a stop mix containing, per RNA sample, 25 μl of 2× Proteinase K buffer plus 20 μg of Proteinase K and 2 μg rRNA in a total volume of 30 μl. Stop the RNase digestion by the addition of 30 μl of this stop mix to each sample. Vortex briefly and incubate for another 30 min at 37 °C.
3. Precipitate the RNA by the addition of 2.5 volumes of 96 % ethanol. Centrifuge for 15 min at maximum speed and discard the supernatant. Since all cellular RNAs have been labeled but the fraction of RNAs bearing a poly(A) tail is only 3–5 %, most of the radioactivity should now be in the supernatant. In contrast, there should be barely any radioactivity in the supernatant of the control RNA. The RNA pellet should be visible because of the rRNA added with the stop mix.
4. Wash the pellet with 200 μl of 70 % ethanol, centrifuge for another 5 min, discard the supernatant, and air-dry the pellet.
5. Dissolve the RNA in 6 μl of formamide loading buffer.
6. Denature the samples for 2 min at 95 °C and separate the poly(A) tails on a denaturing polyacrylamide/urea gel. For normal

poly(A) length distributions, an 8 % gel with a length of 40 cm provides very good resolution. Also load an aliquot of the non-digested control RNA to compare with the digested sample. As a size marker, we routinely use an *Hpa*I digested pBR322 plasmid 5′-labeled with T4-PNK and [γ-^{32}P]-ATP.

7. Visualize the radioactive RNA by phosphoimaging or on an X-ray film.

3.2 PAT Assay to Determine the Poly(A) Tail Length of Specific mRNAs

3.2.1 cDNA Synthesis

1. Mix 3 μg of the RNA preparation to be analyzed with 50 ng of a 5′-phosphorylated $(dT)_{12}$ oligonucleotide in a total volume of 7 μl (*see* **Note 14**). If available, 2 ng of the control RNA should be included.
2. Denature the mixture for 5 min at 65 °C and put the samples at 42 °C; do not put them on ice.
3. In a second tube mix, per RNA sample to be analyzed, 4 μl of 5× M-MLV reverse transcriptase buffer, 2 μl 100 mM DTT, 1 μl 10 mM dNTP mix, 1 μl 10 mM ATP, and 1.5 μl T4 DNA ligase (NEB) in a total volume of 13 μl.
4. Preincubate this mix for 3–5 min at 42 °C and then add 13 μl aliquots to each denatured RNA sample.
5. Incubate the reactions for 30 min at 42 °C.
6. Add 1 μg oligo(dT) anchor primer and transfer the tubes to 12 °C. Incubate for another 2 h.
7. Transfer the tubes back to 42 °C, add 1 μl (~200 u) M-MLV Reverse transcriptase (RNase H minus) and incubate for another hour at 42 °C.
8. Inactivate the enzymes at 70 °C for 15 min.
9. Store the samples on ice.

3.2.2 Primer Labeling Reaction

1. Incubate 150 pmol of the gene-specific upstream PCR primer in a 15 μl reaction in 1× T4 PNK buffer with 10–20 μCi [γ-^{32}P]-ATP [5,000 Ci/mmol] and 1 μl T4-PNK at 37 °C for 1 h.
2. Precipitate the RNA by adding 1.5 μl of 3 M sodium acetate and 55 μl of 96 % ethanol. Centrifuge for 30 min at maximum speed and remove the supernatant.
3. Add 200 μl of 70 % ethanol to the pellet, centrifuge again for 10 min, remove the supernatant, and let the pellet air dry.
4. Dissolve the labeled primers in 50 μl water (~3 μM final concentration) and determine the amount of incorporated radioactivity by scintillation counting. For the subsequent PCR, we always use a mixture of labeled and unlabeled primers.

3.2.3 PCR

1. For each sample to be analyzed, set up a 50 μl reaction mix containing 1 μl of the cDNA synthesis reaction from Subheading 3.2.1, 300 μM of each dNTP, 0.5 μM of oligo(dT)-anchor primer, 0.5 μM of a gene-specific 5′-primer, 100,000–300,000 cpm of the same primer, radioactively labeled as in Subheading 3.2.2, and 1 u *Taq* DNA polymerase in 1× *Taq* buffer containing $MgCl_2$ (*see* **Note 15**). If you have mixed a control RNA with the RNA to be analyzed, set up, for each RNA sample, a second PCR mix containing a 5′ primer specific for the control RNA instead of the gene-specific primer.
2. Perform a PCR under conditions appropriate for the amplification of your gene of interest. The annealing temperature should be set to 58 °C based on the oligo(dT)-anchor primer (*see* **Note 7**). We typically use the following program: 3 min. 94 °C; (30 s 94 °C, 45 s 58 °C; 2 min 72 °C) 35×; 72 °C 5 min; and cooling down to 4 °C.
3. At the end of the PCR, add 50 μl of water and 100 μl of phenol/chloroform. Vortex vigorously, separate phases in a microcentrifuge, and transfer the upper aqueous phase to a fresh tube.
4. Precipitate DNA by adding 1/3 volume ammonium acetate and 2.5 volumes (with respect to the total aqueous volume) of 96 % ethanol. Centrifuge at maximum speed for 30 min and remove the supernatant.
5. Add 200 μl of 70 % ethanol to the pellet, centrifuge again for 5 min, take off the supernatant, and let the pellet air dry for approximately 10 min.
6. Dissolve the pellet in 10 μl of formamide loading buffer.
7. Analyze 5 μl of each sample on a denaturing polyacrylamide/urea gel next to a suitable radioactive standard (*see* **Note 16**). 6 % acrylamide and 20 cm gel length are a good starting point.
8. Apply a phosphoimager screen or an X-ray film overnight to visualize the hot PCR products (*see* **Note 17**).

4 Notes

1. RNase-free BSA can be obtained commercially. Alternatively, it can be prepared by reductive methylation, which inactivates RNases. For a protocol, *see* ref. [24].
2. If you use any other type of poly(A) polymerase, use the buffer conditions recommended by the supplier.
3. DTT is prone to oxidation. Store a 1 M solution in tightly closed containers at −20 °C and dilute small aliquots as needed. Do not store solutions, in particular dilute solutions, for an extended period of time.

4. RNA should be isolated by the Trizol method (Invitrogen) or any other method that delivers clean and undegraded RNA.
5. For both methods of poly(A) tail length measurement, it is useful to have a positive control. An in vitro transcript carrying a poly(A) tail encoded in the plasmid template is convenient. For example, deadenylation substrates as described in the accompanying article [25] can be used.
6. Dilute the RNA from a stock solution. Commercial polynucleotides are often not entirely free of proteins. It may be useful to perform a phenol/chloroform extraction after the RNA has been dissolved, followed by chloroform extraction and ethanol precipitation. As the pellet will be big, wash thoroughly with 70 % ethanol to remove traces of phenol. Also because of the size of the pellet, allow sufficient time for it to dissolve in DEPC-treated water. Measure the concentration in a spectrophotometer, assuming that $A_{260} = 1$ corresponds to 40 μg/ml.
7. Follow the general rules for the design of PCR primers. Theoretical melting temperature should match that of the anchor primer (58 °C). We normally use a primer binding 200–300 nucleotides upstream of the poly(A) tail.
8. The amount of PAP required must be determined empirically. The goal is to incorporate a large fraction (~80 %) of the radioactive label into the RNA. The amount of enzyme needed to achieve this obviously depends on the type of enzyme and on the quality and age of the preparation. For bovine PAP, 0.1 μg is a good starting point.
9. If a polyadenylated control RNA is not available, straight poly(A) of a defined length can also be used. The former is better, as it serves as a twofold control, both for the complete digestion of the non-poly(A) segment and for the absence of RNase cuts in the poly(A) part. Straight poly(A) will only serve as a control against the latter, i.e., overdigestion.
10. In all procedures involving ethanol precipitation of radiolabeled RNA, one can use a Geiger-Müller counter to make sure that the pellet is not lost.
11. Use of ammonium acetate for precipitation reduces the co-precipitation of mononucleotides. Thus, the radioactivity left at this stage should be due to the end-labeling of RNA. If too much non-incorporated nucleotide is left, the precipitation can be repeated or the RNA can be purified over a Sephadex G50 spin column.
12. It is not necessary to determine the efficiency of labeling in molar terms. However, for the subsequent analysis, the amount of radioactivity in the sample should be known.

13. Under the conditions used, RNases A and T1 should cleave specifically on the 3′ side of pyrimidine and guanine nucleotides, respectively. Longer incubation times, larger amounts of enzyme and reduced salt concentration increase the risk of cutting within the poly(A) tail.
14. If the concentration of your RNA preparation is not high enough, ethanol precipitate the RNA and dissolve in water.
15. If the *Taq* buffer does not contain Mg^{2+}, add $MgCl_2$ to 1.5 mM final.
16. Polyacrylamide/urea gels are suitable for PCR products up to at least ~600 nt. PCR products can also be analyzed in non-denaturing agarose gels, but the resolution is lower. If you use an agarose gel, note that they cannot be dried directly on a standard gel dryer for exposure of a film or phosphoimager plate. Instead, place the gel on DEAE paper (Whatman DE81 or similar), which itself is placed on a stack of paper towels. Put a glass plate on top of the gel and a suitable weight on top of the glass plate. Wait for several hours until the gel is only 1 or 2 mm thick. Transfer the gel-DEAE paper sandwich to a gel drier and dry to completion.
17. The PAT assay carried out with the control RNA should yield a PCR product with a length corresponding to the known poly(A) tail length of the RNA. Amounts of this PCR product should be approximately uniform for all reactions independently of the cellular RNA that was mixed with the control RNA.

References

1. Minasaki R, Eckmann CR (2012) Subcellular specialization of multifaceted 3′ end modifying nucleotidyltransferases. Curr Opin Cell Biol 24:1–9
2. Schmidt M-J, Norbury CJ (2010) Polyadenylation and beyond: emerging roles for noncanonical poly(A) polymerases. Wiley Interdiscip Rev RNA 1:142–151
3. Houseley J, Tollervey D (2009) The many pathways of RNA degradation. Cell 136:763–776
4. Wahle E, Rueegsegger U (1999) 3′-End processing of pre-mRNA in eukaryotes. FEMS Microbiol Rev 23:277–295
5. Zhao J, Hyman L, Moore C (1999) Formation of mRNA 3′ ends in eukaryotes: mechanism, regulation and interrelationship with other steps in mRNA synthesis. Microbiol Mol Rev 63:405–445
6. Goldstrohm AC, Wickens M (2008) Multifunctional deadenylase complexes diversify mRNA control. Nature Rev Mol Cell Biol 9:337–344
7. Harnisch C, Moritz B, Rammelt C et al (2012) Eukaryotic RNases and their partners in RNA degradation and biogenesis. In: Chanfreau G, Tamanoi F (eds) The enzymes, vol 31. Academic Press, New York, pp 181–211
8. Bönisch C, Temme C, Moritz B et al (2007) Degradation of hsp70 and other mRNAs in Drosophila via the 5′-3′ pathway and its regulation by heat shock. J Biol Chem 282: 21818–21828
9. Couttet P, Fromont-Racine M, Steel D et al (1997) Messenger RNA deadenylylation precedes decapping in mammalian cells. Proc Natl Acad Sci U S A 94:5628–5633
10. Berndt H et al (2012) Maturation of mammalian H/ACA box snoRNAs: PAPD5-dependent adenylation and PARN-dependent trimming. RNA 18:958–972
11. Lee SY, Mendecki J, Brawerman G (1971) A polynucleotide segment rich in adenylic acid in the rapidly-labeled polyribosomal RNA component of mouse sarcoma 180 ascites cells. Proc Natl Acad Sci U S A 68:1331–1335

12. Hadjivassiliou A, Brawerman G (1966) Polyadenylic acid in the cytoplasm of rat liver. J Mol Biol 20:1–7
13. Herrick D, Parker R, Jacobson A (1990) Identification and comparison of stable and unstable mRNAs in Saccharomyces cerevisiae. Mol Cell Biol 10:2269–2284
14. Grigull J, Mnaimeh S, Pootoolal J et al (2004) Genome-wide analysis of mRNA stability using transcription inhibitors and microarrays reveals posttranscriptional control of ribosome biogenesis factors. Mol Cell Biol 24:5534–5547
15. Coller J (2008) Methods to determine mRNA half-life in Saccharomyces cerevisiae. Meth Enzymol 448:267–284
16. Dölken L et al (2008) High-resolution gene expression profiling for simultaneous kinetic parameter analysis of RNA synthesis and decay. RNA 14:1959–1972
17. Lingner J, Keller W (1993) 3′-End labeling of RNA with recombinant yeast poly(A) polymerase. Nucleic Acids Res 21:2917–2920
18. Martin G, Keller W (1996) Mutational analysis of mammalian poly(A) polymerase identifies a region for primer binding and a catalytic domain, homologous to the family X polymerases, and to other nucleotidyl transferases. EMBO J 15:2593–2603
19. England TE, Uhlenbeck OC (1978) 3′-Terminal labelling of RNA with T4 RNA ligase. Nature 275:560–561
20. Sallés FJ, Strickland S (1995) Rapid and sensitive analysis of mRNA polyadenylation states by PCR. PCR Methods Appl 4:317–321
21. Sallés FJ, Strickland S (1999) In: Haynes S (ed) Methods in molecular biology, vol 118, Humana Press, Totowa, NJ.
22. Jänicke A, Vancuylenberg J, Boag PR et al (2012) ePAT: a simple method to tag adenylated RNA to measure poly(A)-tail length and other 3′ RACE applications. RNA 18: 1289–1295
23. Xu N, Loflin P, Chen C-YA et al (1998) A broader role for AU-rich element-mediated mRNA turnover revealed by a new transcriptional pulse strategy. Nucleic Acids Res 26: 558–565
24. Wahle E (1991) A novel poly(A)-binding protein acts as a specificity factor in the second phase of messenger RNA polyadenylation. Cell 66:759–768
25. Jeske M, Temme C, Wahle E (in press) Assaying mRNA deadenylation in vitro. In: Rorbach J, Bobrowicz AJ (ed) Polyadenylation, methods molecular biology, Humana Press, Totowa, NJ.

Chapter 26

In Vitro Analysis of RNA Degradation Catalyzed by Deadenylase Enzymes

Joel Hrit, Nathan Raynard, Jamie Van Etten, Kamya Sankar, Adam Petterson, and Aaron C. Goldstrohm

Abstract

In this chapter, we describe a method for purification and analysis of the enzymatic activity of deadenylase enzymes. Nearly all eukaryotic messenger RNAs are modified at the 3′ end by the addition of an adenosine polymer: the poly-adenosine tail. The poly(A) tail plays a central role in protein expression and mRNA fate. The poly(A) tail promotes translation of the mRNA. Shortening of the poly(A) tail, referred to as deadenylation, reduces protein synthesis and initiates destruction of the mRNA. A specialized class of exoribonucleases, called deadenylase enzymes, carries out this process. Deadenylases are found throughout eukarya, but their functions remain largely unexplored. We present a detailed protocol to analyze deadenylase activity in vitro. First, recombinant deadenylase enzyme is over-expressed and purified from bacteria. Next, labeled RNA substrate is prepared. Deadenylation reactions are performed, and reaction products are analyzed by denaturing gel electrophoresis. Reaction rates are then determined quantitatively. Crucial controls and experimental parameters are described along with practical tips that promote success.

Key words Poly-adenosine tail, Deadenylase, Exoribonuclease, RNA degradation

1 Introduction

In eukaryotes, newly synthesized mRNAs (with the exception of replication dependent histone mRNAs) are modified by the addition of a poly-Adenosine tail to the 3′ end of the message [1]. The cleavage and poly-adenylation machinery recognizes sequence elements in the nascent mRNA and cleaves the mRNA at a specific site [2]. Subsequently, poly(A) polymerase adds adenosine residues to the 3′ end of the mRNA [2]. Poly-adenosine tails differ in length depending on the species. For instance, in *S. cerevisiae* the nascent tails are 60–80 nucleotides, whereas mammalian tails are 200–250 nucleotides. After the addition of a 5′ cap structure, polyadenylation, and splicing of introns, the mRNA is transported to the cytoplasm where it can be translated. Throughout these events, the mRNA is susceptible to degradation by specialized and highly

Joanna Rorbach and Agnieszka J. Bobrowicz (eds.), *Polyadenylation: Methods and Protocols*, Methods in Molecular Biology, vol. 1125, DOI 10.1007/978-1-62703-971-0_26, © Springer Science+Business Media New York 2014

regulated mRNA decay pathways [3, 4]. Messenger RNAs are often bound by regulatory factors that modulate their decay rates and translation efficiency. Quality control mechanisms also scrutinize mRNAs and destroy aberrant, defective mRNAs.

Decay of mRNAs typically initiates by enzymatic removal of the poly(A) tail by deadenylases [3]. These enzymes are exoribonucleases that act in a 3′–5′ direction on the tail, releasing 5′AMP as a product [5–7]. Deadenylases are hydrolytic, using enzyme-bound divalent cations as essential cofactors that are coordinated by acidic residues in their active sites [3, 4]. Two groups of deadenylases have been identified based on sequence relationships of their catalytic domains [3]. The DEDD family of deadenylases, so-called because of conserved catalytic Aspartate and Glutamate residues, are related to founding members Pan2 and PARN [3, 8, 9]. This group shares similarity to the bacterial RNase D exoribonuclease [4]. The second group are deadenylases that belong to the exonuclease–endonuclease–phosphatase superfamily [3, 10, 11]. The yeast Ccr4 enzyme is the founding member of this class [12, 13].

The number of predicted deadenylases varies among eukaryotes [3, 11, 14]. Many of these enzymes have not been studied and the methods described in this chapter provide a direct, quantitative assay for measuring deadenylase catalytic activity, substrate specificity, and cofactor requirements. In vivo, deadenylation rates vary dramatically, possibly due to the action of specific deadenylase enzymes [5, 6, 15–18]. Deadenylation is also controlled by *cis-acting* regulatory sequences and *trans*-acting factors including RNA binding proteins and small regulatory RNAs called micro-RNAs [3, 19–27]. The methods described here provide a means to study regulation of deadenylase activity in a defined biochemical system.

To begin, the deadenylase of interest must be cloned into the appropriate expression vector with an affinity tag. Multiple affinity tag purification systems have been successfully used to purify deadenylase enzymes. No single tag works best for all proteins; therefore, we recommend starting with several tags and testing protein induction and purification. Choose the tag that gives the highest yield and purity. We have used Glutathione S-Transferase (GST), Halotag, and Maltose Binding Protein (MBP) tags with excellent results [26–29]. We recommend avoiding 6-histidine tags because the metal ions used on the affinity resin (usually Nickel) can inhibit deadenylase activity (our unpublished data). Further, contaminating bacterial nucleases are prone to coelute from the metal affinity resin. Removal of the affinity tag from the deadenylase protein by engineering a site-specific protease cleavage site between tag and deadenylase is often desirable. We have used Tabacco Etch Virus (TEV) protease with excellent results.

In this chapter, we first describe a method to purify recombinant deadenylase enzyme: it is over-expressed as a fusion to Maltose Binding Protein in *E. coli* using conditions optimized to remove bacterial contaminants. Next, a method for preparing radioactively labeled RNA substrates is described. Finally, we describe the procedure for quantitatively measuring deadenylation activity and present instructions and suggestions for data analysis and interpretation, as well as determination of reaction rates.

2 Materials

The method described herein is by design a highly sensitive ribonuclease assay, therefore, the main contributor to experimental problems is RNase contamination. Because bacteria have abundant RNases, this is the most common and likely source of contamination. Reagents, equipment, and the scientists themselves are also possible sources of RNases.

Chemicals must be handled only with gloves and measured with sterile, RNase-free plasticware. All buffers should be prepared using sterile, RNase-free, distilled deionized water. We use water purified with a Millipore Advantage A10 system with a BioPak point of use filter. All buffers should be prepared in baked glassware or RNase-free disposable plasticware. A dedicated set of pipettes should be reserved for working with RNA and aerosol barrier filter pipette tips must be used. RNase Inhibitor (RNaseIn) can be added to deadenylation reactions to specifically inhibit the RNase A family, which are common environmental contaminants. Before beginning this assay, new users should consult instructions on creating and maintaining an RNase-free environment. For additional information and tips for working with RNA, refer to RNA: A Laboratory Manual [30].

2.1 Protein Purification Components

1. Distilled, deionized sterile H_2O (ddH_2O), autoclaved.
2. Protease deficient *E. coli* strain transformed with deadenylase expression vector. We use BL21 Gold (Invitrogen) or KRX (Promega) strains, unless a specialized strain is required.
3. 2xYT broth with 2 mM $MgSO_4$. Autoclave and then add sterile glucose to a final percentage of 0.2 %.
4. 40 % Glucose: Dissolve 40 g into 100 mL of ddH_2O, filter sterilize.
5. Amylose affinity resin (New England Biolabs): supplied as a 50 % slurry.
6. 1 M Isopropyl β-d-thiogalactoside (IPTG): Dissolve in ddH_2O, filter sterilize, store at −20 °C.
7. 1 M Dithiothreitol (DTT): Dissolve in ddH_2O, store at −20 °C.
8. Washing Buffer: 50 mM Tris–HCl pH 8.0, 10 % sucrose. Filter sterilize and store at 4 °C.

9. Binding buffer: 50 mM Tris–HCl pH 8.0, 2 mM $MgCl_2$, 500 mM NaCl. Store at 4 °C. NP-40/IGEPAL at a final concentration of 0.5 % and DTT at a final concentration of 1 mM should be added immediately before use (*see* **Note 1**).
10. Protease inhibitor cocktail (50×): 50 mM phenyl methyl sulfonyl fluoride, 500 μg/mL aprotinin, 500 μg/mL pepstatin, 500 μg/mL leupeptin, in 100 % ethanol (*see* **Note 2**). Store at −20 °C.
11. Lysozyme: 10 mg/mL in 50 % glycerol, store at −20 °C.
12. DNase I (from Bovine pancreas, grade II): 10 mg/mL in 50 % glycerol, store at −20 °C.
13. Wash buffer: 50 mM Tris–HCl pH 8.0, 2 mM $MgCl_2$, and 1,000 mM NaCl. Store at 4 °C. NP-40/IGEPAL should be added to a final of 1 % and DTT at a final concentration of 1 mM should be added immediately before use.
14. Bio-Rad econo-pac plastic chromatography columns.
15. Bio-Rad micro bio-spin columns.
16. Deadenylation buffer: 50 mM Tris–HCl pH 8.0, 1 mM $MgCl_2$, 50 mM NaCl, and 20 % glycerol. Filter sterilize and store at 4 °C. Add 1 mM DTT, final concentration, just before use.
17. Maltose elution buffer: 10 mM maltose in deadenylation buffer with 1 mM DTT. Make fresh just prior to elution.
18. Pierce 660 protein assay reagent and protein standards.
19. SDS-PAGE Loading Dye (5×): 80 mM Tris–HCl pH 6.8, 10 % Glycerol, 2 % SDS, 100 mM DTT, 100 mM beta-mercaptoethanol, 0.02 % Bromophenol Blue.
20. SDS-PAGE running buffer (1×): 25 mM Tris–HCl, 192 mM glycine, 0.1 % SDS, pH 8.3.
21. SDS-PAGE gel: precast 4–20 % gradient gels from Bio-Rad.
22. SDS-PAGE Protein Ladder: PageRuler unstained protein ladder (Pierce) or equivalent.
23. Destain: 40.0 % methanol and 10.0 % acetic acid in water.
24. Coomassie Stain: Final concentration of 0.25 % Coomassie Blue dissolved in destain and filtered.

2.2 RNA Labeling Components

Design of the substrate RNA involves consideration of several variables including length, sequence content, and method of detection. Our standard substrate is poly(A_{20}). Radioactively labeled RNA has several advantages: it is highly sensitive, versatile, and many substrates can be labeled at low cost. Once labeled, the RNA has a limited shelf-life before degradation by radiolysis. Use of radiation requires institutional approval, proper safety equipment, monitoring, and waste disposal. Additional methods of producing labeled RNA substrates have been described elsewhere [27, 30].

1. Chemically synthesized poly(A_{20}) substrate RNAs (Thermo Scientific Dharmacon or Integrated DNA Technologies; *see* **Note 3**): 100 pmol/μL (100 μM) stock RNA in ddH_2O. Store at −80 °C.
2. Radioactive γ-^{32}P ATP: 6,000 Ci/mmol, 150 mCi/mL, 1.7 μM (Perkin-Elmer).
3. T4 polynucleotide kinase.
4. 10× T4 kinase buffer supplied with T4 polynucleotide kinase.
5. G25 sephadex: 1:1 slurry in ddH_2O. Autoclave to sterilize, store at 4 °C.

2.3 Deadenylation Assay Components

1. Deadenylation Buffer: 50 mM Tris pH 8, 1 mM $MgCl_2$, 50 mM NaCl, 20 % Glycerol (*see* **Note 4**).
2. RNA substrate diluted in deadenylation buffer.
3. Purified deadenylase in deadenylation buffer (*see* **Note 5**).
4. Recombinant RNase Inhibitor (RNasin).
5. 10× TBE Buffer: Dissolve 108 g Tris Base, 55 g Boric Acid into ddH_2O and add 20 mL of 0.5 M EDTA pH 8.0. Add H_2O to a final volume of 1 L.
6. Vertical Gel Electrophoresis Apparatus: We use the versatile adjustable system from CBS Scientific with leak proof gaskets and Teflon spacers and combs.
7. High Voltage Electrophoresis power supply.
8. RNA gel loading buffer: 95 % Formamide, 18 mM EDTA, and 0.025 % SDS, Xylene Cyanol, and Bromophenol Blue.
9. Denaturing 7 M Urea, polyacrylamide gel (15 cm × 25 cm × 0.5 mm): 15 % Polyacrylamide 19:1 bisacrylamide (*see* **Note 6**), 7 M Urea, 1× TBE gel mix (for 20 nucleotide RNA).
10. 10 % Ammonium Persulfate (APS) in ddH_2O. Store 1 mL aliquots at −20 °C.
11. *N*,*N*,*N*′,*N*′-Tetramethylethylenediamine (TEMED).

2.4 RNA Ladder Components

1. 50 mM $NaHCO_3/Na_2CO_3$, pH 9.2, 1 mM EDTA.

3 Methods

3.1 Purification of Deadenylase Enzymes

3.1.1 Bacterial Growth and Protein Induction

1. Transform *E. coli* cells with deadenylase expression plasmid.
2. Plate on appropriate selective media, grow at 37 °C.
3. Inoculate a 5 mL liquid culture of LB media containing appropriate antibiotic from a single bacteria colony. Agitate at 37 °C overnight.

4. Dilute 1.0 mL of culture into 1 L of 2xYT with antibiotic, 0.2 % glucose, and 2 mM $MgSO_4$ and shake at 37 °C (*see* **Note 7**).
5. When $OD_{600} = 0.9$, collect 1 mL "Pre-induced" sample, pellet bacteria at 14,000 rpm(18,000 × *g*) for 1 min, remove supernant, and freeze bacterial pellet for subsequent analysis. Samples taken during purification will be analyzed by SDS-PAGE.
6. Induce by adding 300 μL of 1 M IPTG (final concentration 0.3 mM IPTG) for 16 h at 20 °C (*see* **Note 8**).
7. Collect 1 mL "Induced" sample in a 1.7 mL tube, pellet bacteria at 14,000 rpm (18,000 × *g*), remove supernatant, and freeze bacterial pellet for subsequent analysis.

3.1.2 Harvesting Bacteria

1. Transfer bacteria to 1 L centrifuge bottles and centrifuge at 5,000 × *g* in a swinging bucket rotor for 10 min.
2. Resuspend pellet in 30 mL of Washing Buffer by vortexing and/or pipetting up and down. Transfer to oakridge centrifuge tube.
3. Spin bacteria at 8,000 rpm (7,600 × *g*) for 10 min in SS34 rotor. Decant supernatant.
4. Proceed to purification or, as a stopping point, snap freeze pellet in liquid N_2 and store at −80 °C.

3.1.3 Preparation of Protein Extract

1. Perform all steps at 4 °C or on ice to minimize protease activity and denaturation of enzyme.
2. Rapidly thaw each pellet in 25 mL of binding buffer with 1 mM DTT and 1× protease inhibitor cocktail.
3. Thoroughly resuspend pellets by pipetting and/or vortexing the sample (*see* **Note 9**).
4. Once completely thawed, add 1.25 mL of 10 mg/mL lysozyme (final concentration of 0.5 mg/mL), mix by gently inverting tube several times and incubate at 4 °C for 30 min with gentle rocking.
5. Add 140 μL of 1 M $MgCl_2$ (5 mM final concentration) and 26 μL of 10 mg/mL DNase I (10 μg/mL final), incubate at 4 °C with rocking for 20 min.
6. Collect 100 μL "total" sample of cell lysate for subsequent gel analysis.
7. Centrifuge cell lysate at 20,000 rpm (48,000 × *g*) in a SS34 rotor for 30 min at 4 °C.
8. Transfer supernatant, the protein extract, to a fresh 50 mL conical tube. Avoid disrupting the pellet. Reserve 100 μL "protein extract" sample for subsequent gel analysis.

3.1.4 Preparation of Affinity Resin

1. Equilibrate amylose affinity resin by washing 0.5 mL bed volume (1.0 mL of a 50 % slurry) of beads per purification in a 15 mL conical tube with 10 mL of binding buffer.
2. Collect beads by centrifugation at 1,000 × *g* for 5 min. Decant supernatant.
3. Repeat washing step two more times for a total of three washes.
4. Resuspend beads in 1.0 mL of binding buffer.

3.1.5 Affinity Purification

1. Transfer protein extract to an ice-cold 50 mL conical tube containing the pre-washed amylose affinity resin (*see* **Note 10**).
2. Allow protein to bind to the beads for 1 h (or longer) with end-over-end rotation at 4 °C.
3. Set up econo-pac plastic chromatography column on a ring stand.
4. Working at 4 °C, pour sample into column and, using gravity flow, collect the flow-through in a 50 mL conical tube. Reserve a 100 μL sample of the "flow-through" for subsequent gel analysis.
5. Wash beads by adding 25 mL of wash buffer to the column, allowing liquid to drain by gravity.
6. Wash beads two additional times, for a total of three washes.
7. Wash beads three times with 25 mL of deadenylation buffer.
8. To elute protein from beads, cap the bottom of the column and add one bed volume (0.5 mL) of freshly made maltose elution buffer. (If cleaving with TEV protease and not eluting with deadenylation buffer, *see* **Note 11** and proceed accordingly.)
9. Suspend the beads by repeatedly pipetting up and down.
10. Transfer beads to a 1.7 mL tube and incubate at 4 °C for 30 min with end-over-end rotation.
11. Transfer beads and elution buffer to a micro bio-spin column placed inside a 2.0 mL microfuge tube.
12. Collect first eluate by centrifugation at 1,000 × *g* for 5 min at 4 °C in a microfuge tube.
13. Repeat elution twice more with maltose elution buffer, incubating 5 min with end-over-end rotation. Collect second and third eluates.
14. Eluates should be stored at −20 °C (short term) or −80 °C (>1 month).
15. Add 0.5 mL deadenylation buffer to beads left in column, resuspend by pipetting up and down, and transfer slurry to new 1.7 mL tube. Reserve 10 μL of beads after elution for subsequent gel analysis.
16. Measure protein concentration of the eluates using the Pierce 660 protein assay.

3.1.6 Analysis of Samples by Coomassie Blue-Stained SDS-PAGE

1. Prepare samples in 1× SDS-PAGE loading dye. Samples include 10 μL each of pre-induced, induced, total cell extract, protein extract, flow-through, and beads after elution. Prepare 1 μg of total protein of each eluates 1–3 in 1× SDS-PAGE loading dye.
2. Boil samples for 5 min at 95 °C.
3. Set up mini-protean SDS-PAGE 4–20 % gel in mini-protean electrophoresis apparatus with 1× SDS-PAGE running buffer.
4. Load samples into wells of the SDS-PAGE gel. PageRuler unstained protein ladder should be loaded into one well.
5. Perform electrophoresis at 100 V until bromophenol blue dye reaches the bottom of the gel.
6. Remove gel from plates and transfer to clean dish containing enough Coomassie Stain to cover the gel.
7. Incubate at room temperature on a rocking platform for at least 1 h.
8. Drain the Coomassie stain from the gel, and wash the gel in Destain buffer by rocking at room temperature. Wash for up to 1 h until the gel is translucent, leaving the blue-stained protein bands visible.
9. Assess induction and purity of the protein. Confirm that the induced and purified deadenylase fusion protein corresponds to the expected molecular weight (take into account the additional 42 kDa added due to the fusion of MBP). Determine if contaminants are present. A specific band containing the deadenylase fusion protein should be observed in induced sample but not in the pre-induced sample. The deadenylase should be observed in the total and protein extract: if not, the purified protein may be insoluble.

3.2 RNA Substrate 5′ End Labeling

1. Dilute stock RNA (100 pmol/μL) to 10 pmol/μL in ddH$_2$O.
2. Assemble labeling reaction (20 μL final volume):

   9 μL ddH$_2$O, sterile, RNAse free.

   2 μL 10× T4 kinase buffer.

   4 μL 10 μM RNA (40 pmol per reaction, final concentration is 2 pmol/μL).

   1 μL γ-^{32}P ATP, 6,000 Ci/mmol, 150 mCi/mL, 1.7 μM.

   2 μL T4 polynucleotide kinase.
3. Incubate at 37 °C for 1.5 h.
4. Add 2 μL of 0.5 mM EDTA pH 8.0.
5. Heat inactivate at 70 °C for 10 min.
6. Remove free nucleotides and salts by size exclusion spin chromatography with G25 sephadex.

7. Add 1 mL of 1:1 slurry of G25 beads to a micro bio-spin column-placed in a 2 mL microfuge tube.
8. Centrifuge at 2,000 × *g* for 2 min in a microfuge.
9. Decant flow-through and repeat centrifugation to remove the liquid from resin.
10. Place column in a new 1.7 mL microfuge tube.
11. Pipette the kinase reaction directly to the center of G25 beads in the micro bio-spin column. Cap the top of the column.
12. Spin at 2,000 × *g* for 2 min to recover labeled RNA.
13. Measure the final volume of the eluate and calculate the concentration of the labeled RNA (*see* **Note 12**).
14. Store labeled RNA at −20 °C until ready to use.

3.3 RNA Ladder Preparation by Alkaline Hydrolysis

1. Prepare three microfuge tubes containing 20 μL of 10 nM radioactive substrate RNA in 50 mM $NaHCO_3/Na_2CO_3$, pH 9.2 and 1 mM EDTA.
2. Heat samples to 95 °C for 2, 5, and 15 min to hydrolyze the RNA.
3. Place samples on ice and add equal volume of RNA gel loading buffer.
4. Prior to loading the ladder on denaturing 7 M urea/polyacrylamide gel, heat to 95 °C for 5 min.

3.4 Deadenylation Assay

3.4.1 Deadenylation Reactions

The next step is to test the purified enzymes ability to degrade the labeled substrate RNA. Here, we provide a framework for reactions and data analysis. Initial experiments should focus on titrating the enzyme and measuring activity at a single reaction time point. Subsequent experiments should be performed over a time course. A number of variables should be optimized for deadenylation reactions, including divalent metal ion concentration, pH, salt concentration, temperature, and substrate concentration [27]. The size, length, and sequence content of the substrate RNA present additional variables. Negative controls are crucial to interpret the results and should include reactions containing RNA without enzyme, RNA with purified affinity tag (i.e., MBP alone), and RNA with catalytically inactivated, mutant enzyme. An additional negative control is to add EDTA to a reaction, before the enzyme is added, so as to chelate the divalent metals. All known deadenylases require a divalent metal ion, whereas some common contaminant ribonucleases do not [3].

1. Prepare reactions on ice. In a microfuge tube, combine the following reagents in sequential order to a final volume of 20 μL (*see* **Note 13**):
   - Deadenylation buffer.
   - RNasin: 2 Units/microLiter final concentration.

- 200 pmol of RNA substrate (10 nM final concentration).
- Purified Deadenylase (0–1 μM).

2. Mix reactions and collect contents at the bottom of the tube by brief centrifugation.
3. Incubate reaction at 30 °C (*see* **Note 14**).
4. At each time point, terminate the reaction by the addition of 20 μL of RNA gel loading buffer.
5. Heat samples at 95 °C for 5 min and place on ice or at −20 °C until ready to run on a denaturing gel.

3.4.2 Denaturing Polyacrylamide Gel Electrophoresis

1. Thoroughly clean glass plates and comb with alkaline, residue-free detergent. Rinse with ddH_2O and then wipe plates with 70 % ethanol and a lint-free Kimwipe.
2. Treat one side of one of the glass plates with Rain-X to assist subsequent separation of the gel from the plates.
3. Assemble plates with appropriate spacers and clamps.
4. Assemble the following gel components in a 50 mL conical tube and mix gently by inversion (Do not vortex. *See* **Note 15**):

   20 mL denaturing polyacrylamide/urea gel mix.

   20 μL TEMED.

   200 μL of 10 % APS.
5. Using a pipet-aid, transfer gel mix to the assembled plates. Fill the gel assembly to the top. Use a slow but steady flow rate and avoid introduction of bubbles into the mixture.
6. Insert the comb into the top of the gel, avoiding bubbles.
7. Allow the gel to polymerize for 30 min.
8. Carefully remove the comb, being cautious to not tear the fragile walls of each well.
9. Place the gel into the electrophoresis apparatus, clamping a metal heat-sink against one of the glass plates to evenly distribute heat during electrophoresis.
10. Fill the reservoirs of the electrophoresis apparatus with 1× TBE running buffer. Be sure to remove bubbles from the top and bottom of the interface of gel and buffer.
11. Pre-run the gel at 20 mA for 30 min or longer to warm the gel.
12. Flush out each well thoroughly after the pre-run step and before loading samples.
13. Using a fine gauge pipette tip, load each sample and RNA ladder into wells.
14. Electrophorese samples at 20 mA until Bromophenol Blue runs at least halfway through the gel.

15. Remove the glass-plate gel assembly from the electrophoresis apparatus.
16. Gently remove the Rain-X treated plate from the gel. A plastic or metal wedge can be used to pry the plates apart. Use care so as not to distort the gel.
17. Apply 3MM Whatman filter paper to the gel. Press gently so the gel adheres to the paper.
18. Flip the paper–gel–glass sandwich so that the paper is on the bottom.
19. Gently remove the remaining glass plate from the gel, leaving the gel adhered to the Whatman paper.
20. Wrap the gel in plastic cling wrap (*see* **Note 16**).
21. Expose gel to a phosphor screen (*see* **Note 17**).
22. Visualize radioactive RNA bands by scanning the phosphor screen using a phosphorimager.

3.5 Data Analysis

Imaging software provided with the phosphorimager is used to visualize the reaction products and make measurements. We use ImageQuant software (GE). In reactions containing deadenylase enzyme, shortening of the substrate should be observed in a dose-dependent and time-dependent fashion. Controls are absolutely essential so as to avoid false-positive measurements of contaminating exoribonucleases. In negative control reactions lacking enzyme, only full-length substrate should be observed. If shorter products are observed, this indicates that reagents and/or equipment are contaminated with RNase. No shortened products should be observed in control reactions containing MBP tag alone or with purified deadenylase enzyme wherein the catalytic residues have been mutated. If products are observed, this indicates contamination of deadenylase purification with bacterial ribonuclease(s). Also, in control reactions containing EDTA, deadenylation should not be observed. RNA degradation in EDTA treated samples is an indication of contamination by RNase A family enzymes, which unlike deadenylases, do not require divalent metal ions.

Results should be validated using multiple independent experiments and enzyme preparations. Once the deadenylase enzyme's activity has been validated, buffer conditions should be carefully optimized to obtain maximally active enzyme. These variables include the concentration of both RNA substrate and deadenylase, divalent cofactor, salt, pH, and temperature. Substrate specificity using substrates with varying sequence content, length and poly(A) length should also be tested as well. We have observed that individual deadenylases differ in the rates of shortening and in the length of the shortest poly(A) products that form. The hydrolyzed RNA ladder should be used to determine the number of adenosines that were removed by the enzyme. Substrates with different poly(A) lengths can be tested to test whether the deadenylase has a specific requirement for poly(A) length. Competition by the addition of unlabeled nucleotide polymers can also be informative for analysis of deadenylase specificity. Finally important kinetic parameters such as K_m and V_{max} can be determined by measuring reaction rates with varying substrate RNA concentrations.

The rate of deadenylation can be determined by measuring the rate of loss of the full-length RNA or by the accumulation of the deadenylated product over time. Using ImageQuant TL (following the software instructions), measure the intensities within each band across the time course. To do so, draw an object rectangle over each band to be measured within a lane. Carefully consider whether to apply background subtraction parameters. We typically apply a manual background subtraction by drawing an image rectangle in an empty lane, measuring its average background intensity, and apply this to all measurements. Slight variations in sample loading per lane can be normalized by dividing the measured disappearance of substrate band (or appearance of deadenylated product band) by the total radioactive RNA present in each lane.

The methods described here can allow the scientist to identify and characterize deadenylase enzymes, discover their cofactor requirements, and optimize reaction conditions. Further, the techniques can be applied to determine the kinetic parameters of the deadenylase. Importantly, these methods can be used to identify inhibitors or activators that modulate deadenylase activity. This system can also serve as the basis of an in vitro assay system for measuring the effect of trans-acting RNA binding factors that modulate deadenylase function [3, 25, 27].

4 Notes

1. Detergent is easier to pipette accurately if it is first heated to reduce viscosity. If desired, immerse detergent bottle in a warm water bath for a few minutes before use.
2. Do not use protease inhibitor cocktails with chelators such as EDTA as they can inactivate deadenylases. Also be aware that some protease inhibitors are toxic.
3. Polyacrylamide Gel Electrophoresis (PAGE) purification is recommended for long substrates. Consult with manufacturer when ordering substrates.
4. Buffer for enzymatic reaction needs to be optimized for your specific deadenylase. Optimal concentrations of divalent metals, pH, and salt will vary from protein to protein. For previously used conditions, refer to ref. [27].
5. Different affinity tag purifications may affect enzymatic activity. Deadenylases may be inhibited by other metals such as Ni. For this reason we advise against using His-tag purification.

6. The percentage of polyacrylamide will be determined by the size of the RNA substrate and products being analyzed.
7. If desired, more starter culture (up to the full 5 mL) may be diluted into media for faster growth.
8. Optimal induction time and temperature may vary. We frequently test 37 °C for 1–3 h and 20 °C for up to 20 h. Long, low temperature induction can facilitate production of proteins that express poorly.
9. If cell pellet is too large to be resuspended in 25 mL Binding Buffer, more Binding Buffer may be added. Reagents added in subsequent steps should be scaled up accordingly.
10. For more information of MBP purification, consult New England Biolab's pMAL protein fusion purification system user manual.
11. It may be desirable to remove the MBP affinity tag from the deadenylase. To do so, a site-specific protease cleavage site such as TEV site should be cloned between the deadenylase and MBP tag. Instead of elution with maltose, the bead-bound MBP-deadenylase would be cleaved and eluted using TEV protease. We use Invitrogen's AcTEV, which is pure and free of contaminating RNases.
12. This assumes efficient recovery of the labeled RNA. In practice, we typically recover greater than 95 % of RNA.
13. When performing a reaction progress curve with multiple time points, make a master reaction and remove an aliquot at specific time points.
14. Optimum temperature for the deadenylase should be determined experimentally.
15. Acrylamide will begin to harden within 10 min.
16. If long overnight exposures are necessary to visualize radioactive RNA, the gel can be dried in a gel drier before wrapping in plastic wrap. Drying the gel will diminish potential diffusion of the RNA.
17. Phosphorimager is the preferred method for visualization because of its broad range of sensitivity. X-ray film can be used as an alternative.

Acknowledgements

We thank Nathan Blewett, Chase Weidmann, Dr. Trista Schagat, and Dr. Brad Hook for help in developing these methods. Nathan Raynard is supported by the Michigan Predoctoral Training Program in Genetics through NIH National Research Service Award 5T32GM007544-33. Aaron Goldstrohm was supported by a Research Scholar Grant RSG-13-080-01-RMC from the American Cancer Society. This work was also supported by a grant from the Edward Mallinckrodt Jr. Foundation.

References

1. Zhao J, Hyman L, Moore C (1999) Formation of mRNA 3′ ends in eukaryotes: mechanism, regulation, and interrelationships with other steps in mRNA synthesis. Microbiol Mol Biol Rev 63(2):405–445
2. Wahle E, Keller W (1992) The biochemistry of 3′-end cleavage and polyadenylation of messenger RNA precursors. Annu Rev Biochem 61:419–440
3. Goldstrohm AC, Wickens M (2008) Multifunctional deadenylase complexes diversify mRNA control. Nat Rev Mol Cell Biol 9(4):337–344
4. Garneau NL, Wilusz J, Wilusz CJ (2007) The highways and byways of mRNA decay. Nat Rev Mol Cell Biol 8(2):113–126
5. Astrom J, Astrom A, Virtanen A (1992) Properties of a HeLa cell 3′ exonuclease specific for degrading poly(A) tails of mammalian mRNA. J Biol Chem 267(25):18154–18159
6. Astrom J, Astrom A, Virtanen A (1991) In vitro deadenylation of mammalian mRNA by a HeLa cell 3′ exonuclease. EMBO J 10(10): 3067–3071
7. Lowell JE, Rudner DZ, Sachs AB (1992) 3′-UTR-dependent deadenylation by the yeast poly(A) nuclease. Genes Dev 6(11):2088–2099
8. Thore S, Mauxion F, Seraphin B et al (2003) X-ray structure and activity of the yeast Pop2 protein: a nuclease subunit of the mRNA deadenylase complex. EMBO Rep 4(12): 1150–1155
9. Zuo Y, Deutscher MP (2001) Exoribonuclease superfamilies: structural analysis and phylogenetic distribution. Nucleic Acids Res 29(5): 1017–1026
10. Dlakic M (2000) Functionally unrelated signalling proteins contain a fold similar to Mg2+-dependent endonucleases. Trends Biochem Sci 25(6):272–273
11. Dupressoir A, Morel AP, Barbot W et al (2001) Identification of four families of yCCR4- and Mg2+-dependent endonuclease-related proteins in higher eukaryotes, and characterization of orthologs of yCCR4 with a conserved leucine-rich repeat essential for hCAF1/hPOP2 binding. BMC Genomics 2:9
12. Tucker M, Valencia-Sanchez MA, Staples RR et al (2001) The transcription factor associated Ccr4 and Caf1 proteins are components of the major cytoplasmic mRNA deadenylase in Saccharomyces cerevisiae. Cell 104(3):377–386
13. Tucker M, Staples RR, Valencia-Sanchez MA et al (2002) Ccr4p is the catalytic subunit of a Ccr4p/Pop2p/Notp mRNA deadenylase complex in Saccharomyces cerevisiae. EMBO J 21(6):1427–1436
14. Wagner E, Clement SL, Lykke-Andersen J (2007) An unconventional human Ccr4-Caf1 deadenylase complex in nuclear cajal bodies. Mol Cell Biol 27(5):1686–1695
15. Chen J, Chiang YC, Denis CL (2002) CCR4, a 3′-5′ poly(A) RNA and ssDNA exonuclease, is the catalytic component of the cytoplasmic deadenylase. EMBO J 21(6):1414–1426
16. Korner CG, Wahle E (1997) Poly(A) tail shortening by a mammalian poly(A)-specific 3′-exoribonuclease. J Biol Chem 272(16): 10448–10456
17. Martinez J, Ren YG, Thuresson AC et al (2000) A 54-kDa fragment of the Poly(A)-specific ribonuclease is an oligomeric, processive, and cap-interacting Poly(A)-specific 3′ exonuclease. J Biol Chem 275(31):24222–24230
18. Viswanathan P, Chen J, Chiang YC et al (2003) Identification of multiple RNA features that influence CCR4 deadenylation activity. J Biol Chem 278(17):14949–14955
19. Fabian MR, Mathonnet G, Sundermeier T et al (2009) Mammalian miRNA RISC recruits CAF1 and PABP to affect PABP-dependent deadenylation. Mol Cell 35(6):868–880
20. Beilharz TH, Preiss T (2007) Widespread use of poly(A) tail length control to accentuate expression of the yeast transcriptome. RNA 13(7):982–997
21. Grigull J, Mnaimneh S, Pootoolal J et al (2004) Genome-wide analysis of mRNA stability using transcription inhibitors and microarrays reveals posttranscriptional control of ribosome biogenesis factors. Mol Cell Biol 24(12): 5534–5547
22. Lackner DH, Beilharz TH, Marguerat S et al (2007) A network of multiple regulatory layers shapes gene expression in fission yeast. Mol Cell 26(1):145–155
23. Meijer HA, Bushell M, Hill K et al (2007) A novel method for poly(A) fractionation reveals

a large population of mRNAs with a short poly(A) tail in mammalian cells. Nucleic Acids Res 35(19):e132
24. Eulalio A, Huntzinger E, Nishihara T et al (2009) Deadenylation is a widespread effect of miRNA regulation. RNA 15(1):21–32
25. Goldstrohm AC, Hook BA, Seay DJ et al (2006) PUF proteins bind Pop2p to regulate messenger RNAs. Nat Struct Mol Biol 13(6):533–539
26. Hook BA, Goldstrohm AC, Seay DJ et al (2007) Two yeast PUF proteins negatively regulate a single mRNA. J Biol Chem 282(21): 15430–15438
27. Goldstrohm AC, Hook BA, Wickens M (2008) Regulated deadenylation in vitro. Methods Enzymol 448:77–106
28. Goldstrohm AC, Seay DJ, Hook BA et al (2007) PUF protein-mediated deadenylation is catalyzed by Ccr4p. J Biol Chem 282(1): 109–114
29. van Etten J, Schagat TL, Hrit J et al (2012) Human Pumilio proteins recruit multiple deadenylases to efficiently repress messenger RNAs. J Biol Chem 287(43):36370–36383
30. Rio DC, Ares M Jr, Hannon GJ et al (2011) RNA: a labratory manual. Cold Spring Harbor Laboratory Press, Cold Spring Harbor, NY

Chapter 27

MicroRNA-Mediated Deadenylation in a Mammalian Cell-Free System

Motoaki Wakiyama and Shigeyuki Yokoyama

Abstract

MicroRNAs (miRNAs) are 21–22 nucleotide small noncoding RNAs that regulate gene expression posttranscriptionally. The miRNA is incorporated into the miRNP effector complex. The miRNP complex binds to the mRNA containing the target sites, which are partially homologous to the miRNA sequence, and represses protein synthesis. One of the critical functions of miRNP is the recruitment of deadenylase complexes to the target mRNAs. Deadenylation causes translational inhibition as well as mRNA destabilization. In this chapter, we describe our method to recapitulate miRNA-mediated deadenylation in a mammalian cell-free system.

Key words Deadenylation, MicroRNA, miRNA, *let-7*, miRNA repression, miRNP, mRNA stability, Translation, CCR4-NOT deadenylase, Cell-free system

1 Introduction

There is growing evidence that microRNAs (miRNAs) regulate the expression levels of proteins involved in diverse biological processes, including development, cell differentiation, immune responses, and various diseases such as cancer [1–3]. To date, more than 1,000 miRNAs have been found in the human genome (http://www.mirbase.org/). MiRNAs function by base-pairing with partially complementary sequences, which are usually present within the 3′-untranslated regions of the targeted mRNA molecules. The effects of the miRNAs are elicited by several protein factors. The mature ~22 nucleotide miRNAs associate with Argonaute (Ago) proteins and form miRNPs, the effector complexes. Another key component of miRNPs is the GW182 family protein. GW182 interacts with multiple Ago proteins through its amino-terminal half and recruits CCR4-NOT deadenylase complexes [4, 5]. Thereby, miRNPs cause the deadenylation of their target mRNAs. Since the 3′-poly(A) tail plays an important role in both the translation and stability of eukaryotic mRNAs,

Joanna Rorbach and Agnieszka J. Bobrowicz (eds.), *Polyadenylation: Methods and Protocols*, Methods in Molecular Biology, vol. 1125, DOI 10.1007/978-1-62703-971-0_27, © Springer Science+Business Media New York 2014

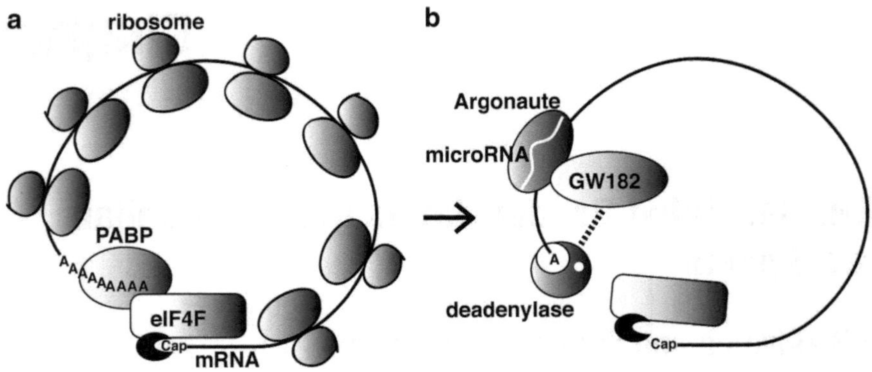

Fig. 1 Model for microRNA-mediated mRNA deadenylation and repression. (**a**) The interaction between the initiation factor eIF4F and PABP, a poly(A)-binding protein, stimulates translation. (**b**) Target mRNAs are deadenylated by the deadenylase complexes recruited by GW182 proteins, a key component of miRNPs. Reproduced and modified from Wakiyama and Yokoyama 2010, with permission from Springer

deadenylation significantly inhibits protein synthesis from the target mRNAs (Fig. 1). Therefore, deadenylation is one of the critical mechanisms of miRNA repression. However, the full picture of miRNA regulation has still not been clarified.

Several groups have developed cell-free systems that recapitulate miRNA repression [6–10]. Cell-free systems offer many advantages over cell-based systems. In cell-free systems, the time courses of translation and mRNA stability can be monitored. Moreover, proteins, RNAs, or chemicals can be added to the system, in order to modify the reaction conditions. We have established a mammalian cell-free system that recapitulates miRNA-mediated deadenylation [9]. In this chapter, we outline the methods to recapitulate *let-7* miRNA-mediated deadenylation, using extracts prepared from suspension-cultured HEK293 cells.

2 Materials

2.1 In Vitro Transcription and Polyadenylation of Reporter mRNAs

1. T7 In Vitro Transcription Kit, such as a mMESSAGEmMACHINE T7 Kit (Ambion).
2. RNA Purification Kit, such as a MEGAclear kit (Ambion).
3. Poly(A) polymerase from Yeast (USB Affymetrix).

2.2 Preparation of Cell Extracts

1. FreeStyle 293-F cells (Invitrogen Life Technologies).
2. FreeStyle 293 Expression Medium (Invitrogen Life Technologies).
3. 293fectin transfection reagent (Invitrogen Life Technologies).
4. Expression vectors encoding human Argonaute 2 (Ago2) and GW182 (*see* **Note 1**).

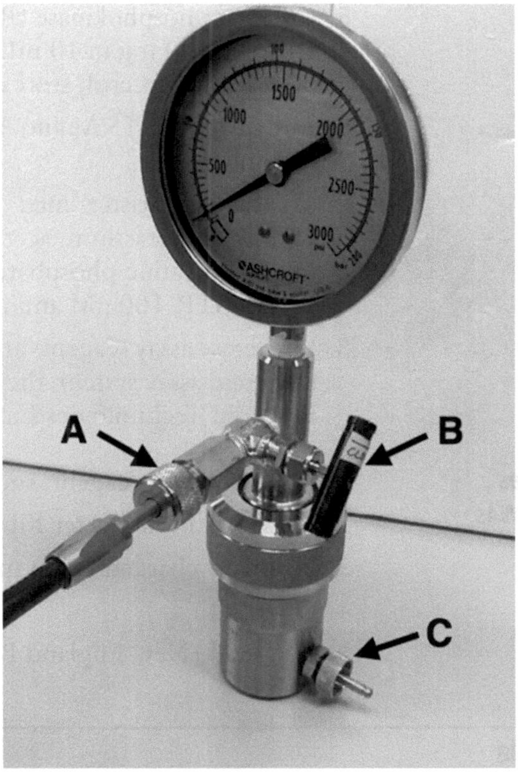

Fig. 2 Cell disruption vessel, model 4639 (Parr Instrument Company). (1) Connect the nitrogen gas hose to valve *A*. (2) Open valve *B*, adjust the nitrogen gas pressure to 150 psi, and then close valve *B*. (3) To retrieve the cell extracts, open the discharge valve *C*

5. 500 mL polycarbonate Erlenmeyer flask.
6. Shaking CO_2 incubator suitable for 125 rpm, 37 °C, 8 % CO_2.
7. Dulbecco's Phosphate-Buffered Saline (D-PBS).
8. Cell lysis solution: 40 mM HEPES–KOH (pH 7.4), 100 mM potassium acetate, 1 mM magnesium acetate, and 1 mM dithiothreitol (DTT).
9. Cell disruption vessel, model 4639 (Parr Instrument Company, USA; *see* Fig. 2).
10. Compressed nitrogen gas (grade 1).

2.3 Assaying let-7 miRNA-Mediated Repression and Deadenylation

1. Chemically synthesized *let-7* pre-miRNA: 5′-UGAGGUAGUAGGUUGUAUAGUAGUAAUUACACAUCAUACUAUACAAUGUGCUAGCUUUCU-3′ (*see* **Note 2**).
2. Chemically synthesized *let-7* siRNA: 5′-UGAGGUAGUAGGUUGUAUAGU-3′.
3. Anti-*let-7* miRNA Inhibitor (Ambion, AM10048).
4. Ribonucleotides: ATP and GTP (100 mM).

5. Creatine phosphokinase (Roche Applied Science): 10 mg/mL. Dissolve 100 mg in 10 mL of 20 mM HEPES–KOH, pH 7.6, 50 % (v/v) glycerol; store at −70 °C in 50–100 μL aliquots.
6. Amino acids: RTS Amino Acid Sampler (5 PRIME, Hamburg), or equivalents.
7. Translation master mix: 30 mM HEPES–KOH, pH 7.4, 100 mM potassium acetate, 4.5 mM magnesium acetate, 40 mM creatine phosphate, 2.5 mM spermidine, 5 mM ATP, 1 mM GTP, 100 μM amino acid mixture (as above).
8. Luciferase assay reagents and a luminometer (e.g., the Promega luciferase assay system, the dual luciferase assay system, and the Berthold Technologies Lumat LB 9506 bioluminometer).

2.4 Analyses of Target mRNAs

1. RNA isolation reagent, such as RNAqueous (Ambion).
2. DIG Northern Starter Kit (Roche Applied Science).
3. Positively charged nylon membrane (Roche Applied Science).
4. Oligo$(dT)_{16}$.
5. RNase H (New England Biolabs).

3 Methods

3.1 In Vitro Transcription and Polyadenylation of Reporter mRNAs

3.1.1 Reporter mRNAs Containing Multiple Target Sites for let-7 miRNA

The reporter construct pSP36T7-Luc-6xT, which encodes firefly luciferase (FLuc) and a 3′-untranslated region (3′-UTR) containing six *let-7* target sequences (6xT), is used to examine the effects of *let-7* miRNAs (Fig. 3). The construction strategies have been described in detail previously [9]. Briefly, the DNA fragment encoding firefly luciferase was excised from pGL3-control (Promega) and inserted into the *Nco*I–*Xba*I sites of the pSP36T7 vector, to create pSP36T7-Luc. A DNA fragment encoding a 6xT sequence was then inserted into the *Xba*I–*Eco*RI sites of pSP36T7-Luc, to create pSP36T7-Luc-6xT. As a control, a 6xTmut6 construct, pSP36T7-Luc-6xTmut6, in which the seeding region of the *let-7* target sequence, UACCUC, was mutated to AUGGAG, was used (Fig. 4). The 3′-UTR sequence of pSP36T7-Luc-6xT is as follows, and the *let-7* target sites are underlined:

5′-TCTAGATAAGGCCACGAGTGCCGATCTTTTTTT
TTAGACCATTTACAAGAGCGAAACCTTAG
CGTAGAGTTTCAAATAACATGACGATTATGA
GGAGGTGCC GAAACCGATTTAATAAGAAAAAATC
TGAAATCATGAAAGTTAGCGAAAATTTT
TAGCTTAAGTGCTACAACGATTACC
TCGATATTGCTTAAGTGCTACAACGATTAC
CTCGTAATTTACAGTGCTACA ACGATTACCTCGG
CGGCCACATCTAGTGCTACAACGATTACCTCG
CTATAGAAGCAGTGCTACAACGATTACCTCG
CATTTACTAGAGTGCTACAACGATTACC
TCGGCGGCCGCGAATTC-3′

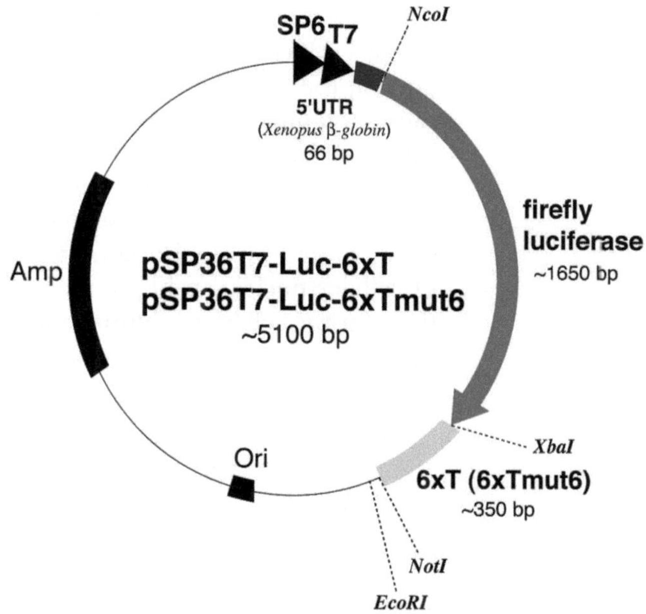

Fig. 3 Map of the pSP36T7-Luc-6xT (6xTmut6) plasmid

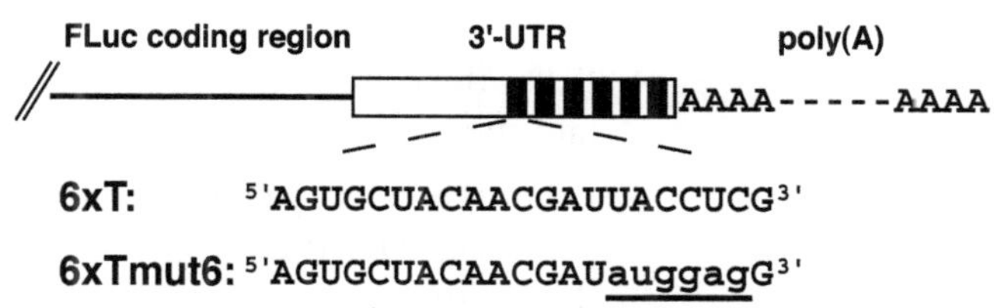

Fig. 4 Schematic representations of the 6xT and 6xTmut6 3′-untranslated regions of the reporter FLuc mRNAs. The sequences of the *let-7* target site and the mutated site are shown below the drawings. Reprinted from Wakiyama et al. 2007, with permission from Cold Spring Harbor Laboratory Press

3.1.2 Polyadenylation of In Vitro Transcribed mRNAs

Capped reporter mRNAs are prepared by in vitro transcription (*see* **Note 3**). The transcripts are purified to eliminate residual cap analogs, NTPs, and template DNAs. The transcripts are then polyadenylated using yeast poly(A) polymerase.

1. For a typical 25 μL reaction, mix 6.6 μg of purified RNA with an appropriate volume of water, 5 μL of 5× reaction buffer (Affymetrix/USB), 0.5 μL of 25 mM ATP, and 2 μL of 600 units/μL yeast poly(A) polymerase (Affymetrix/USB).
2. Incubate at 37 °C for 60 min (*see* **Note 4**).
3. Purify and quantify the RNA by measuring its ultraviolet (UV) absorbance (260 nm).
4. Check the length of the poly(A) tail by denaturing agarose gel electrophoresis.

3.2 Preparing 293-F Cell Extract

The cell disruption method is the key to obtaining good cell extracts. Mechanical shearing of cells with Dounce homogenizer is one of the common techniques. However, this method often causes the nuclei to break, which results in the leakage of chromosomal DNA and nucleases into the cell extracts. We prefer the nitrogen cavitation technique. In this technique, the cells are subjected to pressured nitrogen gas in a vessel for a while, and then are disrupted by rapid decompression from the vessel. Disruption of cellular nuclei can be avoided by choosing appropriate conditions for the cell lysis solution and the pressure of the nitrogen gas.

1. 293-F cells are maintained in suspension, using FreeStyle 293 Expression Medium. The cells are cultured in polycarbonate Erlenmeyer flasks (*see* **Note 5**), using a CO_2 incubator shaker (37 °C, 8 % CO_2, 125 rpm). Prepare three 293-F cell culture flasks.
2. For two culture flasks, transfect the 293-F cells separately. Transfect one culture with the expression vector encoding human Ago2 and the other with the expression vector encoding GW182 (*see* **Note 1**), using the 293fectin transfection reagent according to the manufacturer's instructions.
3. Propagate the cells for 48 h before harvesting (*see* **Note 6**).
4. Pellet the cells by centrifugation (400 × *g*, 10 min at 4 °C), aspirate the medium, and resuspend the cells in ice-cold D-PBS.
5. Pellet the cells by centrifugation as above, aspirate the D-PBS, and resuspend the cells in fresh ice-cold D-PBS. Repeat **step 5** three times.
6. Suspend the cell-pellet in ice-cold lysis solution (at 2×10^8 cells/mL, *see* **Note 7**).
7. Place the cell suspension in an ice-cold Cell Disruption Vessel, model 4639 (Fig. 2, *see* **Note 8**).
8. Pressurize the cell suspension with 150 psi nitrogen gas for 60 min on ice (Fig. 2, *see* **Note 9**).
9. Collect the cell extract in a 50 mL tube placed in ice (Fig. 2, *see* **Note 10**).
10. Centrifuge the cell extract at 1,000 × *g*, 4 °C for 10 min.
11. Transfer the supernatants to fresh micro-tubes.
12. Centrifuge the cell extract at 16,000 × *g*, 4 °C for 10 min.
13. Transfer the supernatants to fresh micro-tubes (*see* **Note 11**).
14. Add creatine phosphokinase to a final concentration of 0.4 mg/mL. Cell extracts with 70–80 OD/mL at 260 nm should be obtained.
15. Dispense the cell extract into micro-tubes, and freeze them in liquid nitrogen. Store the frozen aliquots at −80 °C.

3.3 Reaction Conditions for let-7 miRNA-Mediated Repression and Deadenylation In Vitro

1. Mix the extracts prepared from the 293F cells, 293F cells expressing human Ago2, and 293F cells expressing GW182 at a volume ratio of 80:10:10. This mixed extract is referred to as "the pre-mixed extract" in the following section.
2. Incubate the pre-mixed extract in the presence of *let-7* pre-miRNAs or *let-7* siRNAs, at a final concentration of 0.1 pmol/μL, for 60 min at 37 °C, and then cool the solution on ice. As a control, use the pre-mixed extracts incubated without the addition of *let-7* pre-miRNAs and *let-7* siRNAs.
3. Perform translation reactions in a final volume of 20–50 μL, composed of 60 % volume of the above pre-mixed extracts, 20 % volume of the translation master mix, and 1 ng/μL of FLuc mRNA. As a control, 1 (or 5) ng/μL RLuc mRNA may be added (*see* **Note 12**). To inhibit the effects of the *let7* miRNAs, add the anti-*let7* inhibitor (Ambion) to a final concentration of 0.3 pmol/μL.
4. Incubate the above reaction mixture at 37 °C for up to 60 min.
5. Withdraw 5 μL aliquots at certain time points, and measure the luciferase activities using an appropriate assay system and equipment. The withdrawn 5 μL samples usually must be diluted 10- to 100-fold with an appropriate buffer (e.g., the cell culture lysis reagent from the Promega luciferase assay system), to stop the reaction as well as to attain a light unit level of luminescence below 10^7 (i.e., within the linear range of detection). Figure 5a shows a time-course analysis of luciferase synthesis.

3.4 Analysis of miRNA-Mediated Deadenylation

3.4.1 Northern Blot Analysis

1. Isolate total RNAs from translation reaction mixtures by using an appropriate reagent. A sufficient amount of RNA may be obtained from a 25 μL translation reaction mixture.
2. Resolve 2.5 μg of total RNA on a 0.8 % agarose gel containing formaldehyde and transfer the RNA onto a nylon membrane.
3. Prepare an anti-sense RNA probe by in vitro transcription with T7 RNA polymerase, using the DIG Northern Starter Kit (Roche). For the template, amplify DNA fragments from the pSP36T7-Luc vector by PCR, using the following primer sets (*see* **Note 13**). Forward primer E1-Luc925F: TTTTGAATT CTTACACGAAATTGCTTCTGGTGGC; Reverse primer T7-Luc1524R: TTTTAAGCTTTAATACGACTCACTATAG GGTTGTTACTTGACTGGCGACGTA.
4. Hybridize the blots with the probe at 68 °C overnight. Perform hybridization, washing and detection using the above kit (Roche), according to the manufacturer's instructions. Figure 5b shows time-course analyses of let-7 miRNA-mediated deadenylation.

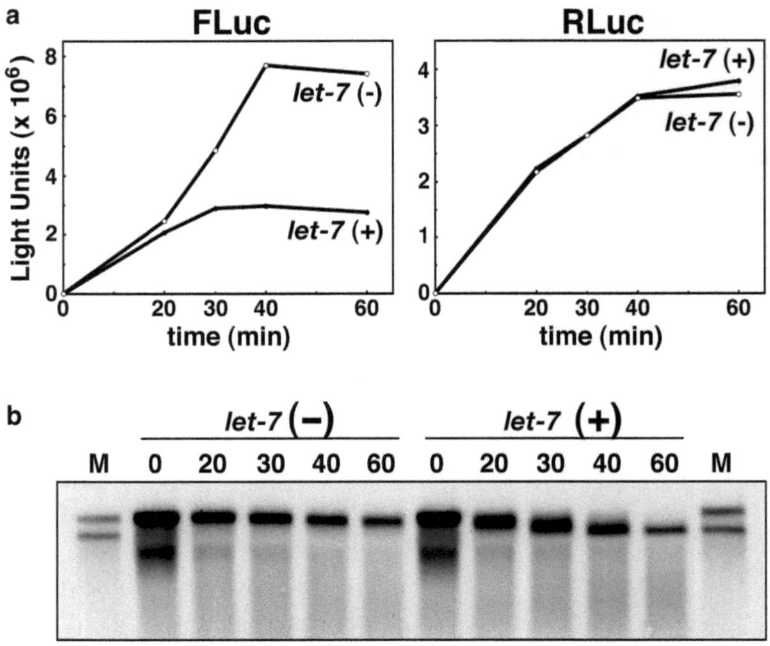

Fig. 5 Target-specific repression and deadenylation mediated by *let-7* miRNA. (**a**) Capped/polyadenylated FLuc-6xT mRNAs and capped RLuc mRNAs were translated in the presence and absence of *let-7* miRNAs. Luciferase activities were assayed for samples taken at time points 0, 20, 30, 40, and 60 min from the start of translation. (**b**) The FLuc mRNA in the RNA samples taken at each time point was analyzed by Northern blotting. In the "M" lanes, capped/polyadenylated and capped/nonadenylated FLuc-6xT mRNAs were loaded. Reprinted from Wakiyama et al. 2007, with permission from Cold Spring Harbor Laboratory Press

3.4.2 RNase H Cleavage Assay

1. Mix 1 μg of total RNA with 50 pmol of oligo$(dT)_{16}$.
2. Denature the samples at 65 °C for 5 min.
3. Anneal the samples at 37 °C for 10 min, and then cool them on ice.
4. Mix the samples with 2.5 units of RNaseH (New England Biolabs) in 1× RNaseH buffer (50 mM Tris–HCl, pH 8.3, 75 mM KCl, 17 mM $MgCl_2$, 10 mM DTT).
5. Incubate the above reaction mixture at 37 °C for 60 min.
6. Subject the above samples to a Northern blot analysis (Fig. 6).

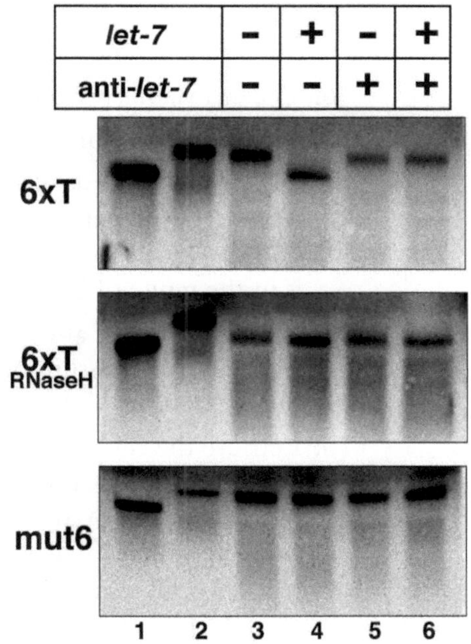

Fig. 6 Sequence-specific deadenylation by *let-7* miRNAs. Capped/polyadenylated mRNAs were translated in the absence (–) or presence (+) of *let-7* and anti-*let-7*. The RNAs were extracted from the reaction mixtures after a 60 min of incubation and were subjected to Northern blotting. *Top panel*: FLuc-6xT mRNA (6xT); *middle panel*: the extracted RNAs, as in the *top panel* (6xT), were annealed with oligo(dT) and subjected to RNaseH digestion; *lower panel*: FLuc-6xTmut6 mRNA (mut6). In lanes *1* and *2*, nonadenylated and polyadenylated transcripts were loaded, respectively, as markers. Reprinted from Wakiyama et al. 2007, with permission from Cold Spring Harbor Laboratory Press

4 Notes

1. We created the human Argonaute2 and GW182 protein expression plasmids by inserting each cDNA into the pCMV-SPORT vector (Invitrogen [9]). Over-expression of Argonaute2 significantly improves the processing efficiency of *let-7* pre-miRNA [9]. Thus, the extra Argonaute protein may be required for the efficient formation of miRNPs incorporating the externally added miRNAs. Over-expression of the GW182 protein enhances the deadenylation of the target mRNAs [9].
2. We use the *Drosophila let-7* pre-miRNA, which is 60 nucleotides in length.
3. The template DNAs, pSP36T7-Luc-6xT and pSP36T7-Luc-6xTmut6, should be linearized with *Eco*RI before in vitro transcription (Fig. 3).
4. Approximately 200 nucleotides of poly(A) are added.

5. Dispense 100–160 mL of the culture medium into each 500 mL Erlenmeyer flask.
6. The cell viability should be more than ~90 % when harvested.
7. Break up cell clumps by careful pipetting with a fine-tipped pipette.
8. Although the internal volume of the cell disruption vessel 4639 is 45 mL, a good working volume of the cell suspension is in the range of 5–10 mL.
9. Several minutes after setting the initial pressurization, the nitrogen gas pressure may drop slightly. In this case, readjust the pressure to 150 psi.
10. Retrieve the cell extract from the valve indicated in Fig. 2. Cell disruption occurs as the homogenate is released to atmospheric pressure through the discharge valve. Adjust the discharge speed to around 1 drop/s.
11. Take the fraction from the clear layer. Centrifuge the fraction again if the lipid layer is contaminated.
12. We created the pSP36T7-RLuc vector as a template for the in vitro transcription of the RLuc mRNA [9]. The pSP36T7-RLuc vector contains the sequences of the 5′- and 3′-untranslated regions from the *Xenopus* beta-globin mRNA.
13. Linearize the PCR fragment with *Eco*RI before using it in in vitro transcription.

Acknowledgments

This work was supported by the RIKEN Structural Genomics/Proteomics Initiative (RSGI), National Project on Protein Structuraland Functional Analysis, Targeted Proteins Research Program(TPRP), from the Ministry of Education, Culture, Sports, Scienceand Technology (MEXT), Japan.

References

1. O'Connell RM, Rao DS, Chaudhuri AA et al (2010) Physiological and pathological roles for microRNAs in the immune system. Nat Rev Immunol 10:111–122
2. Ambros V (2011) MicroRNAs and developmental timing. Curr Opin Genet Dev 21: 511–517
3. van Kouwenhove M, Kedde M, Agami R (2011) MicroRNA regulation by RNA-binding proteins and its implications for cancer. Nat Rev Cancer 11:644–656
4. Wakiyama M, Yokoyama S (2010) MicroRNA-mediated mRNA deadenylation and repression of protein synthesis in a mammalian cell-free system. In: Rhoads RE (ed) Progress in molecular and subcellular biology, miRNA regulation of the translational machinery. Springer, Heidelberg, pp 85–97
5. Fabian M, Sonenberg N (2012) The mechanics of miRNA-mediated gene silencing: a look under the hood of miRISC. Nat Struct Mol Biol 19:586–593

6. Wang B, Love TM, Call ME et al (2006) Recapitulation of short RNA-directed translational gene silencing in vitro. Mol Cell 22:553–560
7. Mathonnet G, Fabian MR, Svitkin SV et al (2007) MicroRNA inhibition of translation initiation in vitro by targeting the cap-binding complex eIF4F. Science 317:1764–1767
8. Thermann R, Hentze MW (2007) Drosophila miR2 induces pseudo-polysomes and inhibits translation initiation. Nature 447:875–878
9. Wakiyama M, Takimoto K, Ohara O et al (2007) Let-7 microRNA-mediated mRNA deadenylation and translational repression in a mammalian cell-free system. Genes Dev 21:1857–1862
10. Iwasaki S, Kawamata T, Tomari Y (2009) Drosophila Argonaute1 and Argonaute2 employ distinct mechanisms for translational repression. Mol Cell 34:58–67

Chapter 28

Quantitative Analysis of Deadenylation-Independent mRNA Decay by a Modified MBRACE Assay

Julie A. Dougherty, Roshan Mascarenhas, and Daniel R. Schoenberg

Abstract

Endonuclease cleavage is the rate-limiting step in the decay of nonsense-containing human β-globin mRNA in erythroid cells. The 5′-truncated intermediates thus generated are polyadenylated and more stable than the parent mRNA. Northern blotting is commonly used to measure the decay rate of full-length mRNA, and S1 nuclease protection is used to assay the fate of decay intermediates. We have adapted the more sensitive and facile MBRACE assay (Lasham et al., Nucleic Acids Res 38: e19, 2010) to quantitatively monitor the decay process by detecting full-length β-globin and its decay intermediates.

Key words β-Globin mRNA, MBRACE assay, Molecular beacons, mRNA decay, Nonsense codon, Endonuclease

1 Introduction

A premature termination codon (PTC) can arise within an mRNA due to mutation, errors in splicing, or via alternative splicing. Translation of PTC-containing mRNAs can cause deleterious effects in cells by producing C-terminally truncated proteins. A surveillance mechanism termed nonsense-mediated mRNA decay (NMD) recognizes and targets these PTC-containing mRNAs for accelerated degradation via two mechanisms. NMD target mRNAs can be internally cleaved by an endonuclease and then rapidly degraded by exonucleases, with Xrn1 catalyzing 5′–3′ degradation of the downstream fragment and the exosome-associated exonucleases catalyzing 3′–5′ degradation of the upstream fragment. Alternatively, the target mRNAs may undergo deadenylation and decapping followed by exonucleolytic decay from the 5′ and 3′ ends (reviewed in ref. 2). In erythroid cells, SMG6 catalyzes endonuclease cleavage of PTC-containing human β-globin mRNA [3], but unlike other forms of NMD this uniquely gives rise to *stable* 5′-truncated products [3–5].

Joanna Rorbach and Agnieszka J. Bobrowicz (eds.), *Polyadenylation: Methods and Protocols*, Methods in Molecular Biology, vol. 1125, DOI 10.1007/978-1-62703-971-0_28, © Springer Science+Business Media New York 2014

The conventional approach to studying the decay of nonsense-containing mRNA is to monitor the disappearance of the full-length mRNA; however, this does not fully illustrate the decay process. Stable decay products of PTC-containing β-globin mRNA were originally detected by S1 nuclease protection and primer extension assays [5]. Each of these approaches is time-consuming, insensitive, difficult to quantify, and impractical for use with large numbers of samples. To monitor changes in both full-length β-globin mRNA and its decay products, we adapted a highly sensitive assay based on *m*olecular *b*eacon *r*apid *a*mplification of *c*DNA *e*nds (MBRACE) [1]. MBRACE involves ligation of an RNA linker onto the 5′ end of monophosphorylated mRNA and detection of the resulting product using a probe specific for the linker-mRNA junction. Molecular beacons are single-stranded, dual-labeled fluorescent probes held in a hairpin structure by complementary stem sequences at the ends of the probe with target-specific sequence in the loop region. While in the hairpin configuration the proximity of the quencher blocks fluorescent emission by the reporter. Hybridization of the molecular beacon to the target sequence opens the hairpin structure, separating the reporter dye and quencher and enabling fluorescent emission. Since only one beacon binds per target molecule the fluorescent signal is proportional to the amount of target DNA. Molecular beacons have several advantages over other quantitative PCR-based assays. They do not interfere with primer binding and extension, they fluoresce only when hybridized to the target sequence, and, unlike Taqman probes, they are not dependent on the exonuclease activity of the polymerase [6].

2 Materials

General Supplies

1. Table top centrifuge.
2. Refrigerated microcentrifuge.
3. 1.7 mL microcentrifuge tubes.
4. Pipette tips with filter.
5. Spectrophotometer to quantify RNA.
6. PCR tubes.
7. Thermocycler.
8. Agarose gel electrophoresis apparatus.
9. Illumina Eco Real-Time PCR system or a comparable instrument.

2.1 Erythroid Cell Harvest, Fractionation, and RNA Extraction

1. Phosphate-buffered saline, PBS.
2. RNase inhibitor, such as RNase OUT (Life Technologies, New York).

3. 0.1 M Dithiothreitol, DTT.
4. Erythroid fractionation lysis buffer (EFLB): 100 mM NaCl, 10 mM Tris–HCl pH 8.0, 2 mM EDTA, 1 % NP-40, 1 % β-mercaptoethanol, 1 mM DTT, RNase OUT (*see* **Note 1**).
5. TRIzol (Life Technologies, New York).
6. Chloroform.
7. Isopropanol.
8. Carrier for nucleic acid precipitation, such as BSA or glycogen.
9. 80 % Ethanol, freshly diluted or stored at −20 °C.
10. RNase-free water.

2.2 5′ RLM-RACE to Identify the Sequence at 5′ Ends

1. Calf intestinal alkaline phosphatase, CIAP.
2. CIAP reaction buffer: 50 mM Tris–HCl pH 8.5, 0.1 mM $ZnCl_2$, 5 % glycerol (v/v) (10× stock provided with enzyme).
3. RNase-free water.
4. Stop solution: 5 M Ammonium acetate, 100 mM EDTA.
5. Acid phenol: chloroform (5:1, pH 4.5).
6. Chloroform.
7. Isopropanol.
8. Carrier for nucleic acid precipitation, such as BSA or glycogen.
9. 70 % Ethanol, cold.
10. Tobacco acid pyrophosphatase, TAP (Epicentre, Wisconsin).
11. TAP buffer: 50 mM sodium acetate pH 6.0, 0.1 mM EDTA, 0.1 % β-mercaptoethanol, 0.01 % Triton-X-100 (10× provided with enzyme).
12. RNase inhibitor, such as RNase OUT (Life Technologies, New York).
13. 100 μM RNA linker, 40–44 bp in length.
14. T4 RNA ligase I.
15. T4 RNA ligase I buffer: 50 mM Tris–HCl pH 7.5, 10 mM $MgCl_2$, 1 mM DTT, 1 mM ATP (10× provided with enzyme, with or without ATP).
16. 10 mM dNTP mix.
17. 25 mM $MgCl_2$.
18. 0.1 M Dithiothreitol, DTT.
19. Reverse transcriptase, SuperScript III (Life Technologies, New York).
20. SuperScript III RT buffer: 20 mM Tris–HCl pH 8.4, 50 mM KCl (10× provided with kit).
21. Taq DNA polymerase (Sigma-Aldrich, Massachusetts).

22. Taq DNA polymerase buffer: 10 mM Tris–HCl pH 8.3, 50 mM KCl, 1.5 mM $MgCl_2$, 0.1 mM DTT, stabilizers, 5 % glycerol (10× provided with enzyme).
23. Forward primer specific to 5′ end of RNA linker.
24. Gene-specific reverse outer primer.
25. Gene-specific reverse nested primer.
26. 1× TBE: 89 mM Tris, 89 mM boric acid, 2 mM EDTA pH 8.0.
27. Gel extraction kit.
28. Cloning vector, such as pGEM-T (Promega, Wisconsin).
29. T4 DNA ligase (Promega, Wisconsin).
30. Rapid ligation buffer: proprietary content including polyethylene glycol (PEG) and ATP (2× provided with pGEM-T vector).
31. 2× Rapid ligation buffer, T4 DNA ligase: 60 mM Tris–HCl pH 7.8, 20 mM $MgCl_2$, 20 mM DTT, 2 mM ATP, 10 % polyethylene glycol (Promega, Wisconsin).
32. Luria broth, LB.
33. LB plus ampicillin at 50 μg/mL.
34. LB agar plates with ampicillin at 50 μg/mL.

2.3 Molecular Beacons

1. In vitro transcription kit—such as MAXIscript T7/SP6 kit (Ambion/Life Technologies, New York).
2. Gene-specific reverse and linker-specific forward qPCR primers.
3. Molecular beacons specific for junction between RNA linker and mRNA sequence.
4. 2× PerfeCTa qPCR FastMix—ROX, UNG (Quanta Biosciences, Maryland).

2.4 Processing Experimental Samples for MBRACE

1. Tobacco acid pyrophosphatase, TAP (Epicentre, Wisconsin).
2. TAP buffer: 50 mM sodium acetate pH 6.0, 0.1 mM EDTA, 0.1 % β-mercaptoethanol, 0.01 % Triton-X-100 (10× provided with enzyme).
3. RNase inhibitor, such as RNase OUT (Life Technologies, New York).
4. RNase-free water.
5. 100 μM RNA linker, 40–44 bp in length.
6. T4 RNA ligase I.
7. T4 RNA ligase I buffer: 50 mM Tris–HCl pH 7.5, 10 mM $MgCl_2$, 1 mM DTT, 1 mM ATP (10× provided with enzyme, with or without ATP).
8. 10 mM ATP.

9. 50 μM oligo$(dT)_{20}$.
10. 10 mM dNTP mix.
11. 25 mM $MgCl_2$.
12. Reverse transcriptase, SuperScript III (Life Technologies, New York).
13. SuperScript III RT buffer: 20 mM Tris–HCl pH 8.4, 50 mM KCl (10× provided with kit).
14. QIAquick PCR purification kit (Qiagen, California).
15. 2× PerfeCTa qPCR FastMix—ROX, UNG (QuantaBiosciences, Maryland).
16. Gene-specific reverse and linker-specific foward qPCR primers.
17. Molecular beacons specific for junction between RNA linker and mRNA sequence.
18. Gene-specific qPCR forward and reverse primers for internal control mRNA.

3 Methods

3.1 Erythroid Cell Harvest, Fractionation, and RNA Extraction (See Note 2)

1. Scrape 1–5 × 10^6 cells into a 15 mL culture tube and centrifuge at 1,000 × *g* for 5 min then remove media.
2. Wash pellet with 1 mL ice-cold PBS and transfer to 1.7 mL microcentrifuge tube. Centrifuge at 1,000 × *g* for 5 min and remove all PBS.
3. Resuspend cell pellet in 3–5 volumes of EFLB by pipetting up and down four to five times, tap the tube, and incubate on ice 5 min. Tap tube and incubate on ice 5 min more, repeat until lysate becomes clear (*see* **Note 3**).
4. Centrifuge at 5,000 × *g* for 10 min at 4 °C to pellet nuclei and debris. Transfer cytoplasmic supernatant to a new tube *pre-chilled on ice* (*see* **Note 4**).
5. Add 1 mL TRIzol, gently invert to mix, and incubate at room temperature for 5 min.
6. Add 200 μL chloroform and shake vigorously by hand for 15 s, incubate at room temperature 10–15 min. Centrifuge at >12,500 × *g* for 15 min at 4 °C to separate phases. Transfer upper aqueous phase to new tube.
7. Add 500 μL isopropanol and 20 μg glycogen, invert to mix and incubate on ice for 10 min. Centrifuge at >7,500 × *g* for 10 min at 4 °C to pellet RNA. Discard supernatant.
8. Wash pellet with 1 mL 80 % ethanol. Centrifuge at >7,500 × *g* for 10 min at 4 °C to pellet RNA. Remove ethanol and air-dry for 5–10 min. Resuspend the pellet in an appropriate amount

of RNase-free water to keep the RNA concentrated. Incubate at 60 °C for 10 min to fully dissolve pellet. Centrifuge briefly to collect and keep on ice (sample may be stored at −80 °C at this point). Quantify RNA using a spectrophotometer, such as the NanoDrop1000.

3.2 5′ RLM-RACE to Identify the Sequence at 5′ Ends

3.2.1 Treatment of Cytoplasmic RNA with CIAP to Remove 5′ Monophosphates from RNA

1. In a 20 μL total reaction volume, combine: 4 μg RNA, 2 μL 10× CIAP buffer, 2 μL (2 U) CIAP, and RNase-free water to 20 μL. Mix gently, centrifuge briefly to collect.
2. Incubate at 37 °C for 1 h.
3. Terminate reaction by adding 115 μL RNase-free water and 15 μL Stop Solution (*see* **Note 5**).
4. Extract RNA by adding 150 μL acid phenol–chloroform. Vortex and centrifuge 5 min at 12,500 × *g* at room temperature. Transfer upper, aqueous phase to a new microcentrifuge tube.
5. Add 150 μL chloroform. Vortex thoroughly. Centrifuge 5 min at 12,500 × *g* at room temperature. Transfer upper, aqueous phase to a new microcentrifuge tube.
6. Add 150 μL isopropanol and 20 μg glycogen. Vortex and incubate on ice 10 min. Centrifuge 20 min at 12,500 × *g* at room temperature. Remove supernatant.
7. Rinse pellet with 500 μL of cold 70 % ethanol. Centrifuge 5 min at 12,500 × *g* at room temperature. Remove ethanol and air-dry for 10–15 min.
8. Resuspend pellet in 10 μL of RNase-free water. Keep on ice (sample may be stored at −80 °C at this point).

Treatment of CIAP-Reacted RNA with TAP to Remove 5′ Cap and Leave a 5′ Monophosphate End (See **Note 6**)

1. In a 16 μL total reaction volume, combine: 5 μL CIAP reaction product (2 μg RNA), 1.6 μL 10× TAP buffer, 0.3 μL TAP (3 U), 0.5 μL RNase OUT, and RNase-free water to 16 μL.
2. Incubate at 37 °C for 2 h, place on ice (sample may be stored at −80 °C at this point).

3.2.2 Ligation of RNA Linker to 5′ Ends of RNA (See **Note 7**)

T4 RNA ligase I ligates the 3′ hydroxyl of the RNA linker to the 5′ monophosphate of RNA molecules.

1. Combine 8 μL of TAP reaction product (1 μg RNA) and 2 μL 100 μM RNA linker. Heat at 65 °C for 5 min and snap chill on ice to relax secondary structures.
2. For a total reaction volume of 16 μL, combine per reaction: 1.6 μL of 10× T4 RNA ligase I buffer, 1.6 μL of 10 mM ATP (if not in 10×), 1 μL of T4 RNA ligase I (10 U), 0.5 μL of RNase OUT, and RNase-free water to 16 μL.
3. Incubate at 37 °C for 2 h, heat-inactivate enzyme by incubating at 65 °C for 15 min, place on ice (sample may be stored at −80 °C at this point).

3.2.3 First-Strand cDNA Synthesis

Perform cDNA synthesis using a gene-specific reverse primer so only β-globin mRNAs are amplified in subsequent PCR reactions.

1. In a PCR tube, combine: 8 μL ligation reaction product (500 ng RNA), 1 μL 10 mM dNTPs, and 1 μL 5 μM reverse primer. Heat contents at 65 °C for 5 min, snap chill on ice.
2. In a separate tube, prepare a 2× reaction mix as follows: 2 μL 10× RT buffer, 4 μL 25 mM $MgCl_2$, 2 μL 0.1 M DTT, 1 μL RNase OUT, and 1 μL SuperScript III RT.
3. Add 10 μL 2× reaction mix to sample tube and incubate at 50 °C for 50 min. Inactivate enzyme by heating reaction at 70 °C for 15 min. Place tubes on ice.
4. Add 1 μL RNase H to tube, gently mix and spin to collect. Incubate at 37 °C for 20 min. RNase H will remove the RNA from the RNA:cDNA hybrids. Place tubes on ice (sample may be stored at −20 °C at this point).

3.2.4 Nested PCR to Amplify and Visualize Full-Length and Truncated Products (See Fig. 1)

1. Perform primary PCR using a forward primer complementary to the 5′ end of the RNA linker and a reverse primer specific to the gene of interest. This will amplify all target gene products with a ligated adapter, full-length and truncated. Prepare the primary PCR in 50 μL total reaction volume under the following conditions: 10 mM Tris–HCl pH 8.3, 50 mM KCl, 1.5 mM $MgCl_2$, 200 μM dNTPs, 400 nM forward primer, 400 nM reverse outer primer, 150 ng cDNA, and 2.5 U *Taq* DNA polymerase (Sigma-Aldrich). Thermal cycle conditions: 94 °C 3 min → [94 °C 30 s → 60 °C 30 s → 72 °C 1 min] × 35 cycles → 72 °C 5 min.
2. The secondary PCR reaction utilizes the same forward primer as the primary and a reverse primer that binds upstream of that used in the primary reaction. Dilute the primary PCR 1:10, 1:100, and 1:1,000 in water for use as template for the secondary reaction. Diluting the primary reaction results in increased detection of the nested product after gel electrophoresis [7].
3. Perform secondary PCR under the following reaction conditions: 1 μL diluted primary PCR reaction, 10 mM Tris–HCl pH 8.3, 50 mM KCl, 1.5 mM $MgCl_2$, 200 μM dNTPs, 400 nM forward primer, 400 nM reverse nested primer, and 2.5 U *Taq* DNA polymerase (Sigma) in 50 μL total reaction. Thermal cycle conditions: 94 °C 3 min → [94 °C 30 s → 60 °C 30 s → 72 °C 45 s] × 35 cycles → 72 °C 5 min.

3.2.5 Identification of 5′ Sequence

1. Electrophorese 8 μL of the secondary PCR reaction on a 3 % low melting point (LMP) agarose gel using 1× TBE buffer. Visualize and document bands with a Bio-Rad GelDoc or similar imager. Size markers allow for identification of products based on size (*see* Fig. 1). The 3 % LMP agarose gel will resolve <500 bp bands very well.

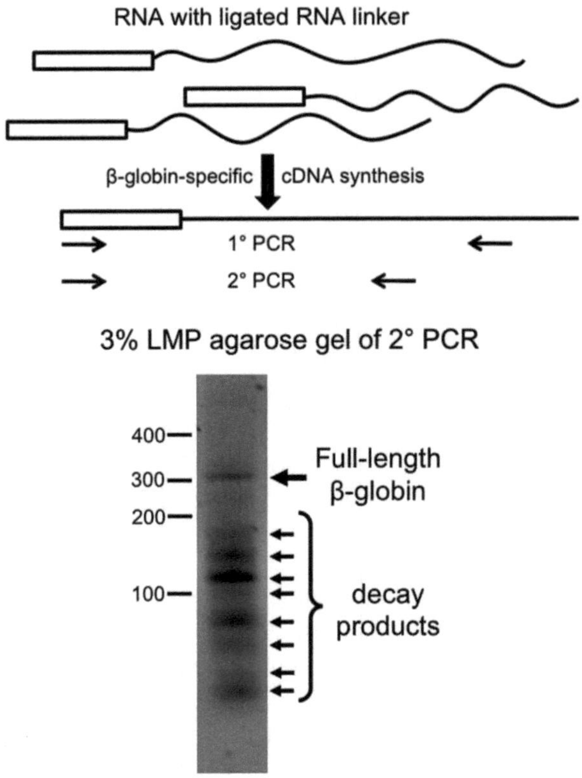

Fig. 1 5′-RLM-RACE to identify 5′ ends of full-length β-globin and its decay products. Cytoplasmic RNA was treated with CIAP to remove the 5′ monophosphates from uncapped RNAs. This is then treated with TAP to convert capped RNAs to RNAs with 5′ monophosphate ends for ligation to the 3′-hydroxyl of an RNA linker. Gene-specific cDNA synthesis produces cDNAs of all β-globin mRNA species. The first-strand PCR amplifies the 515 bp full-length RNA linker-β-globin cDNA and any decay products with a ligated 5′ linker. The second round of PCR amplifies the 312 bp full-length RNA linker-β-globin cDNA and 5′-truncated decay products

2. Excise bands from the gel and purify using a gel extraction kit. Ligate purified bands into pGEM-T cloning vector in a 1.7 mL microcentrifuge tube. For a 10 μL reaction add the following: 20 ng purified DNA, 1 μL pGEM-T vector, 5 μL 2× rapid ligation buffer, 1 μL T4 DNA ligase (3 U), and RNase-free water to 10 μL. Incubate the reaction at 4 °C overnight. Thaw competent *E coli* cells, such as DH5α, on ice and transform by adding 2 μL of the ligation reaction to 50 μL of cells and keeping on ice for 5 min. Heat shock the cells at 42 °C for 45 s and quickly return tubes to ice for another 5 min. Add LB to 1 mL and incubate at 37 °C with shaking for 1 h. Plate 100 μL of the transformed culture on LB agar plates with 50 μg/mL ampicillin, or appropriate antibiotic, and grow overnight in

37 °C incubator. Choose single colonies, verify ligation of target band with PCR, and sequence multiple positive clones using vector-specific forward and reverse primers (*see* **Note 8**).

3. Identify the RNA linker's 3′ end sequence and following it will be the 5′ sequence of the mRNA of interest. Align sequencing results with mRNA of interest using NCBI's specialized alignment engine (*bl2seq*) or other alignment software. Multiple identical sequencing results ensure confidence in the 5′ end of the mRNA of interest.

3.3 Molecular Beacons

Molecular beacons are single-stranded, dual-labeled fluorescent probes held in a hairpin structure by complementary stem sequences at the ends of the probe. The loop region is complementary to the target sequence; hybridization of the molecular beacon to the target sequence opens the hairpin–loop structure and separates the reporter dye and quencher resulting in fluorescent emission. Since only one beacon binds per target molecule the measured fluorescence is quantitatively proportional to the amount of target DNA.

3.3.1 Molecular Beacon and Primer Design

Most importantly, the beacon needs to be designed to efficiently hybridize to and detect the proper target while reducing nonspecific fluorescence. Beacon design may vary as to the purpose of the assay—standard target detection or allele discrimination (refer to ref. [6] for detailed design tips). A qPCR master mix specific for probes is to be used in these reactions, do not use a SYBR green master mix. These experiments utilized QuantaBiosciences PerfeCTa qPCR FastMix—ROX since one of the beacons is detected in the ROX channel.

1. Design molecular beacon target sequences to span the junction between the RNA linker and the target product sequence.
2. Forward and reverse primers should have a T_m of 50–65 °C, with 50–60 % GC content. The forward primer needs to be at least 6 nucleotides upstream of the molecular beacon binding site to provide room for the polymerase. NOTE: this means the molecular beacon cannot be designed to the 5′ most end of a target (*see* **Note 9**). Design the reverse primer so the resulting amplification product is 75–250 nucleotides in length.
3. Optimize molecular beacon conditions empirically by testing varying primer and beacon concentrations. For these experiments, conditions were found to be 250 nM beacon and 500 nM each primer. Illumina technical support is very helpful and a tutorial is available on their website: http://www.eco-qpcr.com/Files/training/Eco-Real_Time_PCR_Assays/Published/index.htm

3.3.2 Determining the Optimal Beacon Annealing Temperature

The annealing temperature of the reaction should be chosen so that the beacon specifically detects the correct target sequence. The optimal annealing temperature can be determined using a thermal denaturation profile of 1) beacon with correct target, 2) beacon with incorrect target, and 3) beacon alone. Choose the annealing temperature that provides maximal detection of the target sequence and minimal nonspecific detection.

1. Measure molecular beacon fluorescence in 1) the absence of target, 2) the presence of perfectly complementary target, and 3) the presence of mismatched target. The target oligonucleotides are created by in vitro transcription and subsequent gene-specific cDNA synthesis (*see* **Note 10**).
2. Prepare three reactions containing 200 nM molecular beacon in 1× final qPCR buffer. Add to one reaction a fivefold molar excess of perfectly complementary cDNA, to the next add a fivefold molar excess of mismatched cDNA, and to the last add only buffer (*see* **Note 11**).
3. Using the Illumina Eco Real-Time PCR machine (Eco) choose the "other" detection chemistry and the high-resolution melt option. Select to detect the fluorescence for each preparation while the temperature decreases from 80 to 30 °C.
4. Normalize the data to the buffer control to get the denaturation profiles. The annealing temperature chosen provides the largest window of discrimination between matched and mismatched target detection.
5. This can be completed on other instruments by running the same samples and measuring fluorescence of the sample while decreasing temperature from 80 to 30 °C in 1 °C increments for 1 min each and comparing detection curves.

3.3.3 Standard Curves

Standard curves allow for absolute quantification of data and ensure reliable detection by the beacon. Standard curves also determine the efficiency of the qPCR, which is needed to perform relative quantification calculations.

1. Using the cDNA from the thermal denaturation profiles, create a tenfold dilution series of standards. The dilution series needs to cover a broad range to ensure a proper curve is generated.
2. In the Eco software, choose the "other" detection chemistry and the standardization option. Prepare a master mix of 2× qPCR buffer, primers, and beacon at the optimized conditions and aliquot into wells. Add template, running each standard in triplicate. Be sure to include a no template control for background detection.
3. The Eco software will generate a standard curve graph and provide values for the standard curve line. For tenfold dilutions

the ideal slope is −3.32, which indicates an efficiency of 100 %. Acceptable curves will have slopes that range from −3.10 to −3.58, indicating an efficiency from 90 to 110 %, with an R^2 value ≥0.98 [8]. If the standard curve is not acceptable, then try adjusting the reaction conditions. Always use a freshly diluted series of standards and run multiple standard curves on different days.

3.3.4 Beacon Specificity

Determining the degree of nonspecific detection by the beacon is essential to assure confidence in positive readings.

1. To determine the specificity of the molecular beacons, a template containing mutations only at the 3′ binding side of the beacon is needed along with the correct template. For these experiments, full-length beacon was reacted with decay product target and the decay product beacon was reacted with full-length target, as the beacons differ only in their 3′ sequence.
2. Run triplicate qPCR reactions of a tenfold dilution series using a similar molar dilution series. Compare Ct values to those for samples of similar molarity, if the Ct value for nonspecific template is 7 cycles higher than the Ct for correct template, then nonspecific detection is less than 1 %, according to the $2^{-\Delta\Delta Ct}$ method of calculation [9]. In our experiments, nonspecific detection was less than 1 %.

3.4 Processing Experimental Samples for MBRACE (See Fig. 2)

3.4.1 TAP Treatment to Remove 5′ Cap

A reaction with 2 μg RNA will be used as an example.

1. Thaw RNA and 10× TAP buffer on ice.
2. Incubate RNA at 65 °C for 5 min and snap chill on ice to relax secondary structures.
3. Prepare 10 μL total volume reactions as follows: 2 μg RNA, 1 μL 10× TAP buffer, 0.3 μL TAP (3 U), 0.5 μL RNase OUT (20 U), and RNase-free water to 10 μL.
4. Incubate at 37 °C for 1 h.
5. Return tubes to ice (samples may be stored at −80 °C at this point).

3.4.2 RNA Linker Ligation

1. Thaw RNA, 100 μM RNA linker, 10× T4 RNA Ligase I buffer, and 10 mM ATP on ice.
2. Combine 5 μL TAP reaction (1 μg RNA) with 2 μL 100 μM RNA linker and incubate at 65 °C for 5 min then snap chill on ice.
3. Prepare reaction master mix, using the following per reaction: 1.6 μL 10× buffer, 1.6 μL 10 mM ATP (if not in 10×), 1 μL T4 RNA ligase I (10 U), 0.5 μL RNase OUT, and RNase-free water to 16 μL total reaction volume.

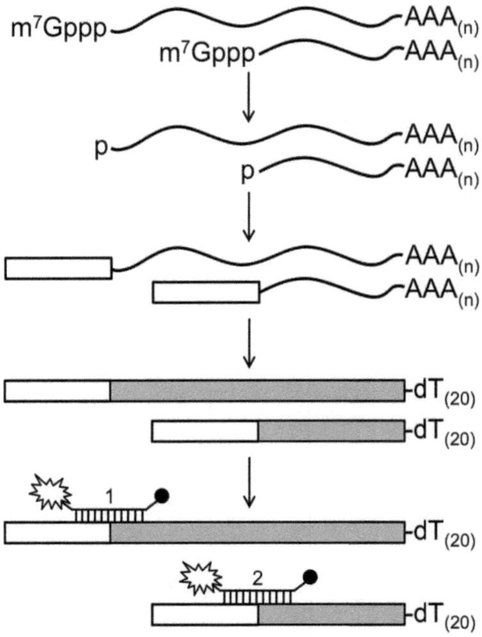

Fig. 2 MBRACE schematic. Cytoplasmic RNA is treated with TAP to convert capped 5′ ends to 5′ monophosphates. Next, the RNA linker is added to the RNA via ligation of its 3′ hydroxyl to the 5′ monophosphates of the RNA. Oligo(dt)-specific cDNA synthesis creates templates only from RNAs with an intact poly(A) tail. The MBRACE assays uses beacons to detect the junction between the RNA linker and [1] full-length β-globin and [2] a decay product by fluorescing upon hybridization. SYBR green qPCR may be used as an alternative detection method by designing the forward primer to be specific to the linker-mRNA junction sequence

4. Add reaction mix to tubes and incubate at 37 °C for 1 h.
5. Heat-inactivate enzyme at 65 °C for 15 min. Return tubes to ice (samples may be stored at −80 °C at this point).

3.4.3 First-Strand cDNA Synthesis

cDNA synthesis will now be performed using an oligo$(dT)_{20}$ primer, which will bind to RNA with an intact poly(A) tail. This allows for detection of an internal control mRNA species for normalization of data later.

1. Thaw ligation reaction, oligo$(dT)_{20}$ primer, dNTP mix, 10× RT buffer, $MgCl_2$, and DTT on ice.
2. For a 20 μL total volume reaction, prepare a master mix of 1 μL 50 μM oligo$(dT)_{20}$ primer and 1 μL 10 mM dNTP mix per reaction. Aliquot 2 μL master mix into 0.2 mL PCR tubes and add 8 μL ligation reaction (500 ng RNA). Denature secondary structure by heating at 65 °C for 5 min and snap chilling on ice.

3. Prepare a 2× reaction master mix on ice, containing the following per reaction: 2 μL 10× RT buffer, 4 μL 25 mM $MgCl_2$, 2 μL 0.1 M DTT, 1 μL SuperScript III reverse transcriptase, and 1 μL RNase OUT. Add 10 μL 2× reaction master mix to each tube.
4. Incubate at 50 °C 50 min. Heat-inactivate the enzyme at 70 °C for 15 min, place tubes on ice.
5. Add 1 μL RNase H to each reaction, mix gently, and centrifuge briefly to collect. Incubate at 37 °C for 20 min.
6. Place tubes on ice (samples may be stored at −20 °C at this point).

*3.4.4 Column Purification of cDNA (See **Note 12**)*

Qiagen QIAquick PCR purification kit columns were chosen as they demonstrated the greatest recovery of cDNA [10].

1. Mix cDNA reaction with 5 volumes (100 μL) of buffer PB. Load onto column and, using a tabletop centrifuge, spin at 12,500 × *g* for 1 min. Discard the flow through and return columns to collection tubes.
2. Add 750 μL buffer PE to each column. Spin at 12,500 × *g* for 1 min, discard flow through, and return columns to tubes.
3. Remove residual ethanol by spinning at 12,500 × *g* for 2 min. Discard collection tubes and place columns in new 1.7 mL elution tubes.
4. Add 52 μL of elution buffer (2 μL will be lost to the column) directly to column and incubate at room temperature for 5 min.
5. Elute by centrifuging at 12,500 × *g* for 2 min. Current concentration of cDNA is approximately 10 ng/μL. Keep tubes on ice (samples may be stored at −20 °C at this point).

3.4.5 MBRACE qPCR

1. Thaw purified cDNA, forward and reverse primers, beacons, and qPCR reagent on ice; keep qPCR reagent and beacons under foil to limit light exposure.
2. For a 20 μL total reaction volume, prepare a master mix as follows: 10 μL 2× PerfeCTa qPCR FastMix, 1 μL 10 μM forward primer, 1 μL 10 μM reverse primer, 1 μL 5 μM beacon, and 5 μL RNase-free water per reaction.
3. Aliquot 18 μL master mix per well, including wells for a no template control and a positive control, such as a standard from curves generated earlier.
4. Add 2 μL cDNA (~20 ng) per well and apply cover to plate. Run experimental samples in triplicate. Agitate plate to mix and centrifuge to collect.
5. Using Eco software, choose "other" detection chemistry. Set the thermal profile: 95 °C 3 min → [95 °C 10 s → 61 °C 30 s* → 72 °C 15 s] × 40 cycles (* denotes fluorescence detection).

Remember to set the fluorescent detection for the annealing step as this is when the beacons will fluoresce.

6. Once the run is completed, set the threshold and base line to "Auto." Export the data for analysis.
7. Set a separate qPCR reaction for an internal control mRNA, such as β-actin or GAPDH, using SYBR green chemistry. Run each sample in triplicate for the control mRNA. Again, set the Eco software to automatically determine baseline and threshold. Export the data for analysis.

3.4.6 Analyzing qPCR Data

Absolute Quantification

1. Absolute quantification calculates the input amount of the target by relating the qRT-PCR signal to a standard curve of known quantity. The quantity of the gene of interest is *interpolated* along the linear regression line generated by the standard curve. The range of standards must be broad enough to generate a curve containing amounts relevant to the experiment—quantity may NOT be extrapolated from the standard curve, as the relationship may no longer be linear outside of the standard curve range.
2. The equation for the linear regression line will allow for absolute quantification of unknowns. The equation will be given in the $y = mx + b$ format—where y is the Ct, x is the quantity, m is the slope, and b is the y-intercept. The unknown Ct values substitute in for y and the equation is solved for x to determine the quantity.
3. Absolute quantification assumes the qPCR reactions for each target have *equal amplification efficiencies.* If efficiencies are not equal then relative quantification should be used. Additionally, absolute quantification may not be effective when small changes in mRNA are observed, thus relative quantification may prove more fruitful.

Relative Quantification

1. Relative quantification reports the fold change between the target transcript of one sample versus another. Changes in fold expression across treatments or groups may prove more useful in some studies. An internal reference transcript, a housekeeping gene, is to be used to account for experimental variation. A proper internal reference must not be affected by the treatment; in our study we used β-actin. If a housekeeping gene is not the desired normalization control then an exogenous RNA, such as luciferase, may be spiked into the cell lysate prior to RNA extraction to act as a calibrator.
2. If the amplification efficiencies of the target and reference gene are *similar* then the Livak method is used for calculations [11].

In this method of calculation the relative expression ratio, R, is given as

$$R = 2^{-\Delta\Delta Ct}$$

To determine the value of ΔΔCt, ΔCt values are first calculated for both the target and reference mRNA using

$$\Delta Ct = Ct_{sample} - Ct_{control}$$

where control refers to the housekeeping gene. This step is also called normalization of the data as the reference control may differ due to experimental variations. Calculate the standard deviation for ΔCt using the equation

$$s = \sqrt{s_1^2 + s_2^2}$$

where s_1 and s_2 represent the standard deviations of the Ct values. The reference sample needs to be chosen, to which the other samples will be compared—untreated control or wildtype is a good choice for the reference sample. Next, ΔΔCt is calculated by the equation

$$\Delta\Delta Ct = \Delta Ct_{sample} - \Delta Ct_{ref}$$

The resulting values are substituted into the relative expression ration equation. Note that the reference sample will always have a fold expression equal to 1. The standard deviation for ΔΔCt is the same as the standard deviation of the ΔCt_{sample} value. In our experiments, β-actin acted as the control and the wildtype cell line acted as the reference (*see* Fig. 3).

3. If the amplification efficiencies *differ*, then the Pfaff efficiency-corrected calculation model based on multiple samples is used [12]. This model states the relative expression ratio, R, is calculated by the equation

$$R = \frac{\left(E_{ref}\right)^{Ct_{sample}}}{\left(E_{tar}\right)^{Ct_{sample}}} \div \frac{\left(E_{ref}\right)^{Ct_{control}}}{\left(E_{tar}\right)^{Ct_{control}}}$$

Control refers to the housekeeping gene or a calibrator. E refers to the efficiency of amplification of a standard curve for those particular qPCR reaction conditions; the slope of the linear regression line determines the amplification efficiency as:

$$E = 10^{(-1/slope)}$$

E_{ref} refers to the qPCR efficiency of the reference sample. Likewise, E_{tar} refers to the qPCR efficiency of the target sample. $Ct_{control}$ refers to the Ct value of the internal normalization control mRNA, the housekeeping gene or calibrator.

Sample	full-length β-globin Average Ct	β-actin Average Ct	ΔCt full-length - β-actin	ΔΔCt ΔCt(sample) -ΔCt(WT)	Fold expression relative to WT 2^-ΔΔCt
WT β-globin	23.79±0.21	18.60±0.22	5.19±0.30	0.00±0.30	1.00 (0.81-1.23)
PTC β-globin	26.32±0.21	19.04±0.21	7.27±0.30	2.08±0.30	0.24 (0.19-0.29)

Sample	β-globin decay product Average Ct	β-actin Average Ct	ΔCt decay product - β-actin	ΔΔCt ΔCt(sample) -ΔCt(WT)	Fold expression relative to WT 2^-ΔΔCt
WT β-globin	26.09±0.13	18.60±0.22	7.49±0.26	0.00±0.26	1.00 (0.84-1.20)
PTC β-globin	22.26±0.07	19.04±0.21	3.21±0.22	-4.28±0.22	19.38 (16.68-22.63)

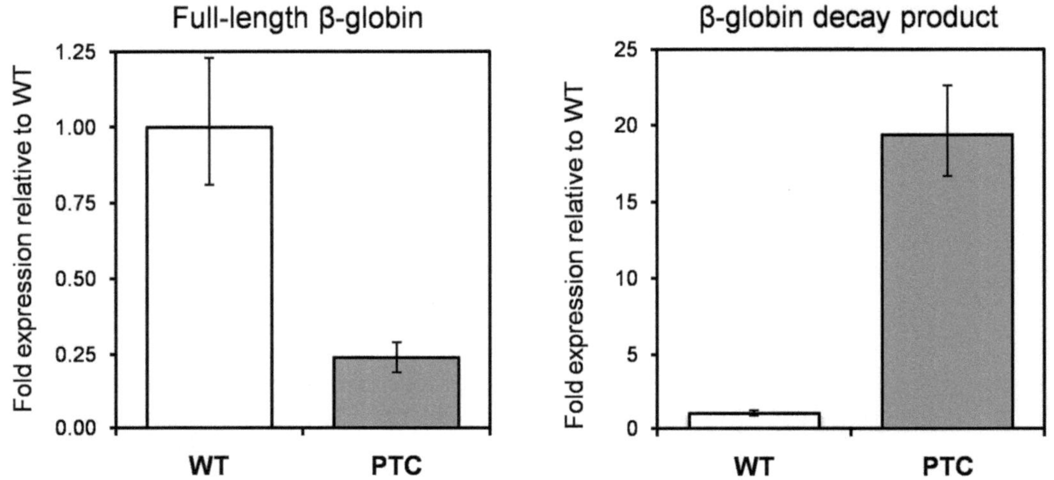

Fig. 3 Effect of PTC on β-globin expression. Data analysis of MBRACE for WT and PTC-containing β-globin in MEL cells. Values and standard deviations are shown for Ct, ΔCt, and ΔΔCt. Value and range shown for $R=2^{-\Delta\Delta Ct}$. The presence of a PTC significantly ($p \ll 0.01$) reduces full-length β-globin and significantly ($p \ll 0.01$) stabilizes the decay product

4 Notes

1. Prepare an incomplete stock of EFLB composed of all components except DTT and the RNase inhibitor; this can be stored at −20 °C. Prior to use, thaw it on ice and complete buffer by adding DTT and RNase inhibitor then keep it on ice. Discard any remaining unused completed buffer.
2. Erythroid cells are difficult to lyse so this method will not be appropriate for all cell types. Use an established fractionation protocol for the cell type being used.
3. The lysate should become clear but not viscous as this denotes lysis of the nuclear membrane. Three cycles of tapping/incubation have been adequate for erythroid cells. RNA can be assessed for proper cyto-nuclear fractionation by denaturing

gel electrophoresis and observation of the mature 18S and 28S rRNAs highly concentrated in the cytoplasm and two larger pre-cursor rRNA bands only seen in the nucleus.

4. To lyse the nucleus, resuspend the pellet in 3–5 volumes of EFLB and process as with cytoplasm—tapping and incubating on ice. Centrifuge at 5,000 × *g* to pellet cell debris. Transfer nuclear supernatant to a new tube *pre-chilled on ice.*
5. CIAP *must* be inactivated/removed or it will continue to dephosphorylate RNA after subsequent TAP-mediated cleavage of the 5′ cap, which would leave all the RNA without a 5′ monophosphate.
6. As of this writing, TAP is exclusively sold by Epicentre and is quite expensive. To economize costs, scale up the reactions and use an aliquot of the TAP reaction for subsequent work. Save the remainder of the TAP reaction at −80 °C.
7. Order an RNA linker without a 5′ monophosphate, the lack of a 5′ monophosphate will prevent concatamerization of the linker molecules. Excess adapter needs to be used in the reaction to ensure as many molecules as possible are ligated. Aliquot the RNA linker for storage at −80 °C to avoid freeze–thaw cycles. The RNA linker sequence was adapted from GeneRacer® protocol by Invitrogen. To complete the 5′ RACE experiments, a 25mer RNA linker was used, which was later changed for the MBRACE qPCR to a 44mer RNA linker to provide room for hybridization of the forward primer and the beacon.
8. Use of vector-specific forward and reverse primers will allow for sequencing the entire insert. Using forward and reverse sequencing primers also lends greater assurance of junction sequence as it is determined by two separate reactions. Sequencing several independent colonies provides verification of junction sequence.
9. Avoid selecting a molecular beacon target sequence in an area of strong secondary structure.
10. Specific target cDNAs are needed to perform the denaturation profiles, standard curves, and verify specificity of the beacons since the linker-specific sequence is shared. Synthetic 90mer DNA oligonucleotides of reverse complementarity to T7 promoter sequence—RNA linker—5′ end mRNA sequence were ordered through Sigma-Aldrich. 1 μg of 90mer DNA was combined with 2 μL of 50 μM T7 promoter-specific primer in a 0.65 mL tube. This tube was placed in a 1.7 mL tube filled with RNase-free water and the set was heated to 95 °C on a heat block for 10 min. The set of tubes was then removed and placed at room temperature to cool slowly and allow the primer to properly hybridize to the 90mer and form a double-stranded

T7 promoter. T7-directed in vitro transcription was completed using the Ambion MAXIscript T7/SP6 kit in a 30 μL reaction volume per manufacturer's instructions. The reaction was then treated with Turbo DNase to degrade the DNA template. Resulting RNA was stored at −80 °C. The full-length mRNA product served as proper target for the full-length beacon and improper target for the decay product beacon. Alternatively the decay product served as proper target for the decay beacon and improper target for the full-length beacon.

11. For detection with the beacons, the original RNA linker sequence needed to be altered to increase the annealing temperature for the beacon. Additionally, the RNA linker length was increased to 44 bp to accommodate binding of the forward primer and beacon.
12. cDNA column purification was necessary in our experiments to yield proper detection amplification curves with the molecular beacons.
13. The conditions are for these particular constructs and experiments and may not be the optimal conditions for other constructs. These conditions could be used as a starting point for optimization. Parameters to optimize include primer concentration, beacon concentration, template concentration (sometimes less is better), annealing temperature, and qPCR reagent. Also, due to the sensitivity of the MBRACE assay, use well-maintained and calibrated pipettes, as small deviations can become large differences.
14. An alternative to using molecular beacons would be to use SYBR green chemistry—if it can be proven to be highly specific. Designing the forward primer to hybridize to the RNA linker–mRNA junction sequence would serve to differentiate the products. Standard curves and specificity testing would be necessary to guarantee-specific detection. qPCR products should be electrophoresed on an agarose gel and the 5′ end sequences of the resulting product(s) verified by 5′ RACE as above to additionally ensure specificity. Bioline's SensiFAST SYBR reagents have shown superior sensitivity.

Acknowledgments

We wish to thank members of the Schoenberg lab for their helpful comments in developing this methodology. This work was supported by the National Institute of General Medical Science of the National Institutes of Health under award number GM079707.

References

1. Lasham A, et al (2010) A rapid and sensitive method to detect siRNA-mediated mRNA cleavage in vivo using 5′ RACE and a molecular beacon probe. Nucleic Acids Res 38:e19
2. Schoenberg DR, Maquat LE (2012) Regulation of cytoplasmic mRNA decay. Nat Rev Genet 13:246–259
3. Mascarenhas, R, Dougherty JA, Schoenberg DR (2013) SMG6 cleavage generates metastable decay intermediates from nonsense-containing β-globin mRNA. PLoS One 8: e74791
4. Stevens A, et al (2002) Beta-Globin mRNA decay in erythroid cells: UG site-preferred endonucleolytic cleavage that is augmented by a premature termination codon. Proc Natl Acad Sci U S A 99:12741–12746
5. Lim S-K, et al (1992) Nonsense codons in human beta-globin mRNA result in the production of mRNA degradation products. Mol Cell Biol 12:1149–1161
6. Vet JA, Marras SA (2005) Design and optimization of molecular beacon real-time polymerase chain reaction assays. In: Herdewijn P (ed) Methods in molecular biology: oligonucleotide synthesis. Humana Press, Totowa, NJ, pp 273–290
7. Youseff BH, Dougherty JA, Rappleye CA (2009) Reverse genetics through random mutagenesis in Histoplasma capsulatum. BMC Microbiol 9:236
8. Bio-Rad Laboratories (2012) Hallmarks of an optimized qPCR assay. Bio-Rad Laboratories, Berkeley, CA
9. Pfaffl MW (2001) A new mathematical model for relative quantification in real-time RT-PCR. Nucleic Acids Res 29:e45
10. Molas ML, Kiss JZ (2007) The effect of column purification on cDNA indirect labelling for microarrays. Plant Methods 3:9
11. Livak KJ, Schmittgen TD (2001) Analysis of relative gene expression data using real-time quantitative PCR and the 2(−Delta Delta C(T)) Method. Methods 25:402–408
12. Pfaffl SL (2006) Relative quantification. In: Dorak MT (ed) Real-time PCR (advanced methods). Taylor & Francis Group, New York, NY, pp 63–82

ERRATUM TO

Cytoplasmic Polyadenylation Assays

Olga Coll, Ana Villalba, and Fátima Gebauer

Joanna Rorbach and Agnieszka J. Bobrowicz (eds.) *Polyadenylation: Methods and Protocols*, Methods in Molecular Biology, vol. 1125, DOI 10.1007/978-1-62703-971-0,

DOI 10.1007/978-1-62703-971-0_29

The incorrect measuring unit was given in Chapter 5 page 58, of the print and online versions of this title. The correction is listed below.

Assemble the polyadenylation reaction on ice. Mix 0.01 pmol of the substrate RNA (*see* **Note 5**) in a final volume of 12.5 μL containing 16.8 mM creatine phosphate, 80 ng/μL creatine kinase, 60 μM amino acids, 24 mM HEPES pH 7.4, 0.4 mM $Mg(OAc)_2$, 48 mM KOAc, and 40 % embryo extract.

The online version of the original chapter can be found at http://dx.doi.org/10.1007/978-1-62703-971-0_5

ERRATUM

Retracted: Poly(A) Polymerase-Based Poly(A) Length Assay

Deepak P. Patil, Baskar Bakthavachalu, and Daniel R. Schoenberg

Joanna Rorbach and Agnieszka J. Bobrowicz (eds.) *Polyadenylation: Methods and Protocols*, Methods in Molecular Biology, vol. 1125, DOI 10.1007/978-1-62703-971-0, © Springer Science+Business Media New York 2014

DOI 10.1007/978-1-62703-971-0_30

The author withdraws this paper due to errors in referencing.

The online version of the original chapter can be found at http://dx.doi.org/10.1007/978-1-62703-971-0_2

INDEX

Joanna Rorbach and Agnieszka J. Bobrowicz (eds.), *Polyadenylation: Methods and Protocols*, Methods in Molecular Biology, vol. 1125, DOI 10.1007/978-1-62703-971-0, © Springer Science+Business Media New York 2014

Q

R

S

T

U

V

X

Y

MIX
Papier aus verantwortungsvollen Quellen
Paper from responsible sources
FSC® C105338

If you have any concerns about our products,
you can contact us on
ProductSafety@springernature.com

In case Publisher is established outside the EU,
the EU authorized representative is:
Springer Nature Customer Service Center GmbH
Europaplatz 3, 69115 Heidelberg, Germany

Printed by Libri Plureos GmbH
in Hamburg, Germany